THE ALEX STUDIES

THE ALEX STUDIES

Cognitive and Communicative

Abilities of Grey Parrots

IRENE MAXINE PEPPERBERG

HARVARD UNIVERSITY PRESS
Cambridge, Massachusetts, and London, England

First Harvard University Press paperback edition, 2002

Library of Congress Cataloging-in-Publication Data

Pepperberg, Irene M. (Irene Maxine)
  The Alex studies : cognitive and communicative
abilities of grey parrots / Irene Maxine Pepperberg.
    p.   cm.
Includes bibliographical references.
ISBN 0-674-00051-X (cloth)
ISBN 0-674-00806-5 (pbk.)
1. African grey parrot—Behavior.   2. Cognition
in animals.   3. Animal communication.   I. Title.
QL696.P7P46   1999
598.7'1—dc21   99-29630

Designed by Gwen Nefsky Frankfeldt

*To my father, Robert Platzblatt, for his love and encouragement,
to Alex, the raison d'être, and to Griffin, Kyaaro, and Alo,
for their supporting roles*

CONTENTS

PREFACE

"Bye. I'm gonna go eat dinner. I'll see you tomorrow." I hear these words, or variations, each night as I leave my laboratory. Exactly what one expects to hear from the typical graduate or undergraduate student—but these words do not emanate from human lips; rather they come from a beak—the beak of my research subject, a Grey parrot.

Not many scientists can describe such interactions with the subject of their studies, but for the past two decades I have been examining the cognitive and communicative abilities of this parrot, Alex, and the dialogue quoted above is one outcome of my work. In recent years, Alex has been joined by Kyaaro, Griffin, and, briefly, Alo. When I began my studies, the capacities of Grey parrots were unknown. Yes, everyone knew parrots could mimic human speech sounds, often to an uncanny extent. And in articles in the popular press, pet owners claimed that their birds used speech in meaningful ways. Was there a kernel of truth in all this anecdotal evidence? How much *did* these birds really understand? How much *could* these birds *learn* to understand? Given an appropriately enriched environment, might parrots turn out to be the great apes of the bird world? I set out to find answers to these questions.

That undertaking has generated an immense amount of data, and in this book I attempt to integrate these data into a coherent whole. In the course of my research, I have had to delve into comparative and cognitive psychology, animal learning, ethology, evolutionary biology, linguistics, and even neurobiology, and have published journal articles and book chapters related to those areas. These publications, however, fail to provide an overall picture: They provide momentary snapshots, rather than a cohesive album, of the capacities of Alex and my other birds. I hope that this book provides the appropriate panoramic view of my data. This book is not, however, meant to be a detailed review of the past 20 years of research in animal behavior, experimental psychology, or even comparative cognition, and I apologize to all my colleagues whose work could not be included in a book of this size. Also, in many instances my intention has been to present to the reader the information available to me in a particular area at the time I began a set of experiments, so that the reasons why I proceeded as I did are clear. Such an approach, however, means that I often specifically omit mention of subsequent, more advanced studies. Again, I apologize to colleagues whose work I therefore appeared to have ignored.

Like every long-term research endeavor, this project could not have been completed without the help of others. Foremost are the dedicated students who have worked in the laboratory throughout the years; the list would run on for pages. Of special importance have been the colleagues and friends who supported me and my ideas during the early days when anyone who studied cognitive processes in animals (and particularly in birds) was not considered a serious scientist. Included in this group are also my critics, who forced me to refine my ideas and methods. Many thanks are due to the National Science Foundation (and especially Fred Stollnitz), the Harry Frank Guggenheim Foundation, the Whitehall Foundation, and donors to The Alex Foundation, who have kept the laboratory afloat these many years, and to the John Simon Guggenheim Foundation for sabbatical support so that this book could be written. Not to be forgotten are the editors who have kept my prose clean, simple, and intelligible, and the many friends whose homes provided ideal (read secluded!) writing environments, and who patiently supported me through difficult times; special thanks go to Misty and Larry, for helping me through a particularly rough patch. I thank Russ and Madonna LaPell of VIP Aviaries for donating Kyaaro and Alo, and Terry Clyne of Apalachee River Aviary for donating Griffin. But most of all, thanks must be given to Alex, for putting up with 20 years of training and testing.

THE ALEX STUDIES

1

In Search of King

Solomon's Ring

As Holy Scripture tells us, the wise King Solomon, the son of David "spake also of beasts, and of fowl, and of creeping things, and of fishes" (I Kings IV.33). A slight misreading of this text . . . has given rise to the legend that the king was able to talk the language of animals, which was hidden from all other men . . . I am quite ready to believe that Solomon really could do so, even without the help of the magic ring which is attributed to him by the legend.

Konrad Lorenz, *King Solomon's Ring* (1952:xiii)

In the very earliest times, when both people and animals lived on earth, a person could become an animal if he wanted to, and an animal could become a human being. Sometimes they were people and sometimes animals and there was no difference. All spoke the same language. That was the time when words were like magic.

Nalunglaq, a Netsilik Eskimo (Rasmussen 1972:45)

The wish to "talk" with animals and understand their lives is not a recent phenomenon. According to legend, King Solomon possessed a ring that enabled him to communicate at will with all the birds and beasts in his realm. Similarly, some Native Americans supposedly had the ability to change into various animals and thus share their lives. And our literature—although primarily that for children—abounds with stories in which the protagonists, human and animal, freely "speak" with each other (e.g., Lofting 1948; Sleigh 1955; Clark 1963; King-Smith 1984). As with most legends and stories, these represent deeply held human dreams and desires.

In the twentieth century, researchers began to test whether these dreams and stories might be turned into reality. Scientists resolved to teach animals to communicate using human speech. Some researchers raised infant apes in their homes, thus providing their subjects with the same experiences as those of a human infant, including a language-rich environment (e.g., Furness 1916; Yerkes 1929; Hayes and Nissen 1956/1971; Kellogg 1968; note Laidler 1978). Despite the scientists' hard work and dedication, the apes never learned to use more than a few words of human speech. The apes' failures caused emphasis to shift toward studies designed to understand an animal's natural communicative behavior. Researchers examined various species, particularly birds, vervet monkeys *(Cercopithecus aethiops)*, chimpanzees *(Pan troglodytes)*, and marine mammals. Thus scientists, lacking access to Solomon's ring or Native American shapeshifting, developed experimental techniques to achieve insight into nonhuman communication. And after decades of work, they did indeed gain some knowledge about why birds sing (reviews in Kroodsma and Miller 1982, 1996), how vervets use alarm calls (Cheney and Seyfarth 1990), the differences that exist between affiliative and aggressive vocalizations in chimpanzees (Goodall 1986), and about dolphins' use of their whistles *(Tursiops truncatus;* Tyack 1986; Sayigh et al. 1990; McCowan and Reiss 1995; review in McCowan and Reiss 1997). Interestingly, at about the same time as these studies began, the field of psychology was shaken by the aptly named "cognitive revolution"—the radical notion that levels and types of intelligence in nonhumans formed a continuum with those of humans. This notion inspired researchers to study a wide range of behavior—including communication—in various species (e.g., Hulse, Fowler, and Honig 1968). Nevertheless, at that time most scientists believed that truly interactive interspecies communication was no more a reality than King Solomon's ring (e.g., Chomsky 1966). Even so, several dedicated researchers set out to prove not only that such communication was possible, but that it could also be used to examine intelligent behavior. They took the then-unacceptable stance that previous failures were simply a consequence of requiring apes to learn to communicate vocally, and developed new training paradigms (Gardner and Gardner 1969; Premack 1971; Rumbaugh et al. 1973). These pioneers set the stage for an entirely new field, and their successes led other researchers to begin working with marine mammals (e.g., Herman, Richards, and Wolz 1984; Schusterman and Krieger 1984). Their paradigms often challenged not only the accepted procedures for training, but also conventional techniques for testing animals and the existent doctrines of what should be tested. A very brief review of the history of several traditions in the scientific community at that time will help clarify why these new techniques constituted such a paradigm shift.

In the 1970s, the behaviorist tradition, best exemplified by Skinner (1974), still represented mainstream, laboratory-based behavioral science. In reaction against studies centered on anecdotal evidence and the introspective, "mentalist" approach that characterized the late 1800s (e.g., Romanes 1883/1977; review in Burghardt 1984), behaviorists emphasized experimental controls and eschewed discussions of thought or mental representations, information processing, or intentional actions. Scientists canonized the cautionary tenets of Morgan (1894) and the strictures of Watson (e.g., 1929). Models of learning were based on "associationist" principles: Simple laws were formulated to explain how external sensory input caused observable behavior (e.g., stimulus-response associations, reviews in Macphail 1982; Rescorla 1985; Roitblat 1987). Thus, for example, an animal saw an environmental stimulus, such as a green-lit button in one location; at some point the animal's subsequent random behavior included an action such as pressing a nearby green-lit button and that particular action was rewarded. A weak association was thus formed between stimulus and action; reward controlled the formation of the association. If the animal instead hit a nearby red-lit button, no reward was given and the action would be, in the parlance of the behaviorists, "extinguished." The next time the lit-button scenario appeared, the correct action was more likely, because it had been strengthened by the previous presence of a reward; thus behaviorists explained "match-to-sample" learning. Behaviorist researchers posited that even complex behavior patterns could be explained by breaking down these patterns into simpler, component parts that each have as their basis the same type of paired association. They argued that all behavior could be reduced to such laws, that the number of these laws was few, and that there was no need to examine cognitive skills, which, in part, include "learning, remembering, problem solving, rule and concept formation, perception, recognition" (Roitblat 1987:2). And, according to Skinner (1938), one needn't study a wide variety of animals, because none would react any differently from a pigeon or a rat: The rules of learning were universal.

Despite initial successes, behaviorists found themselves faced with situations their laws could not explain. At first their general reaction was not unlike that of the classical physicists at the turn of the twentieth century, who were able to explain all but a few physical phenomena and who believed that their current paradigms would eventually provide appropriate answers; later, such physicists realized that explaining these phenomena required the entirely new paradigm of quantum physics (Kane and Sternheim 1988). Similarly, the apparently anomalous activities of animals (for examples, see Breland and Breland 1961; Garcia and Koelling 1966; Smith and Roll 1967; Williams and

Williams 1969; reviewed by Roitblat 1987), whose natural responses to stimuli could not be reshaped by behavioristic training, forced at least some researchers to consider a new paradigm in which animals were seen as active processors of information (Kamil 1984, 1988; e.g., Pepperberg 1990c). This need was made even clearer by behavioral ecologists, whose data could be explained only by positing mechanisms such as selective attention and long-term memory (e.g., Schoener 1991; Pyke, Pulliam, and Charnov 1977; Kamil and Sargent 1981; reviewed by Roitblat 1987).

But even this cognitive stance, which argued that behavior is best explained in terms of mental representations and information processing, still often retained the physical techniques of behaviorism. Researchers in many cases continued to examine animals in isolation, in sterile laboratory conditions, and using operant conditioning techniques. In operant procedures, the researcher typically holds an animal at 80% of free-feeding weight and places it in a small box (an "operant chamber") containing devices that deliver a stimulus (such as a colored light), provides a response mechanism (such as a lever to press), records these responses (via a computer link), and delivers food rewards (perhaps via an automatic pellet dispenser). The animal is isolated from almost all stimuli other than those controlled by the experimenter; this isolation extends to all interactions with a natural environment, including interactions with other organisms. The experimental stimuli, usually tones, colored lights, or line patterns, generally have little to do with the animal's natural experiences. Also, an animal cannot communicate directly with the experimenter, so no explicit transfer of information exists about the task to be learned. Thus the animal must determine, through trial and error, the question that the experimenter wishes to ask and to which it must respond—the specific nature of the task (Pepperberg 1992)—as well as the appropriate response. The rationale is that such procedures examine learning proclivities so basic that they exist in the absence of environmental influences.

Researchers determined, using such procedures, that animals did indeed actively process information and use mental representations to solve problems (see, e.g., review in Roitblat 1987; Roitblat and von Fersen 1992); why, therefore, might one need to engage an animal in direct conversation to determine its cognitive capacities? Two major reasons exist. One is basically procedural, the other philosophical; they are, however, closely intertwined.

The procedural reason for using interspecies communication becomes apparent if one compares the experimental design of a representative task—the study of concepts of color, shape, similarity and difference—from operant and communicative perspectives. In a typical operant setting, a subject is given a red triangle, a red square, and a blue square, and is rewarded for choosing blue when the objects are backed in white and for choosing triangle when the backing is black (see Thomas 1980). A successful subject may learn something about similarity and difference (choose "odd color" versus "odd shape"), but only with respect to the *specific* objects rather than the *concept* of same versus different (Premack 1978; Pepperberg 1987a). A concept, in contrast, can be extended to any modality. In general, reports of how well the animal transfers its knowledge to novel operant situations (which would demonstrate some understanding of the concept) do not include its responses to the very first presentation of novel objects; rather, researchers report "learning curves" that show merely that the animal takes less time to learn to respond correctly on the novel task than to learn the task in the first place (Thomas and Noble 1988). Moreover, an animal in such a study need not determine specifically *what,* if anything, is same or different.[1] Contrast this study with one in which an animal learns to communicate with humans by using some form of symbolic code (e.g., vocal labels, American Sign Language, abstract computer symbols). Here the animal first acquires labels for various colors, shapes, and materials, category labels for these attributes, and comprehension of the phrases "What's same?" and "What's different?" The animal is then shown a red wooden triangle and a red wooden square, and asked, "What's different?" The animal must, based on its understanding of the question, first determine whether the experimenter is interested in similarity or difference. The animal must then observe the two objects, determine what attribute they do not have in common (the "odd" attribute), encode that information, and tell the experimenter the category label of this *attribute,* "shape." If the experimenter then asks "What's same?" the animal can respond either "color" or "matter." The animal can be given any set of objects, familiar or novel, and still respond appropriately on the very first trial (see Chapter 5). Clearly, both techniques explore an animal's cognitive abilities, but the communicative approach enables researchers to ask more complex and advanced questions—which is the philosophical reason for using interspecies communication.

Specifically, possession of (or the ability to acquire) human language or at least some form of a human-based communication code has been posited as a necessary precondition for an organism to organize and process information for certain complex cognitive tasks (for details, see, e.g., Macphail 1982, 1985, 1987; Premack 1983, 1986).[2] A brief summary of the position is that without some form of language-like code, an animal is capable only of remembering and working from images (such as mental pictures of the location of items in space), which restricts its capacities. An animal that has been taught a human-based code, in contrast, can supposedly work from abstract representations (e.g., not simply learn that "red" and "green" are associated with particular hues, but understand that these labels form a category known as "color" and that this knowledge of the category can be used to solve problems). Thus, according to this argument, a language-trained animal can understand and use a concept it would otherwise be unable to learn. One refutation of this position is that an animal trained in this manner simply has more experience in categorizing its world according to the experimenter's specific demands (see, e.g., Bronowski and Bellugi 1970; Marshark 1983; Miller 1983). The most interesting refutations, however, come from studies of human children. Reese (1972), for example, showed that labeling seems not to be a prerequisite for understanding concepts of same and difference because even children with normal language skills (grades K-2) had somewhat more difficulty labeling objects than denoting if pairs are same or different. Rice (1980), moreover, showed that language training does not improve a child's ability to learn concepts such as "color" (as opposed to the labels for individual colors) if the child has not already reached a certain stage of development. Subsequent studies (review in Mandler 1990) support the idea that children form many concepts well before they acquire language.[3] What is not easily resolved, however, is how complex a concept must be before its comprehension might indeed require a form of language learning: Animals that have been trained to use a human-based language code do better than untrained animals on some tests, such as those requiring understanding of analogies, whereas little difference exists between trained and untrained animals on tasks such as those involving spatial relations (Premack 1983).

Evaluating whether language training enables animals to succeed on complex tasks is also tricky because it is almost impossible to separate procedural and philosophical effects. The reason is that interspecies communication is a particularly powerful means of studying animal cognition. As I have previously noted (e.g., Pepperberg 1987c), interspecies communication (1) enables researchers to communicate directly to their subjects the precise nature of the questions being asked (the animals need not determine the nature of the question through trial and error); (2) takes into account research showing that social animals respond more readily and often with greater accuracy when placed within a social context at least somewhat similar to their field situation (see Menzel and Juno 1982, 1985); and (3) enables researchers to compare data not only between humans and other animals but among various nonhuman species. Interspecies communication also provides an open, arbitrary code that can create an enormous variety of signals; such variety allows researchers to examine the nature, and not just the extent, of the information perceived by an animal. Two-way communication also allows rigorous testing because animals can be required to choose a response from their entire repertoire rather than from a subset relevant only to the topic of a particular question. Moreover, an animal that learns such a code may also respond in novel and possibly innovative ways that may demonstrate even greater competence than that required by the responses of operant paradigms. Interspecies communication may therefore simply allow researchers to demonstrate more readily the inherent capacities of their animal subjects, rather than enable the animals to learn more complex tasks. Of course, an animal must first be taught the communication code, and here again major contrasts exist between an operant and an interactive protocol.

In the late 1960s and early 1970s, the several research laboratories studying interspecies communication used three quite different approaches. Several reviews of these projects are available (e.g., Ristau and Robbins 1982; Roitblat, Herman, and Nachtigall 1993; Hill 1995), and thus a detailed description of the projects and their results are not necessary here. I will instead focus on studies with chimpanzees, and briefly describe how the different training approaches led to strikingly different results (for details, see the review in Pepperberg 1997).

Although the extent to which chimpanzees can acquire a communication code isomorphic with that of human language is unresolved (see, e.g., Terrace et al. 1979; Savage-Rumbaugh et al. 1980a,b; Greenfield and Savage-Rumbaugh 1991), the animals studied did learn to label, request, refuse, and categorize objects and some actions, and often learned to comprehend some form of human speech (see, e.g., Fouts, Chown, and Goodin 1976; Savage-Rumbaugh et al. 1985; see Williams, Brakke, and Savage-Rumbaugh 1997 for a recent review). The conditions under which chimpanzees were taught human-based codes ranged from computer-driven systems with limited social interaction (Rumbaugh 1977), to a system in which the ape was trained to use magnet-backed plastic chips to communicate with humans in a laboratory setting (Premack 1971), to cross-fostering programs in which the ape was raised much like a human child (Gardner and Gardner 1969, 1989). Even when the actual code being taught in two laboratories—American Sign Language (ASL)—was identical, techniques used for training often differed radically (Gardner and Gardner 1969; Terrace 1979b), and the different projects obtained different results. All these projects subsequently were harshly critiqued (primarily with respect to claims of language abilities; see, e.g., Lenneberg 1971; Terrace et al. 1979), but the severest criticism was reserved for projects with the most naturalistic settings; the arguments were that these projects lacked experimental controls (Sebeok and Umiker-Sebeok 1980). Interestingly, animals in programs that relied most heavily on training related to operant conditioning (and often controls, e.g., Premack 1976; Rumbaugh 1977; Terrace 1979a) demonstrated behavior that was far less flexible and less "language-like" (note Terrace 1979b) than that of apes trained in ways that more closely resembled the experiences of young children (Gardner and Gardner 1969, 1989).[4] Those who criticized the more naturalistic, communicative approach eventually adopted the primary elements of this paradigm (Savage-Rumbaugh 1991; Williams et al. 1997), but only the criticisms were voiced in the late 1970s.

In any case, the accepted dogma in the mid-1970s was that interspecies communication studies were possible, but that subjects must be animals with a close phylogenetic relationship to humans, such as chimpanzees (see, e.g., Sarich and Cronin 1977),[5] or at least with large brains, like dolphins (see, e.g., Russell 1979; note, however, Morgane, Jacobs, and Galaburda 1986). Thus when I decided to study parrots, colleagues generally expressed extreme skepticism. Numerous studies had already shown that birds—represented by the laboratory pigeon—were greatly inferior to mammals on standard tests of complex behavior, for example, on tasks such as reversal learning (Bitterman 1975) and insight detour problems (Krushinskii 1960). Researchers had also shown that birds lacked, to any great extent, cerebral cortex—the so-called mammalian organ of intelligence (see, e.g., Jerison 1973). Finally, studies using mynahs and various parrot species had attempted—and failed—to teach birds to communicate with humans (see, e.g., Mowrer 1950; see Chapter 2). The overall perception was that birds were capable, at best, of nothing more than mindless mimicry (see, e.g., Lenneberg 1967), and certainly not capable of any kind of complex information processing.

My desire to work with parrots, however, stemmed from both scientific knowledge and an awareness of the folklore surrounding these birds. Reports of intelligent speech derive from at least the time of Aristotle (Sundevall 1863; Ilyichev and Silayeva 1992); parrots, if not common, were considered exceptional creatures by the British in the

1500s (Boosey 1947), and the popular press of the past centuries abound with stories of their intelligence (e.g., Temple 1692; Menault 1875; Stevens 1888; von Lucanus 1923; Amsler 1947; Boosey 1947). But anecdotal evidence was not enough: I had to find scientific evidence that parrots not only were good imitators of human speech, but also had the requisite intelligence to understand, acquire, and use English vocalizations meaningfully. It wouldn't hurt if I could also find evidence that their natural squawks, whistles, and calls were used for intentional communication: Teaching animals a human-based task (or any task) is always easier if the task corresponds to some natural predisposition (Seligman and Hager 1972; Hinde and Stevenson-Hinde 1973; Kamil 1988). Fortunately, a detailed literature search uncovered tantalizing bits of information from studies on neurobiology, animal learning, and avian song learning that suggested that my ideas were not quite as far-fetched as they at first seemed.

Neurobiological studies on parrots date back to the beginning of the twentieth century. Researchers at the time suggested that mammalian standards for correlations between brain structure—absolute brain size and particularly relative cortical size (Jerison 1973; Lenneberg 1973)—and intelligence might not hold for birds. Kalischer (1901), using what were clearly rather primitive techniques, found that striatal rather than cortical areas might be involved in avian intelligence. The metaphor I like to use involves looking at avian and mammalian brains as early Macintosh versus IBM-style computers: These different information-processing machines use the same wires, and when you enter the same data into their programs you get the same results—but the wires are organized differently and you must use programs designed for their differently configured systems. Although the work of the early researchers was essentially ignored for several decades, later elegant experiments drew more convincing parallels between avian learning and memory and these striatal areas (e.g., Portmann 1950; Cobb 1960; Portmann and Stingelin 1961; Stettner 1974). Of course, even more is known today about avian brains (see, e.g., deVoogd, Krebs, Healy, and Purvis 1993), but such was the state of knowledge when I wished to begin my studies.

Of particular interest was research that suggested a link between striatal development and "intelligence." On studies of reversal learning (Gossette, Gossette, and Riddell 1966; Gossette and Gossette 1967; Matyniak, Wheeler, and Stettner 1971), set learning (Stettner and Matyniak 1968; Kamil and Hunter 1970; Dücker 1976), oddity problems (Pastore 1955), number-related problems (Koehler 1943, 1950, 1953; Braun 1952; Lögler 1959; Pastore 1961; Rensch and Dücker

1973), and insight detour problems (Krushinskii 1960; Dücker and Rensch 1977; Zorina 1982), birds with the greatest striatal development—such as crows, parrots, and mynahs (Portmann 1950; Portmann and Stingelin 1961)—performed more accurately than birds with lesser striatal development—such as pigeons and domestic fowl—and were often superior to some monkeys (see, e.g., Stettner and Matyniak 1968; Hodos 1982; Zorina 1982). Moreover, lesions in these areas appeared to interfere with learning (Stettner 1967, 1974). Parrots had also performed at high levels on problems involving simple labeling and intermodal associations (Koehler 1950, 1953, 1972; Mowrer 1950, 1954; Lögler 1959; Thorpe 1964, 1974), and Grey parrots *(Psittacus eritha-cus)* had demonstrated the ability to respond as accurately on new problems as on related training problems (on "transfer tests": Koehler 1943; Lögler 1959). This ability to transfer information between problems is generally considered evidence for advanced cognitive capacities (see, e.g., Rozin 1976; Premack 1978). Such findings suggested that birds did not need an extensive cerebral cortex to perform complex cognitive tasks, and that the extent of avian intelligence, based primarily on studies on pigeons, might be markedly underrated.

Serendipitously, again at about the same time that I was contemplating working with parrots, researchers studying how birds learned their songs found intriguing parallels between avian song acquisition and human language learning. Some studies suggested that these parallels might also apply to parrots, and thus that psittacine vocal learning abilities might be channeled into acquiring meaningful communication with humans. Four aspects appeared common to humans, songbirds, and parrots. First was the role of imitative learning (Amsler 1947; Lorenz 1952; Mowrer 1950; Marler 1967, 1970, 1974; Cazden 1972; Nottebohm and Nottebohm 1969; Todt 1975a). Second was the existence of a "babbling," or practice, period, during which individuals experiment with sounds that ultimately become part of their repertoire (Baldwin 1914; Mowrer 1950; Thorpe 1964; Lenneberg 1967; Marler 1970, 1973; Nottebohm 1970, 1976; Brown 1973). Third was the presence of repetition and imitation, which, at least in the early stages, appear to be self-rewarding and self-reinforcing (Mowrer 1950, 1953, 1958; Mattick 1972; Marler 1974). Fourth was that specific brain areas seemed to be involved in the learning, storage, and production of vocalizations (see, e.g., Nauta and Karten 1971; Nottebohm 1980; Patton, Manogue, and Nottebohm 1981).[6]

In addition, some research suggested that parrot vocalizations might be used in meaningful, if simple ways. Although very little information existed on natural parrot vocalizations in the 1970s (see Nottebohm and Nottebohm 1969; Nottebohm 1970), limited data suggested that communication within psittacine flocks might be mediated by complex vocalizations that were probably learned through social interaction with parents, flock members, or other organisms (Dilger 1960; Power 1966a; Nottebohm 1970; Busnel and Mebes 1975). Specifically, other data suggested that parrots have intrapair duets that are distinct from interpair or other interparrot vocalizations (e.g., Power 1966b; Mebes 1978), and several researchers proposed that such duets might mediate interactions among flock members (Gwinner and Kneutgen 1962; Thorpe 1972; Wickler 1972). Interestingly, shortly after I began my studies in 1977, several researchers demonstrated that some parrot species used vocalizations to recognize specific individuals in a flock (Rowley 1980; Saunders 1983); later work provided evidence that some species had specific calls to alert flock members to the presence of predators (sentinel behavior: Lawson and Lanning 1980; Levinson 1980). These studies suggested that human failure to document complex psittacine communication in the wild might be more a consequence of our lack of detailed field work than the lack of such capacities in parrots.

Finally, in the late 1970s, serious thought was given to using interspecies communication as "a possible window on the minds of animals," and not only for great apes (Griffin 1976, ch. 7). The arguments are complex, and engendered strong disagreements that continue to this day (see Yoerg and Kamil 1991; Kamil 1998), but the basic idea is that, according to Griffin (1976), researchers might benefit from studying animals the way that anthropologists study a previously undiscovered primitive tribe: by attempting to establish two-way communication and using this communication to determine how they process information and interpret the events in their world. Granted, the researcher will never be sure that an absolute correlation exists between the way that the animal and researcher define the elements of their mutually derived code (Quine 1960; Premack 1986). But researchers studying great apes by using various levels of two-way communication had demonstrated behavior patterns far more complex than were expressed in standard operant paradigms (in addition to the studies noted above, see Fouts 1973a; Miles 1978).[7] Why not parrots? Given the material discussed above, parrots might be capable of learning to communicate with humans. These birds apparently acquired their vocalizations in ways that had some parallels with human language learning, had some cognitive capacities (and underlying neurological structures) that suggested that such learning was possible, and used their natural vocalizations in meaningful ways; all these findings intimated that human vocalizations might be mapped in a 1:1 correspondence with natural parrot behavior. Previous studies in which these birds failed to learn to communicate with humans used basically an operant paradigm (e.g., Mowrer 1950). How might these birds fare if I used instead a system involving interactive communication? Quite possibly, the general perception of birds as intellectually inferior to mammals and the previous failures to teach these birds to use English speech had more to do with inappropriate training and testing procedures than the birds' actual abilities.

Thus I decided to obtain a Grey parrot and begin its training in a laboratory setting; this species had been the subject of many of the previously described studies on psittacine intelligence and was known for the clarity of its speech (see, e.g., Boosey 1947). My goal, however, was not to examine the extent to which a bird could learn to use a human language, but rather to establish some form of avian-human communication and then use this communication to examine further avian *cognitive* capacities. The next chapters provide the details of the training and testing procedures, the rationale for these procedures, and the results that my students and I obtained. The work has also branched out into studies concerning cognitive capacities that do not appear to involve two-way communication (e.g., object permanence). Finally, I discuss both the acoustic properties and production mechanisms of avian speech. Such analyses might seem quite removed from cognition and learning but, to produce human speech, a bird must discriminate among and appropriately categorize speech sounds despite individual speaker variation; to use such speech referentially, the bird must understand that minor acoustic variations such as "want corn" versus "want cork" create major differences in meaning. Such abilities require considerable cognitive processing, including learning of flexible, voluntary control of portions of the vocal tract: Differences between "cork" and "corn" must be real and measurable. My students and I thus needed to determine exactly what a parrot could produce—and how.

2 Can We Really Communicate with a Bird?

> Let a man decide upon his favorite animal and make a study
> of it, learning its innocent ways. Let him learn to understand its
> sounds and motions. The animals want to communicate with man,
> but Wakan-Tanka does not intend they shall do so directly—man
> must do the greater part in securing an understanding.
>
> Brave Buffalo, a Teton Sioux (Densmore 1918:172)

In this chapter I describe the training and testing techniques I developed to study cognition and communication in Grey parrots. I first provide a brief history of the projects and failures that preceded my research and what I deduced were the possible reasons for these failures. I describe the one project to succeed in training birds to acquire human speech, albeit without comprehension, and the likely reasons for its success. I explain how my techniques were derived from a theoretical framework that examined the effect of social interaction on learning; I describe this theoretical framework in detail. I conclude with a somewhat technical description of the testing procedures and rigorous controls I used to determine what my parrot had learned and to ensure that his responses were based on his understanding of the questions and concepts and not on extraneous cues.

Previous Failures to Establish Two-Way Communication with "Talking" Birds

I was not the only scientist to reason that the vocal ability of mimetic birds, coupled with their considerable intelligence, should enable them to learn to communicate with humans using speechlike sounds. Mowrer (1950, 1952, 1958) was one of the first to investigate this possibility. He studied mynahs *(Gracula religiosa)*, magpies *(Pica pica)*, crows (species not reported), and several psittacids (budgerigars, *Melopsittacus undulatus;* a Yellow-headed parrot, *Amazona ochrocephala;* and a Grey parrot), and used standard psychological techniques—basically the operant methodology described in the previous chapter—prevalent at the time. Interestingly, although his birds were trained according to principles of association and reward, they were neither socially isolated nor placed in operant chambers for their sessions.

Mowrer (1952, 1969) used procedures such as the following to teach a bird to produce "Hello" upon the appearance of its trainers. The first step was to make the bird totally dependent upon its trainers for food, water, and social attention—that is, a bird received food and water only in the presence of its trainers. Next, trainers produced "certain characteristic noises" (p. 264), for example, "Hello," followed by a positive action such as a trainer appearing from behind a screen or uncovering the cage. The noise would then be considered a signal (in the parlance of the behaviorists, a "conditioned stimulus") for subsequent appearance of the positive action (a reward, or "primary reinforcer"). According to the behaviorists, moreover, the sound itself becomes reinforcing, if only secondarily (the "secondary reinforcer"), because of what it predicts (Mowrer 1954). If a bird began to produce the sound itself, it received this secondary reinforcement. Production of the sound would, theoretically, increase more quickly if an additional reward occurred for vocalizing (another "primary reinforcer," such as a favorite food). Thus another association would be formed, between vocalizing and a lessening of hunger, and a bird would be expected to produce the sound with increasing frequency. Mowrer introduced several different words and phrases, but the reward for all vocalizations was food. The idea was that, after a bird emitted vocalizations with some frequency, it could be trained to produce the utterance only in the original, appropriate context (on the appearance of the trainer) by providing the food only when the vocalization was emitted in such a situation.

Mowrer's birds acquired few vocalizations. His use of food rewards that directly related neither to the task being taught nor the skill being targeted (such as saying "Hello" when the trainer appeared) probably delayed or possibly prevented learning: Most likely, birds confounded the label of the object or action to be taught with that of the unrelated food reward (Pepperberg 1978, 1981; see Bruner 1978; Greenfield 1978; Miles 1983). That is, his birds apparently connected reproduction of human sounds with the inevitable appearance of food (a salient object to a hungry bird) rather than with their actual referents, for example, "Hello" and the appearance of the trainer. Birds clearly did not realize that a trainer's appearance was the relevant stimulus for producing "Hello." Attempts to obtain food by producing "Hello" when a trainer was already in place would eventually fail: The trainer would consider that the vocalization had been used inappropriately and provide no reward; a bird's production of the strange sound ("Hello") would consequently lessen (in behaviorist terms, "be extinguished"). Moreover, because a bird received food for whatever it produced, it may have stopped learning after acquiring one or two utterances that were sufficient for decreasing its hunger.[1]

Some researchers, possibly believing that Mowrer's social setting was responsible for his failure, attempted to train mimetic birds under more rigorous operant conditions. Ginsburg (1960) managed to place some natural vocalizations of budgerigars under stimulus control—had birds respond with a particular sound after experiencing a particular stimulus—and Gramza (1970) trained budgerigars to mimic tones and musical phrases, but neither tried to replicate Mowrer's goal of achieving interactive communication. Other scientists placed mynahs in soundproof boxes and played recordings of human vocalizations followed by presentation of food pellets; their birds also failed to acquire much in the way of trained vocalizations, although Ginsburg's birds learned to produce two simple utterances in two distinct contexts (Ginsburg 1963; Gossette 1969; Grosslight, Zaynor, and Lively 1964; Grosslight and Zaynor 1967). The puzzling aspect of these failures was why these birds, which were so vocal in the wild and which learned allospecific vocalizations so readily in the informal setting of a home (Amsler 1947; Boosey 1947; Hensley 1980), were incapable of significant vocal learning under well-controlled laboratory conditions.

The answer was not immediately forthcoming and required three breakthroughs, one in data collection and two in research methodology (Pepperberg 1988c). First, researchers had to document the experiences and behavior of wild mimetic birds. Second, laboratory scientists had to be shown how their training paradigms differed from mimetic birds' natural learning situations and how adapting laboratory conditions to reproduce a natural environment might affect their results. Third, a theoretical framework was needed to integrate data from the wild into a valid training procedure.

Learning by Mimetic Birds in Natural Environments and in Certain Laboratory Settings

Until the late 1960s, little evidence existed for imitation or for *any* vocal learning for wild mimetic birds. Researchers thus wondered if vocal learning in captivity was an artifact of unnatural housing conditions (Thorpe 1964). At that time, however, Nottebohm found that Orange-winged Amazon parrots *(Amazona amazonica)* nesting within audible range of one another in Trinidad often shared calls, whereas colonies of the same species living in separate habitats had completely different dialects. Dialects were so different as to suggest initially (and erroneously) that birds inhabiting the various areas were different species (Nottebohm and Nottebohm 1969). These data made sense only if parrots learned their vocalizations through interactions with parents, flock members, and other organisms (Nottebohm 1970)—that is, if parrots mimicked, but generally other birds of the same species.[2] At roughly the same time, Bertram (1970) showed that wild mynahs predominantly shared calls with and appeared to learn vocalizations from neighboring mynahs and, on occasion, other animals in their habitat (Tenaza 1976). Wickler (1976, 1980) subsequently suggested that the extensive duetting often observed in mated parrot pairs came about through a complex socialization process involving learning: To create a duet unique to the pair, both individuals synergistically adapt and adjust their repertoire (Mebes 1978).

Before too long, some laboratories incorporated social interactions into their experimental designs. If Nottebohm (1970) was correct, some interaction between birds—or at least observation—should be involved in the acquisition process. To determine optimal conditions for allospecific learning, Todt (1975a) used Grey parrots to investigate what might happen if training involved social interaction. He developed the model/rival (M/R) technique in which humans assume roles played by psittacine peers in the wild. Humans thus demonstrate to the parrots the types of interactive vocalizations to be learned. In Todt's procedure, one human is exclusively the principal trainer of each parrot, asking questions and providing increased visual and vocal attention for appropriate responses. Another human is exclusively the *model* for the parrot's behavior and simultaneously the parrot's *rival* for the attention of the principal trainer. So, for example, the trainer says, "What's your name?" and the human model/rival responds, "My name is Lora." Such human interchanges are similar to duets observed between parrots in large aviaries (Mebes 1978). Todt's parrots learned the model/rival's response often in less than a day, in striking contrast to the slow and sparse acquisition in operant paradigms (compare Grosslight et al. 1964; Grosslight and Zaynor 1967). The rapidity with which Todt's birds acquired human speech was impressive, but the phrases he used did not allow him to show if a bird understood their meaning. That is, words and phrases did not refer to specific objects or actions, such as "tickle," to which an experimenter could respond by scratching the bird's head. Thus Todt's birds may have learned a human-imposed form of antiphonal duetting (an elaborate form of contact calling for interacting with social peers; Thorpe and North 1964; Thorpe 1974) or a simple conditioned response (e.g., Lenneberg 1971, 1973). Also, Todt's parrots vocally interacted solely with their particular trainer and learned only the phrase or sentence spoken by the model/rival, never that of the principal trainer. Todt's intent, however, had been not to train birds to communicate meaningfully with humans, but only to determine optimal learning conditions. Subsequent studies showed that another mimid, the starling *(Sturnus vulgaris),* acquires human utterances only in conjunction with social interaction (West, Straud, and King 1983), and that white-crowned sparrows *(Zono-trichia leucophyrs)* learn allospecific song (e.g., of a strawberry finch, *Amandava amadava*) from live but not taped tutors (Baptista and Petrinovich 1984, 1986).

Social interaction, therefore, seemed important for learning, but none of these studies had established actual communication between humans and birds. In addition, scientists wondered about the *extent* of social interaction necessary for learning communication skills. At present, some researchers argue that only minimal input is needed even for children to learn to communicate appropriately (Goldin-Meadow 1997, but see Menyuk, Liebergott, and Schultz 1995; Locke and Snow 1997); and when I began my research, isolated white-crowned sparrows had been shown to learn conspecific song after hearing only a limited number of taped renditions (12 songs/day for 21 days; Petrinovich 1985). Why then should we be concerned with social interaction? The reason is that social interaction is a part of social context and we need to determine its specific role in learning, particularly *allospecific* learning. Context includes, for example, the quality and quantity of input during interaction, the relative status of individuals who are interacting, the environment in which interaction occurs, and the feedback a learner receives for attempting the behavior it is learning (Pepperberg 1993b). Interaction thus may not provide all components necessary for learning to communicate, but may provide components that facilitate such learning. Two examples illustrate this point (Pepperberg 1997).

First, birds tutored in a laboratory may not understand how to use their learned vocalizations. We now know that birds learn not only what to sing but also when to sing and the appropriate context for specific songs (King and West 1983, 1989; Kroodsma 1988; Spector, McKim, and Kroodsma 1989); such learning occurs when birds are taught in a social context (review in Brown and Farabaugh 1997). We also know that wild birds respond less vigorously to songs of laboratory-reared conspecific isolates than to songs of wild conspecifics (see, e.g., Thielcke 1973; Shiovitz 1975; Searcy, Marler, and Peters 1985). Something appears to be missing in the overall singing behavior when it is learned in isolation in a laboratory.

Second, social interaction may not be the reason learning occurs but rather may facilitate learning or modify the speed and amount of learning; moreover, the type and amount of interaction may be important (see, e.g., Pepperberg 1985). For example, positive correlations often exist between high model status and the rate and amount learned in both humans (Mischel and Liebert 1967; Bandura 1977) and birds (Payne 1978, 1981, 1982, 1983; Mundinger 1979; Baptista and Morton 1982; Snow and Snow 1983; McGregor and Krebs 1984).[3] Boys imitated actions of camp mates regarded as having high status by their peers far more often than actions of other group members (Lippitt, Polansky, and Rosen 1952; cf. Dollinger and Thelen 1978); birds may react similarly to different song tutors (Payne and Groschupf 1984). And if no single model seems "superior" and modeled patterns are diverse, both humans (Bandura 1977) and birds (see, e.g., Laws 1994) may *combine* elements of modeled patterns to develop a repertoire. Furthermore, a live tutor may simply be more effective (Pepperberg 1997): Certain birds learn more from live tutors than from tapes; some birds choose to learn songs of live tutors with whom they can interact rather than songs presented by tape or noninteractive tutors (Waser and Marler 1977; Kroodsma 1978; Todt, Hultsch, and Heike 1979; Payne 1981; Kroodsma and Pickert 1984b; note Brown and Farabaugh 1997). Data showing, for example, that juvenile Bewick's wrens *(Thryomanes bewickii)*, ostensibly before learning their final adult song pattern, countersing with adults using "nearly perfect renditions of the adult songs" (Kroodsma 1974:360), suggest that the effect of social interaction on song acquisition is not a laboratory artifact (see Baptista 1983 for comparable data on white-crowned sparrows).

Clearly, social interaction is important for learning, but is not the only factor involved, and we must return to issues of social context and type of input. Todt (1975a), for example, varied neither types of input nor contexts in which his birds were taught. Might these factors also be important for learning a communication code? Researchers as diverse as Piaget (1952), Vygotsky (1962), and Bandura (1971a) had described how intellectual, environmental, and social contexts affect human learning; might context affect a bird's ability to learn to communicate with humans?

Contextual Effects on Learning

For Piaget (1952), all intellectual development arises from a subject's continuous interaction with its specific environment. Development proceeds as novel experiences are assimilated into and modify pre-existing patterns, which themselves were formed during the subject's reactions to prior experiences. A crucial factor is that the subject's environment provides experiences that encourage such "assimilation" and "accommodation" (see, e.g., Doré and Dumas 1987). Though not directly stated by Piaget, an idea inferred from his writings is that a concept or behavior is more likely to be assimilated if it has functional value for the student, particularly if this functionality is demonstrated explicitly (Wadsworth 1978; see Bandura 1971a; Merrill et al. 1996). For example, teaching the concept of color might proceed more quickly if a subject is shown that different colors represent different flavors of candy: The subject might be motivated to distinguish among differently colored candies and to learn to label red to obtain the cherry flavor, and yellow to refuse the lemon.

A somewhat different but complementary idea has been inferred from Vygotsky's writings (1962, 1978): that "assimilation" and "accommodation" involve use of intellectual skills developed in one context to solve problems in a new one (e.g., see Rozin 1976; Rogoff 1984; Wertsch 1985). A crucial factor is then the existence of appropriate contextual support, or "scaffolding" (Bruner 1977): a bridge showing *how* to transfer skills between familiar and novel situations by explicit demonstration or emphasis of features shared by the two contexts (Rogoff and Gardner 1984). Scaffolding may involve comments or questions such as "What did we say when we saw Grandma yesterday?" when a child sees a neighbor and a parent is trying to instill appropriate greeting behavior. Similarly, Bandura (1971a) suggests that learning is most effective when subjects observe demonstration of—and themselves practice—novel, targeted behavior in a familiar context. Thus a child might learn a new shape most efficiently if the teacher uses a familiar object of that shape and presents the task in the context of previously learned shapes and familiar objects.

This idea makes a strong case for involving environmental context in the learning process, particularly with respect to communication: Communication is, after all, a social process, and thus acquisition, as noted above, should occur most readily in a social situation. Consequently, researchers conducting studies to determine how organisms learn to communicate must delineate the exact role (or roles) played by social input *and* its context, and must search for specific mechanisms through which such input can influence learning. In particular, the material discussed above suggests that learning to communicate—acquiring the ability to convey intent and respond to the intent of other individuals (see Smith 1977)—whether the communication is between humans, between birds, or between humans and animals, requires more than simple interaction with a tutor. I have previously argued that for true communication to develop, tutoring must expose a student to three components of the desired communication code (Pepperberg 1991): (1) the code's *semantics* (the meaning of its individual elements), (2) its *syntax* (the appropriate rules for combining its elements), and (3) its *pragmatics* (how the code is used and how its use affects other individuals). How do these components affect learning and what input provides all three components? Determining how different input affects learning can be simplified if a conceptual framework already exists that identifies the critical factors to be studied and provides a prescription for teaching. I believe that human social psychology, in the form of social modeling theory (see, e.g., Bandura 1971a, 1977), provides just such a framework for studies of communication (Pepperberg 1986a,b, 1988c).

Social Modeling Theory and Its Application to Birds

Social modeling theory was the outcome of social psychologists' successful attempts to determine the underlying mechanisms of what they considered real-world learning, in contrast to the behavioristic associationist paradigm. Social psychologists, for example, reasoned that an animal whose survival depended upon differential reinforcement of trial-and-error learning (the basis for the operant paradigm) would rarely live long enough to learn the task in question (Bandura 1971a); they posited similar bases for cultural learning in humans. They proposed that "provision of models not only serves to accelerate the learning process, but also, in cases where errors are dangerous or costly, becomes an essential means of transmitting behavior patterns" (Bandura 1963:54). Other aspects of modeling theory devolved from these same researchers' efforts to devise procedures to enable humans to overcome strong inhibitions or phobias; the idea was that by determining how the toughest learning problems were resolved, scientists could make inferences about general learning mechanisms (Bandura 1971b). Eventually these ideas were used to teach aspects of communication to humans (see, e.g., Brown 1976; Snow and Hoefnagle-Höhle 1978; Pepperberg 1981, 1985; Fey 1986).[4]

Social modeling theory systematically identifies specific contributions to the learning process that may not otherwise be easily distinguished. The theory encompasses several levels of interaction (see, e.g., Pepperberg 1993b). One level separates learning situations into three functionally distinct categories: two types of active input—social modeling and social interaction—and the passive phenomenon of observational learning. A second level involves determining the optimal form of input for a given type of learning. For the purposes of this chapter, the modeling situation provides the best examples for explaining the theory.

Within every aspect of the modeling paradigm, the learner is an active force in the learning process, and trainers motivate as well as model. Unlike a subject in an operant paradigm, who learns to respond to a few simple stimuli (e.g., a ball, a block) with actions acquired by trial and error (e.g., point to one item rather than to another) and is rewarded with an unrelated item (e.g., food), a learner in a modeling paradigm uses the modeled act and the modelers to direct the course of action. Optimally, modelers signal which environmental aspects should be noted, emphasize common attributes—and thus possible underlying rules—of diverse acts, provide contextual reasons for the acts, and demonstrate consequences of the acts (note the parallels with the work of Piaget and Vygotsky). An example is a modeled interaction for a child who lacks language, such as an autistic individual. The child is shown a tray of objects and observes the following interactions between two trainers: Trainer 1: (points to a block) "What's this?" Trainer 2: *"Block."* Trainer 1: "Yes, that's a *block.* Here's the *block.*" The block is given to Trainer 2, who plays with it for a few moments. The roles of questioner and receiver are then exchanged; later, a ball becomes the targeted object in the session. The interactions clearly depict speaker-listener-respondent relationships, in which both an action and the consequences of the action are clearly demonstrated by an interactive tutor and model.

Social modeling theory thus emphasizes how attention, comprehension, and motivation affect learning. The theory consists of a set of principles—the second level referred to above—that describe the optimal form of social input for any type of learning (Bandura 1971a, 1977). Because some principles presume the existence of cognitive abilities (such as symbolic coding, cognitive organization, and mental rehearsal) that cannot be unquestionably attributed to all nonhumans (cf. Griffin 1985), not all principles may be relevant for animal studies. I believe, however, that a subset of four principles is clearly applicable to avian vocal learning (see, e.g., Pepperberg 1986a,b).

One principle states that the student's level of competence must be taken into account (note Jouanjean-L'Antoëne 1997). Human children, for example, most easily and most often imitate whatever is just slightly beyond their current abilities (Piaget 1954a; Ryan 1973; Scollon 1976; Nelson 1978, cited in Krashen 1983; Krashen 1980, 1982; Kuczaj 1982a,b, 1983; Masur 1988). Interactions that model a new behavior that differs only slightly from an existing behavior or that encode only slightly novel information are most easily learned. Thus children generally acquire additional color labels more quickly than their first color labels. Some songbirds seem to respond similarly, in that they more easily acquire allospecific vocalizations that resemble their natural songs in, for example, tonal pattern or syntax (see Pepperberg 1997 for detailed examples).

A corollary is that for learning to continue, tutor/models must constantly adjust their demonstration to take into account—and continue to challenge—a student's increasing knowledge. Finding the appropriate level of input can be difficult: Input that is too simple may be ignored because the learner loses interest; input that is too advanced and not understood may similarly be ignored. Thus trainers and student must work in concert. Trainers must interact with the student, determine the student's current level, and make upward or downward adjustments; the process is recursive and continuous. Humans do learn most effectively from tutors who actively adjust their input (see, e.g., Bruner 1983; Dunham, Dunham, and Curwin 1993; Tomasello 1992; Menyuk, Liebergott, and Schultz 1995; Peters 1996; Landry et al. 1997). For songbirds, evidence is circumstantial, because we do not know if adults intentionally teach song to juveniles.[5] A female cowbird *(Molothrus ater ater)* does, however, actively and continually direct male singing by a wingstroke display (West and King 1988), but the extent to which she adjusts her input is not easily analyzed (see, e.g., Caro and Hauser 1992). Researchers have also studied whether different types of adult-juvenile interactions and different models presented by adult male-female pairs during the juveniles' song acquisition period affect learning in birds like zebra finches *(Taeniopygia guttata;* see, e.g., Williams 1990; Williams, Kilander, and Sotanski 1993; Mann and Slater 1994, 1995; see also Zann 1997), but much remains to be learned.

A second principle of social modeling theory states that modeling must help the student understand how new material relates to current problems and what advantage is conferred by learning new material (again, note the parallels with Piaget and Vygotsky). Thus training is most effective when two conditions are met: (1) the student sees and then practices the targeted behavior under conditions similar to those in its regular environment, and (2) the appropriate use and consequence of the behavior are explicitly demonstrated (Bandura 1971, 1977; Brown 1976; Harris et al. 1986; Pierce and Schreibman 1995; note Schwartz and Terrell 1983; Baptista and Petrinovich 1984, 1986; Payne, Payne, and Doehlert 1984; Lock 1991). Thus, for example, a child who is shown exactly how to request a toy, or a bird that sees exactly how a particular vocalization is used in an aggressive encounter, are both likely to learn more readily than if they were in situations without such demonstrations.

A third principle states that the more intense the interaction between a student and its models, the more effective is the training. Intensity—the extent to which tutors arouse a response in a student (Bandura 1977; Locke and Snow 1997)—is determined from direct observation of the interactants (e.g., by recording emotional reactions) or from indirect measures (e.g., blood pressure or hormone levels). One implication, supported by data reviewed in Pepperberg and Neapolitan (1988), is that, for both humans and birds, intense interaction requires one or more tutors per student. Of course, increasing the intensity of interaction may not always increase learning: Overly nurturant models may inhibit learning by preventing a student from experimenting on his or her own (see Rogoff 1990; note the tie-in to the corollary concerning challenging a student intellectually); overly aggressive models may arouse fear or counter-aggression strong enough to block processing of any input (see Casey and Baker 1993 for a possible example for white-crowned sparrows). For some birds, such as zebra finches, intense interaction may affect which tutor is chosen, not necessarily how much is learned (Jones and Slater 1996).

The fourth principle states that if inhibition or resistance exists toward learning, the first three principles are even more important. The fourth principle is particularly relevant to what I call *exceptional* communication (Pepperberg 1985): communication characterized by vocal learning that is unlikely to occur in the normal course of development. For birds, exceptional learning can take two forms. One is use of allospecific vocalizations by subjects generally expected to acquire functional use of only conspecific vocalizations (e.g., contextual use of the song of unrelated species; see Baptista 1988 and Baptista et al. 1981 for data on exceptional song learning in, respectively, a wild song sparrow, *Melospiza melodia,* and a Lincoln sparrow, *Melospiza lincolnii*). The second is age-independent acquisition of vocalizations in species generally recognized as having a limited "sensitive phase" for vocal learning (for laboratory studies on white-crowned sparrows, see Baptista and Petrinovich 1984, 1986; Petrinovich 1988; Jones, ten Cate, and Slater 1996; cf. Marler 1970; for possible parallels in the wild, see Baptista 1985; Baptista and Morton 1988; review in Baptista and Gaunt 1994; cf. Nelson, Marler, and Morton 1996; Nelson 1998). The term "exceptional" implies some resistance toward acquiring the targeted behavior. Thus, for exceptional learning to occur, social modeling theory predicts that tutor/models be even more attuned to the student's level, that interactions be even more intense, and that demonstrations be even more explicit as to real-world uses and consequences of a targeted behavior than during normal learning. This principle has been important for analyzing exceptional learning in some oscine birds (see, e.g., Neapolitan and Pepperberg 1988; Pepperberg and Schinke-Llano 1991), and we'll see how it is similarly critical for understanding how I adapted social modeling theory to teach a Grey parrot to communicate with humans and to use this communication to examine his cognitive abilities (e.g., Pepperberg 1990d).

Social Modeling Theory and Real-World Learning

Social modeling theory thus provides a framework for devising optimal training procedures, but does not delineate how its tenets are to be put into practice. For example, the theory does not define the specific factors that characterize optimal input. Based on the principles described above and the ideas of Piaget and Vygotsky, optimal input should (1) correlate well with specific aspects of an individual's environment (be "referential"; Smith 1991), (2) have functional meaning relevant to the individual's environment (also known as "contextual applicability"), and (3) be socially interactive. Although I have described these factors in some detail previously (e.g., Pepperberg 1993b, 1998), they bear reexamination here.

Reference. Reference is, for the most part, what signals "are about" (Smith 1991). Reference concerns the direct relationship between a signal and an object or action. Reference is not always easily determined. For example, "ape" generally refers to one of a small set of nonhuman primates (what Smith [1991] calls an "external referent"), but may also refer to an action, as in to "ape," or imitate, a behavior. Similarly, a bird that emits an alarm call may refer either to the predator or to the action it is about to take (or both). Thus not all information contained in a signal involves a single referent. The more explicit the referent of a signal, however, the more easily the signal appears to be learned.

Functionality. Functionality (contextual applicability) involves the pragmatics of signal use: when a signal is to be used and the effects of using information in the signal. Explicit demonstration of functionality shows when using a signal is advantageous and the specific advantage gained by its use. The way a signal is used and its effect on recipients may depend upon environmental context; for example, a comment such as "My, don't we look nice today" will have one meaning and effect for a little girl in a party dress and a different meaning and effect for a hungover friend. Functionality also helps define reference; that is, context defines "ape" as a noun or verb. The more explicit a signal's functionality, the more readily the signal appears to be learned.

Social interaction. Social interaction has, as noted above, three major functions that can be clarified by examples. First, social interaction can highlight which environmental components should be noted; a subject can be directed to an object's color to learn color labels ("Look at the blocks. The *color* of this one is *blue;* the *color* of that one is *green.*"). Second, social interaction can emphasize common attributes—and thus possible underlying rules—of diverse actions ("*Give* me the *ball*" versus "*Give* me the *block*"). Third, social interaction allows input to be continuously adjusted to match the receiver's level ("Yes, you found a block in this pile of toys! Now can you find the *green* block?"). Interaction may also provide a contextual explanation for an action and demonstrate its consequences ("I don't know which toy you want . . . do you want the *ball* or the *block?* Tell me which one you want, and you can have it."). Interactive input thus facilitates learning.

In sum, reference and functionality refer to real-world use of input,[6] social interaction highlights various components of input, and all are necessary for meaningful learning. I thus reasoned that to teach a bird to communicate with humans, my training procedure needed to take these factors into account. The critical point, however, was my hypothesis that a parrot's acquisition of a human-based code was a form of exceptional learning: I believed that despite these birds' abilities to reproduce all sorts of sounds, some strong inhibition existed toward learning to use allospecific sounds in a functional manner; I further believed that, to overcome this inhibition, training would have to be carefully adjusted to the parrot's abilities and include intense interactions and extremely clear demonstrations of reference and functionality. I decided that the best approach would be to modify Todt's technique. He had demonstrated the effectiveness of social interaction; what if I adapted his method to incorporate referentiality and functionality?

A New Version of the M/R Technique

My training system, because of its similarity to Todt's, is also called the M/R technique. In my procedure, however, an interaction is not only modeled; it also involves three-way interactions among two human speakers and the avian student. Because almost all the data in the following chapters were obtained by using M/R training, I provide details of the procedure, although the material is available elsewhere (e.g., Pepperberg 1981, 1990d).

During M/R training, humans demonstrate to the bird the types of interactive responses that are to be learned. In a typical interaction, the bird is on a jungle gym, its cage, or the back of a chair, and observes two humans handling some objects in which he has already demonstrated interest (perhaps has used them as preening implements). While the bird watches, one human "trains" the second human. The trainer presents an object, asks questions about the object (e.g., "What's here?" "What color?" "What shape?"), and gives praise and the object itself as a reward for a correct answer. Unlike Todt's procedure, my technique thus demonstrates referential and contextual use of labels for observable objects, qualifiers, quantifiers, and, on occasion, actions.[7] As in Todt's procedure, the second human is a *model* for the bird's responses and a *rival* for the trainer's attention. The model/rival occasionally errs (produces garbled utterances, partial identifications, etc., that are similar to mistakes being made by the bird at the time). Disapproval for an incorrect response is demonstrated by scolding and temporarily removing the object from sight. Because the human model/rival is, however, encouraged to try again or talk more clearly (e.g., "You're close; say better"), the procedure also allows the bird to observe "corrective feedback," which also apparently assists label acquisition (see Goldstein 1984; Vanayan, Robertson, and Biederman 1985; Moerk 1994).

As part of the demonstration of functionality and relevance, the M/R protocol that my students and I use also repeats the interaction while *reversing* the roles of trainer and model/rival, and occasionally includes the bird in interactions. Thus, unlike Todt's subjects (note Goldstein 1984), our birds do not simply hear stepwise vocal duets, but rather observe a communicative process that involves reciprocity: We show that interaction is indeed a two-way street in that one person is not always the questioner and the other always the respondent, and how the process can be used to effect environment change. An excerpt from a training session on the label "five" is presented in Table 2.1, and the physical arrangement for a different session is depicted in

Table 2.1. Excerpt of an M/R training session, 1/20/81. The aim of the session was to review and improve pronunciation of the label "five." B refers to Bruce Rosen, one of the secondary trainers, I to the principal trainer (me), and A to the parrot Alex. This portion of the session lasted approximately 5 minutes. (After Pepperberg 1988a.)

I: (Acting as trainer): Bruce, what's this?

B: (Acting as model/rival): *Five* wood.

I: That's right, *five* wood. Here you are . . . *five* wood. (Hands over five wooden craft sticks. B begins to break one apart, much as Alex would.)

A: 'ii wood.

B: (Now acting as trainer, quickly replaces broken stick and presents the five sticks to Alex): Better . . . (briefly turns away, then repositions himself in visual contact with Alex) . . . how many?

A: No!

B: (Turns from Alex to establish visual contact with the PI.) Irene, what's this? (Presents sticks.)

I: (Now acting as model/rival): 'ii wood.

B: Better . . . (turns, then resumes eye contact) . . . how many?

I: *FIVE* wood (takes wooden sticks) . . . *five* wood. (Now acts as trainer, directs gaze to Alex and presents sticks to him) . . . how many wood?

A: Fife wood.

I: OK, Alex, close enough . . . *fivvvvve* wood . . . here's *five* wood. (Places one stick in the bird's beak and the others within his reach.)

Figure 2.1. I surmised, on the basis of Bandura's (1971a,b, 1977) studies of interactive modeling, that Todt's failure to demonstrate role reversal between trainer and model/rival caused some of the problems (at least for my goals) with his results, and explained why his birds could not transfer their responses to anyone other than the particular human who posed the questions, and why his birds never learned both parts of the interaction. In contrast, my birds respond to, interact with, and learn from all their trainers.

Three actions by trainers ensure that our birds indeed attend to sessions; these actions are consistent with the principles of social modeling theory (Pepperberg 1992b). First, trainers adjust the level of modeling to match a bird's current capacities. If, for example, a label being trained ("wool") resembles one already in the repertoire ("wood"), trainers praise but do not reward the bird's probable initial use of the existent label and clearly demonstrate how the two labels differ with respect to both sound and referent. Trainers then adjust rewards as a bird practices its utterances to challenge it to achieve correct pronunciation. Second, a bird must be motivated to obtain items used in train-

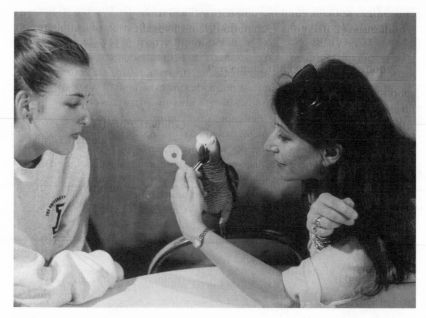

Figure 2.1. Typical arrangement of two trainers and a parrot for an M/R session.

ing. Trainers working on a numerical task who choose, for example, corks rather than keys are more likely to engage one bird's attention, whereas another bird might prefer keys. Third, trainers must act as though they themselves find the task interesting. A bird is less likely to ignore the session and begin to preen if the emotional content of the trainers' interactions suggests that there is real relevance to the task, and trainers who actively engage the bird in the task are more likely to be successful.

The M/R technique is the primary method for introducing new labels and concepts and for shaping correct pronunciation, but another procedure helps clarify pronunciation (Pepperberg 1981). Because this technique does encourage some imitation, it is used only after a bird begins to attempt a new label in the presence of a new object—after the bird makes some connection between sound and object. We present the new object along with a string of "sentence frames"—phrases like "Here's *paper!*" and "Such a big piece of *paper!*" The target label, "paper," is consistently stressed and is the one most frequently heard, but *not* as a single, repetitive utterance. The target label is also consistently at the end of the phrase; conceivably parrots, like humans (see, e.g., Lenneberg 1973) most easily remember ends of word strings. This combination of nonidentical but consistent vocal repetition and physical presentation of the object resembles parental behavior for introducing labels for new items to very young children (Berko-Gleason 1977; de Villiers and de Villiers 1978; review in Moerk 1996), and appears to have two effects: a bird (1) hears the label employed as it is to be used in normal, productive speech; and (2) learns to reproduce the emphasized, targeted label without associating word-for-word imitation of its trainers with reward.

On occasion, birds experiment with labels in their repertoire and produce novel vocalizations. To encourage such recombinations of, or variants upon, parts of labels and enlarge the referential repertoire, we reward these utterances (when possible) with appropriate objects and use a variant of the M/R technique to associate the novel vocalizations and objects. The technique, called "referential mapping" (Pepperberg 1990c), is discussed in detail in Chapter 13.

Thus by integrating social modeling theory with the ideas and results of scientists as diverse as Piaget, Vygotsky, and Todt, I devised training procedures for teaching a parrot to communicate with humans. I now had to ensure that my bird was using English speech in a truly communicative manner and was not simply responding to external cues, such as how a trainer held an object, to a specific object, or to other irrelevant circumstances. I thus devised testing procedures to avoid such cuing. As with the training procedures, this material is described in detail elsewhere (e.g., in Pepperberg 1981, 1987a,b,c), but needs to be repeated so that the findings I subsequently report will be seen as truly reflecting the abilities of Grey parrots.

Testing a Parrot for Communicative and Cognitive Skills

Before we begin testing, we must be able to identify a bird's vocalizations with some degree of confidence. While learning a label, a bird will often produce combinations of human phoneme-like sounds that resemble the targeted label in cadence and length but that are not completely recognizable (Pepperberg 1981). The length of time needed to develop a recognizable utterance from these initial vocalizations indicates only how long it takes a bird to learn to manipulate its vocal tract appropriately (see Chapter 16); sometimes comprehension precedes clarity and sometimes clarity precedes comprehension. The criterion for initiating testing (and generally for ending M/R training) is thus based on the clarity of a bird's speech, and not accuracy of labeling during training. For testing to begin, a vocalization must be recognized by multiple trainers in blind trials (ones in which listeners do not know what the bird is being asked to identify) with more than

90% interobserver agreement. Blind trials ensure that listeners do not recognize a vocalization merely because it is "expected" in the presence of the object (e.g., do not mistake "wood" for "wool"). Interobserver agreement ensures that vocalizations are clear enough to be understood by anyone. Note that a bird can misidentify an object and still satisfy this criterion; if it says "wood" to a woolen pompon and everyone agrees that the label is clearly "wood," we begin testing on "wood." Thus we separate the effect of our procedures on a bird's ability physically to emit a label from the effect on its ability to associate label and referent. Only when the former skill is considered satisfactory is the latter skill tested (Pepperberg 1981).

Test situations include precautions to avoid trainer-induced cuing, defined as any behavior by a trainer that allows birds to guess a correct answer without really understanding what the test is about (Pepperberg 1981). One precaution is a design that prevents either bird or examiners from predicting which questions (and thus which answers) will appear on a given day. Tests are constructed as follows: I list all object labels to be examined and a student not engaged in testing randomly orders the list.[8] The list includes all current labels in the bird's repertoire in addition to those being tested. Two to five questions are then presented intermittently during training on current (and thus unrelated) topics for several days until all questions are asked.[9] While targeted object labels are tested, for example, a bird can be trained on sequential number recognition and color labels. Training questions are thus as likely during test sessions as a test question, and a specific test object may appear only once or twice per session and its appearance cannot be predicted. A second precaution against cuing is to ensure that tests on a label are conducted by a student who never trained that label. Students question a bird on several different labels that they have not trained, so that the presence of a specific student cannot cue a bird as to which label is to be tested.

I also had to avoid something known as "expectation cuing": the possibility that the limited context of single-topic tests (a homogeneous set of questions) might lead a subject to ignore all but a small subset of responses and thus facilitate a better performance than expected based on its actual knowledge of a topic (see, e.g., Terrace 1979a; Pepperberg 1983a). An example may clarify this problem. Remember being in high school and preparing for a geometry test? The night before the test, did you even look at French or chemistry homework? The day of the test, could you conjugate a French verb or write equations for chemical reactions? Not likely—you had focused on one set of questions, had crammed for the test, and were likely to do okay—

even though you really might not be very good at geometry. I must ensure that my birds can't do the same thing. Thus I intermingle different test questions (e.g., "How many?" "What matter [material]?") during training on other topics so that my birds are never queried on a single topic (such as object labels) in a session, and more important, are never tested successively in one session on similar questions ("What's here?") or on more than three questions that have a particular correct response (e.g., "cork"). Only novel objects are used for testing, and identical items are never used for similar questions (e.g., birds are asked about material and colors of wool and paper that differ from training items and from question to question). A question is repeated in a session only if the initial answer is incorrect. Thus, though the range of correct responses to, for example, "What's here?" or "What matter?" was limited initially to a few object or material labels, in any session a bird still had to choose from among several possible responses to object and material questions to be correct (Pepperberg 1981).

Concurrent work on several tasks is also necessary because birds become restless during sessions on a single task. They cease to work, begin preening, or interrupt with requests (such as "Want tickle," "Want cork"); such behavior impedes testing.[10] Averting inattention to a presented object is particularly important if a bird has not yet learned to separate identifications from requests by appending "want" appropriately to object labels. In the absence of "want," trainers cannot distinguish misidentifications from requests for more desirable objects. In such instances, test scores might decline for reasons unrelated to a bird's competence (Pepperberg 1988a, 1990d).

Once a bird understands and uses "Want X" (prefaces requests with "I want . . ." or "Wanna . . .") and uses object (and attribute) labels, such as "blue wood," for identifications (Pepperberg 1988a), we modify our procedures slightly so that a bird will work with objects in which it has little interest. A bird that uses "want" can request alternative objects for its reward (Pepperberg 1987a,b), which ensures that inappropriate responses are not requests for preferred items. About 75% of the time a bird does indeed want the requested object: The bird eats the walnut, uses the key to scratch its neck or the cork to clean its beak, and refuses substitute items, usually by saying "no" and repeating the original request. Thus, during testing, the response "cork" to a nail is considered an error, whereas "Want cork" is taken as a valid (if interruptive) request rather than a mistake. The procedure still stresses referentiality, because a bird never receives a cork directly after identifying a nail; only after a bird produces the correct response to a targeted object and rejects that object will trainers accommodate requests for an alternative item.

The actual testing protocol is fairly straightforward and involves a correction procedure. The number of times objects are presented to a bird depends upon its accuracy, which is determined as follows. The examiner (a student trainer) presents a targeted item to the bird. The principal trainer (me) sits so that she cannot see the bird, examiner, or object being presented, and thus cannot cue the bird as to the correct response. The examiner (who, remember, has never trained the bird on the test item) asks a test question, to which the bird responds. The principal trainer then repeats what she heard the bird say. This repetition prevents the examiner from accepting an indistinct, incorrect utterance that is similar to an expected, correct response (such as "or" for "cork"). Interpretation of a bird's response is not likely to be influenced by hearing the type of question. Post-test transcriptions of contextless tapes of a bird's responses agree to within 98.2% of original evaluations (Pepperberg 1992a). If what the principal trainer heard is correct (e.g., the appropriate object label), the bird is given praise and the object, and no additional presentations occur—only a "first trial" exists. If a response is incorrect or indistinct, the examiner removes the object, turns his or her head (a brief "time-out"), and says, "No!" The examiner then implements a correction procedure in which the misnamed object is immediately (re-)presented up to three times or until a correct identification is made; errors are recorded.

Birds thus find that an incorrect response, such as substituting the label of a more desired item for the one presented, is fruitless; instead, correct responses allow a bird to proceed to a preferred item. Because immediate re-presentation of objects during a test occurs only when response to the initial presentation is incorrect, the protocol penalizes a win-stay strategy. Incorrect repetition of a previously correct response (e.g., the name of the previous item) elicits no reward.

Interestingly, sometimes an examiner must be corrected. In about 5% (1 in 20) of test trials (particularly during student exam periods), an examiner will err and scold a bird for a correct response. One of our birds will repeat its correct response, despite our procedures, which encourage a lose-shift strategy. The examiner then usually recognizes his or her error, and the bird gets rewarded. Although this is not a formal blind test, it produces the same results.

The issue of cuing is particularly tricky (Sebeok and Umiker-Sebeok 1980), and I am often asked why I don't prevent examiners from knowing the content of the test. I have, in fact, hidden test objects inside a box with a movable flap (Pepperberg and Brezinsky 1991); the closed box was given to the examiner, who could then lift the flap so that only the bird could see the test item (for details and a diagram, see Chapter 12). We eventually obtained data via this method, but trials were incredibly time-consuming. The bird ignored the box, or attempted to chew it apart and ignore its contents. When examiners hold test objects right at the birds' faces, in contrast, we obtain data most efficiently, because the bird cannot ignore what is being presented. Note, too, that the physical look of certain objects (e.g., Play-Doh and rawhide items) may not provide sufficient information for a bird to discriminate their material. Thus the bird must be allowed to touch the item with its tongue but not obtain the item, because the item is the reward for correct labeling. Only by having a trainer manipulate the item can we provide the appropriate level of interaction with these objects and, to avoid cuing the parrot about these specific objects, we must allow similar manipulations on a random basis with all objects. Finally, young birds do not necessarily sit quietly on perches during tests; they attempt to climb down, jump to the trainers for preening, and so forth. Trying to get these birds to attend to material in a box under such circumstances is next to impossible, but gaining their attention with a hand-held object works very well.

Note, too, that our oldest subject does indeed correctly label items even if the tester does not know the answer: We have several idiosyncratic labels (e.g., "cork nut" for almond, "banerry" for apple, "truck" for a toy car, "four-corner" for square), and have had repeated instances in which television crew members, unaware of the meaning of these labels, have asked our bird to identify these items; the bird has been correct on 7 of 8 trials. In such trials, the testers cannot possibly cue the bird because they do not know the appropriate label.

Questions also arise as to whether I am truly "blind" during testing: Could I remember answers to specific questions on a given test? Note that to remember the correct answers, I would have to store in memory, usually over the course of 5 to 14 days, not only the answers to each of the numerous (10–12) questions of the type given to one bird on a specific test, but also the answers to test questions that are simultaneously given to three parrots. At present, I would also have to remember such answers for a set of more than 50 different possible arrays. This memory feat would also have to occur in the midst of fast-paced training sessions on sets of completely different topics in which notes were constantly being made of correct and incorrect responses. Such considerations do not include other demands on memory load during these 5 to 14 days. Given how many of us can't remember where we have put our car keys on a given day, I believe that such memory is an unlikely confound in the testing procedure (Pepperberg 1994a).

I report test results two ways, as "first" and "all trials." First trial data are the percentage of first trials that are correct and are used for statistical analyses. For comparison, I report "all trials" scores for each task.[11] The "all trials" score is the total number of correct identifications (i.e., the predetermined number of tested items) divided by the total number of presentations required to elicit a correct response. If a bird cannot produce a correct response, I report only first trial data. "All trials" scores are important because they often suggest trends in the parrot's behavior: A bird who is consistently correct on second trials may, for example, be making what we call generic errors, such as labeling an object's material rather than its color (see Chapter 4 for a discussion).

Using Communication to Study Cognition: Pitfalls and Benefits

I have described how I developed procedures to train a Grey parrot to communicate with humans and the test procedures used to determine if the bird actually understands this human-based code and uses it appropriately. In the next chapters I describe how I've used this code to examine the bird's cognitive abilities, to compare its abilities to those of chimpanzees and marine mammals, and how my students and I are determining how our birds produce human sounds. Studies with juvenile birds are clarifying and confirming the need for reference, functionality, and interaction in training. I end this chapter, however, with a cautionary note about using interspecies communication to study cognition.

Researchers who train animals to use a human-based code in order to study nonhuman cognition often lose sight of the code as an investigative tool and argue instead about the extent to which the code is equivalent to human language (Pepperberg 1993a); that is, they compare human and nonhuman *linguistic* rather than *cognitive* abilities. Thus data and discussions about relative cognitive capacities are lost amidst open-ended debates about the extent to which a particular animal has acquired "language" or how well particular studies or procedures demonstrate linguistic competence (see, e.g., Premack 1986; K. Nelson 1987; Savage-Rumbaugh 1987; Seidenberg and Petitto 1987; Herman 1989; Schusterman and Gisiner 1989; Kako 1999). Such debates are based on the assumption that the extent to which an animal learns and uses such a code indicates its cognitive capacities (Premack

1976; Terrace 1979a). But training techniques and codes themselves vary considerably with respect to the subject species and vary even among projects studying the same species; such variation in experimental design invalidates any comparison of cognitive capacity based solely on facility in using a particular interspecies communication code.[12] It would, for example, be foolish to claim that a parrot that uses a vocal code but does not read symbols exhibits, respectively, a greater cognitive capacity than a dolphin and a lesser capacity than a chimpanzee (Pepperberg 1993a). Nevertheless, even widely divergent codes may, in properly designed experiments, be used to obtain data on cognitive tasks that can be directly compared across species. We will see that although the parrot vocalizes and some chimpanzees manipulate plastic symbols, both have comparable understanding of concepts of similarity and difference (Chapter 5; Premack 1976, 1983). In all the material to follow, we must remember that the power of interspecies communication is best exploited by focusing on how an animal's use of a vocal code facilitates investigations of its cognitive abilities, rather than on what the animal's use of the code itself implies about such abilities.

3 Can a Parrot

Learn Referential Use

of English Speech?

It is characteristic of the behavioristic Zeitgeist of the mid-twentieth century that Mowrer's work has since been cited primarily as evidence that because birds cannot be trained to imitate speech by means of standard operant conditioning methods, their imitation cannot entail any understanding of the meanings conveyed by the words they mimic.

Donald Griffin, *Animal Minds* (1992:171)

With all the training and testing procedures in place, I could begin working with a parrot. The subject, a Grey, came from a pet store in the Chicago area. It was one of eight birds in a cage, and had had no formal training since hatching, approximately 12–13 months earlier. The bird was chosen by the store manager; thus it was a matter of chance that this particular individual became my subject. Although I believe that parrots should be bought from reputable breeders rather than from stores that might be selling smuggled animals or potential disease-carriers, I specifically chose to obtain a bird in this manner because I wanted to show that any success I achieved was not related to my choice of subject; I didn't want anyone to claim that my subject was particularly smart or had specifically been bred for vocal ability. Thus, on June 15, 1977, I acquired Alex, who was to change our perception of the term "birdbrain."

The Environment

Alex was brought to a laboratory in the biological sciences building at Purdue University. The room was sound-deadened so that his input would be relatively free of reverberations. I designed a stimulating environment, but one that discouraged simple mimicry or duetting. He was kept in a cage (62 × 62 × 73 cm) during sleeping hours, but was allowed to explore the laboratory room and the desk on which his cage was kept while trainers were present (~8 hours a day) and formal training and testing were not in progress. Basic food (at the time, a standard seed mix; later, pellets) and water were always accessible. Toys were initially available only during training. At various times throughout the day, Alex received foods (e.g., fruits, vegetables, nuts) that the trainers appropriately labeled, but without the repetition that accompanied formal instruction. He was also trained to respond to a limited number of commands, such as "climb" and "go perch."[1]

Alex initially received one or two formal training sessions per day. Sessions were scheduled for an hour, but were usually broken into short segments based on his attention span. Training was interrupted to give him food or water, to allow him to preen, and so forth. Also, because Alex's rewards were the objects he identified, we had to let him chew, tear, or otherwise manipulate these items for several minutes after a correct identification to maintain his interest in the task.

Note, too, that Alex received additional information about the functional use of labels during modeling sessions: He witnessed how trainers directed each other, for example, to pick up a key rather than a cork from a pile of objects; such interactions occurred either at the beginning of a session or during sessions when we switched to a different label. These interactions demonstrated the effect of comprehending, as well as producing, the labels.

I also monitored Alex's vocalizations between training sessions. Initially, I kept a daily record of all our interactions, tabulating and sequentially numbering his utterances. For each label in his repertoire, I noted if he produced it spontaneously, subsequently correctly identified the corresponding item and received it as a reward, if an attempted identification was unclear or an error, and what happened when I showed him an object. Table 3.1 describes the morning of December 1, 1977. After Alex's repertoire increased beyond about a dozen labels, I could not maintain this form of recording and substituted weekly taping sessions, which I transcribed to chart his progress.

The First Goal: Training Object Labels

My initial goal was to learn whether Alex could label items. My rationale was twofold. First, no one had ever shown, in a controlled laboratory setting, this ability in a bird. Labeling was considered the sole province of humans (Chomsky 1966; Lenneberg 1967), although studies suggested that this skill could be acquired by great apes (Gardner and Gardner 1969; Premack 1971; Rumbaugh 1977). But by a bird?

Table 3.1. Typical daily record for human-parrot interactions when neither training nor testing was in progress. Each number represents an utterance or an action and the order in which the events occurred. (After Pepperberg 1981.)

Object	Object shown[a]	Object named[b]	Correct i.d.		"Close" i.d.[e]	Incorrect i.d.[f]	Spontaneous vocalization[g]
			Reward[c]	No reward[d]			
Paper	2,17,40		3,18,41				1,16
Key	38		39				37
Wood	11,19,35		12,20,28,36				10,28
Hide	22,30	24,26	27,31		23,25		21,29
Peg wood	4,8,14,33	6	9,15,34	7			13,32
Green key	42	44,45				43U	
Other[h]							

a. The object is presented but not named by the trainer; the parrot is asked, "What's this?"

b. The object is named by the trainer during presentation.

c. The parrot correctly identifies the presented object and receives the item as a reward, or a spontaneous vocalization is rewarded. In the latter case, the same number is assigned to "reward" as to "spontaneous vocalization."

d. The parrot correctly identifies the object, but is asked to repeat the identification for emphasis.

e. The vocal response of the parrot to the presented object is a variant of the correct label, e.g., "peh wood" instead of "peg wood" (a clothes pin).

f. The vocal response of the parrot to the presented object is either incorrect or unrecognizable and thus designated U.

g. The parrot names the object in the absence of any action by the trainer.

h. Random noises or unrecognizable attempts at labels.

Second, if Alex acquired labels, certain cognitive capacities could be studied. Might he, like children, over- or underextend categories (Bloom 1973; Brown 1973; Clark 1973; Nelson 1974)? Might he initially overextend "banana" to encompass all fruit? Or underextend "paper" to refer only to the initial exemplars of index card pieces? Such data would provide insights into how another species might categorize the elements in their world (see, e.g., Bates, Thal, and Marchman 1991).

In achieving my goal, however, I faced a problem more basic than simply overcoming the preconceived notions of avian cognitive and communicative inferiority discussed in Chapter 1. The real problem was ensuring that what Alex learned was, indeed, labeling and not some other behavior. Re-analysis of data from the great ape projects suggested that some animals had acquired a skill far simpler than labeling (see, e.g., Terrace 1979a). Thus before examining the results of my study with Alex, I describe possible ways that labels can be used that differ from true labeling. My discussion is not exhaustive; entire books have been written on the subject (e.g., Savage-Rumbaugh 1986). Moreover, unlike other analyses, mine was not to determine the extent to which an animal's labeling behavior is identical to human language, but rather to determine exactly what an animal has learned.

Defining Levels of Reference

We can begin to examine labeling by returning to the discussion of reference, based on research by Smith (1977). As stated, reference is what signals "are about," and the general notion is that a label is the signal that refers to an object (ball), action (hit), or attribute (red). Unfortunately, defining labeling is not quite that simple. A serious problem exists with the study of referential communication—and particularly labeling—in animals: Although scientists may clearly describe an animal's behavior and call that behavior "referential labeling," different researchers mean very different things by the terms "referential" and "labeling." The definition of "referential" is particularly variable, and usually depends upon the discipline of the scientist who is doing the defining. Unfortunately, for each discipline, the definition is strikingly different, and these differences cause considerable miscommunication among scientists.

Let me start with a simple example to demonstrate the subjective nature of a definition. I state, "I'm a native New Yorker." To someone from Manhattan, that sentence defines me as cultured and stylish, as someone who understands the finer points of art, music, and theater, who appreciates fine food and wines. To someone from the Heartland, however, I'm more likely to be defined as boorish, mean-spirited, rude, and as having a dreadful accent. The example is somewhat trivial, but the problem with defining referentiality is not all that dissimilar.

Figure 3.1 shows some terms commonly used as synonyms for referential and the group of people most likely to use these terms to mean referential. The terms are organized top to bottom such that they go from describing simple to complex forms of behavior, and each subsequent term excludes the behavior described in the terms above. Let's look at each of these behavior patterns.

The first entry refers to mimicry, whether physical or auditory. The layperson—say, a pet owner—uses the term referential to describe mimetic actions that, some small percentage of the time, actually do correspond to a sensible action or event. Thus a dog that has learned to move its paw in a human-like manner may perform that action when its owner says, "Wave goodbye," and a parrot may take and eat a cracker after saying, "Polly want a cracker." Describing such behavior as referential is, by almost any imaginable scientific standard, unacceptable, but I bring it up for both completeness and comparison. Note that we say a mockingbird mimics songs of other species—maybe, based on the information I have just conveyed, we should be careful to say "reproduces" these songs instead of "mimics," because we have little understanding of the use the bird makes of these allospecific sounds. If, for example, a mockingbird *consistently* uses robin song to defend its territory against male robins, we cannot classify the mockingbird's communicatory behavior as mimicry; its behavior would fit into one of the other categories I subsequently examine. We must therefore account for context when we classify a behavior, and this point will arise several times.

The differences inherent in the next set of definitions of terms are anything but trivial, and have caused endless confusion in the literature and in discussions among scientists (Terrace 1979a; Savage-Rumbaugh 1986; Greenfield and Savage-Rumbaugh 1991; Savage-Rumbaugh et al. 1993). The problem is that "referential" is used by behaviorists to define what is really associative use, by biologists to define what is really contextual and conceptual use, and by psychologists to define what I term "peri-referential" use; only linguists seem to use the term appropriately. Describing each of these uses in some detail will, I hope, clarify their differences.

Behaviorists (mostly those adhering to Skinnerian precepts) often use "referential" to describe what is merely an *association* between one particular item and one particular action. Association is a widespread phenomenon throughout the animal kingdom; some research-

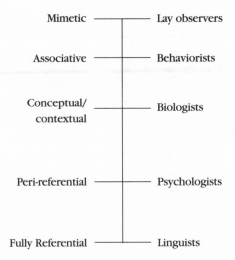

Figure 3.1. Depiction of various levels of "referentiality" and the people most likely to use the terms.

ers (e.g., Macphail 1987) suggest that all learning (including that of humans) can be explained in terms of paired associations. Unfortunately, behaviorists may use the terms "labeling" or "naming," which imply referential use, to describe paired associations. But association is merely the first step toward referential labeling (K. Nelson 1985). Association does not require communicative intent: It is generally performed as a means to an end (obtaining a reward such as a bit of food) rather than as a means to transfer information about the item being "labeled" (Terrace 1979a; Savage-Rumbaugh 1986). Association also almost always lacks generalization; that is, the action often relates to a particular item rather than a class of items (see, e.g., Clark 1973; Carey 1978, 1982; Lock 1980). Let's take a specific example: In an experiment by Manabe, Kawashima, and Staddon (1995), budgerigars learned to make a particular pitched sound to a specific red light and a different pitched sound to a specific green light. The authors claim the birds "named" the colors. The birds had, however, simply associated a specific light hue with a specific sound. Unless the birds could make the same sounds, respectively, to red balls and flowers, and green blocks and lettuce, the birds cannot be said to have "named" the colors, nor can their sounds be said to imply concepts of "redness" or "greenness."[2] Moreover, any communicative intent in the birds' behavior is independent of the concept: The birds make the sound not to convey information about the light, but to receive food.[3] The conceptual understanding and communicative intent missing from simple associations are, as we will see next, some of the defining characteristics *leading up to* referential communication.

The next term, conceptual or contextual use, is generally what biologists mean by referential communication. An animal uses a symbol to indicate the presence of, or to request, a particular type of item or a class of such items. This category of communication probably has the most subcategories. To keep this discussion concise, I will not discuss all possible subcategories but give only a few examples. Certain subcategories, for instance, involve particular responses to foods of different qualities (Elowson, Tannenbaum, and Snowdon 1991; Roush and Snowdon 1994; Caine, Addington, and Windfelder 1995). Other subcategories involve probably the most famous example, vervet alarm calls. Seyfarth, Cheney, and Marler (1980) have shown that a vervet monkey, in the context of being in a group and sighting a particular type of bird flying above, utters a particular call and she or he and other members of the group flee into cover, in contrast to other actions taken in response to other calls. Is the caller specifically referring to a martial eagle? To the action the troop should take? To the outcome if they fail to flee? We have some hypotheses (Cheney and Seyfarth 1988, 1990), but cannot completely rule out multiple explanations (see Dennett 1988; Bickerton 1990; Seyfarth and Cheney 1997). Each call refers to the concept of some kind of danger, in a way more specific (although not entirely dissimilar) to the way, for example, that food calls refer to some hierarchy of something good to eat. Furthermore, aspects of the alarm call can be modified by, for example, the intensity of danger and type of audience to which it is directed (see, e.g., Cheney and Seyfarth 1990). Clearly, a close correlation exists between context and concept. The vervets' communication does not refer to a specific eagle, or even to a particular action (although it does refer to a particular *type* of action); therefore, the connection between call and object is not a simple association, but involves something far more complex. I suggest that the term contextual/conceptual communication be used in any situation in which an animal produces specific forms of communication in particular contexts—for instance, to describe the behavior of warblers that use one song type for what seems to be intrasexual communication and another for what seems to be intersexual communication (Kroodsma 1988; Spector et al. 1989). Based on our current level of understanding—and I stress "current," for what we currently know is conceivably quite limited—such communication is not truly referential. I also suggest that some spontaneous combinations of signing chimpanzees—such as use of "water-bird" on the appearance of a swan (Fouts and Rigby 1977)—fit into the contextual/

conceptual category. The chimpanzee has some concept of what constitutes a bird (wings, beak, etc.), of what constitutes water (wetness, etc.), and the context (interaction with humans) in which labeling occurs. Without further information about how such a term is used, however, we cannot designate it as referential.

The importance of context in the attempt to define "referential" cannot be overstressed. Often, two (or more) somewhat or very disparate notions may be indicated by a single label. As noted in Chapter 2, when humans say "key," for example, they usually mean a noun: a specific metal object used to open a lock. The term may, however, be used as a verb, as to "key in" data. Thus information contained in a signal does not necessarily involve a single referent. In the absence of additional contextual information, we have no idea what is meant by the isolated term.

Most cognitive psychologists are closer to appropriate use of the term referential. They use the term to describe behavior that, however, I believe should be called peri-referential, or proto-referential.[4] Here animals use a sign as a symbol in a way that suggests that the symbol functions as a *mental representation* of an item. Most great apes (and, as I will show in future chapters, parrots) that have been trained to communicate with humans have peri-referential communication systems. Savage-Rumbaugh and colleagues (e.g., Savage-Rumbaugh et al. 1991) and the Gardners (1978) demonstrated that apes use particular signs as symbols for types of objects (e.g., for all apples) and actions (e.g., requests), and different signs as symbols for categories that include the specific objects (e.g., food) or actions. Moreover, these animals do not use symbols only for objects and actions in the here and now. They, like humans, will request an absent object, or an action not currently being performed, and accept that object or action and no other. They may use symbols dispassionately—label a picture of an eagle without fleeing—or separate identification of an object from a request for that item. They generally demonstrate label comprehension as well as production; thus, upon being given the label, the animal can indicate the object in some manner (either by pointing or by telling the questioner something about the indicated item). Such animals use labels to refer to similar but nonidentical items; for example, they can identify the material of any colored or shaped piece of wood without additional training. Similarly, they understand that the label "green" refers to a concept of "greenness"—to beans as well as to training items—and how the label "green" is subsumed into a category whose label is "color." These animals use labels to describe concepts of "same" and "different," "large" and "small" (Premack 1971). Although almost all "animal language researchers" describe such behavior in their animals as referential—and will probably continue to do so—such communication is not fully referential.

Referentiality, as defined by linguists (e.g., Bickerton 1990), requires full abstract use of a symbol. Let's look at an example. Money is a concrete item in one sense, but is useless paper by itself; its worth is in what it represents to the world: what can be done with it, what it means in terms of power and influence. Thus referential use is the ability to talk about qualities of the item, to talk about how you think about the item—the referent—in its absence, to talk about it in future and past tense—but *not* simply in the sense of a request for something not yet present. As far as we currently know—and, again, I stress "currently"—no animal has demonstrated the communication immortalized in a Gary Larson comic, in which two baboons, during a grooming bout, discuss a set of interactions that have happened earlier between two other baboons. Quite possibly, our inability to demonstrate such use has more to do with our incompetence than with animal capacities. Most likely, much of an animal's total signaling goes unperceived, and thus uninterpreted, by humans who study behavior in the wild.[5] In the laboratory, we have yet to develop appropriate means of training and testing such behavior. The point, however, is that until we devise ways to *show* clear referential use in animals, we must be extremely careful as to how we define various types of communication. To call a behavior referential when it probably fits instead into a simpler category not merely causes confusion and miscommunication among scientists, but also may prevent us from determining the animals' actual abilities. We may, for example, settle for describing the simplest behavior if we give such behavior the most complex of labels, and thus miss the full level of competence the animal might demonstrate.

I emphasize, however, that by providing these definitions, by using a human standard for the term "referential," and thus classifying only the most complex behavior as referential, I am *not* claiming that animal communication, whether natural or trained, cannot be referential. I do not accept the argument, stated years ago by a colleague, that language, and by implication, referential communication, will continue to be redefined by linguists as whatever animals cannot be shown to do (Fouts 1973b). As I've said twice already, lack of evidence for truly referential communication in animals is most likely a consequence of *our* incompetence, rather than that of the animals. We face a task not unlike that of Quine's (1960) anthropological linguist, briefly alluded to in Chapter 1, who wonders at the meaning of the label "gavagi" that is shouted as a rabbit runs into view in a tribal village. Does the label refer to the small furry creature itself? To a future meal? To the animal as a symbol of good or bad fortune? Quine argues that the level of reference can never thoroughly be elucidated, even when humans try to decipher other *human* communication systems. To some extent, our everyday experiences, which often include miscommunication with other humans, suggest that Quine's point has some validity. Nevertheless, I believe we can examine certain levels of reference, and what I hope to do by providing these definitions is to clarify which cognitive capacities can be demonstrated when we teach an animal to communicate with humans.

A Parrot's Functional Use of English Labels

After 26 months of training with the procedures described in Chapter 2, Alex acquired a fairly extensive vocabulary (Pepperberg 1981). He used the following English labels to identify 9 physical items: paper, key, wood, hide (rawhide chips), peg wood (clothes pins), cork, corn, nut, pasta (bowtie macaroni). He labeled the colors rose (red), green, and blue. He used the phrases "three-corner" and "four-corner," respectively, to identify triangles and squares, and had functional, if limited use of "nuh" for "no." He used these vocalizations singly or in combinations to identify, request, or refuse more than 30 objects. He identified objects that differed somewhat from training items; for example, although all paper initially consisted of fairly similar pieces of white, unlined index cards, he could identify pieces that varied considerably in size and shape, and, without training, transferred to items such as computer and notebook paper. Hide, wood, cork, and clothes pins ("peg wood") were consistently correctly identified even if these items were considerably chewed and barely resembled their original forms. Moreover, shades of the colored objects differed somewhat as they were, with the exception of the keys, hand-dyed with food colors.[6]

Alex correctly identified, on first try, 80% of all objects presented in over 200 tests; he scored 78% when all responses to all presentations were included. Mastery of a label was defined as identification of its corresponding object at that 80% accuracy. Given our criteria for beginning testing, Alex's label use sometimes improved with time. Table 3.2 presents results of 16 tests, given between October 14, 1977, and November 8, 1977, fairly soon after we began testing "hide." Although Alex's eventual accuracy for "hide" reached 100%, on these tests his accuracy was only 69.6%. Note that accuracy on the earlier labels ranged from 78.7–82.8% (Pepperberg 1981).

Table 3.2. Test results from 10/14/77 to 11/8/77 of all vocabulary items in Alex's vocabulary at that time. (After Pepperberg 1981.)

Object label	Total number of presentations	Responses elicited by presentation of object (number of occurrences)			
		Paper	Key	Wood	Hide
Paper	58	48	2	4	4
Key	61	3	48	5	5
Wood	60	2	2	48	8
Hide	69	12	6	3	48

Acquisition of Color Labels: A Test of "Segmentation" and Transfer

A crucial test of an animal's ability to progress beyond associative use of labels is its ability both to generalize these labels to novel situations and to view the labels as interchangeable units that can be combined in novel, untutored ways to identify novel objects. This latter ability (which subsumes the former) is known as "segmentation" (see Marler 1973; Fromkin and Rodman 1974). Alex's segmentation was first investigated when I introduced color labels (Pepperberg 1981).

Although Alex's segmentation was not initially spontaneous, such behavior appeared with limited instruction, and increased consistently with little or no additional training. Alex was first trained on green key and green wood and then shown a novel piece of green hide. He initially called it "hide," but identified it as "green hide" 60% of the time after only 5 minutes of M/R training; that is, in seven presentations immediately following training, he used "green hide" four times and "hide" three times. His accuracy improved over the next few days. He was then shown a green clothes pin (peg wood) and asked "What's this?" He said "green wood, peg wood," and repeated the combination several times. He paused briefly between the combination of phrases, so that we heard two two-word phrases rather than four words or one four-word phrase. True segmentation would have been a response of "green peg wood." Alex, however, never previously needed to combine more than two labels at one time; we suspected that he did not quite understand what we were asking him to do. Nevertheless, he had attempted to identify this new object by linking phrases already in his vocabulary (Pepperberg 1981).

Subsequently, we elicited apparent segmentation during presentation of red (rose) paper. Alex first called it "rose hide," which was one of the training items (the other being rose key). The object did somewhat resemble hide, and, as noted in Table 3.2, Alex initially confounded paper and rawhide. After chewing the paper for a few seconds, Alex said "paper." We asked "What color?" and he replied "rose." Our further query, "Rose what?" elicited "rose . . . paper," with a second or so interval between the two labels.

We soon collected additional data (Pepperberg 1981). Green cork, blue peg wood, and blue hide were correctly identified when first introduced, but rose cork elicited distress calls.[7] A particularly interesting, if anecdotal, result occurred when we unintentionally asked Alex to label a square piece of paper. At the time, he was being trained on, but had not yet mastered, labels for three- and four-corner wood; he was still being rewarded for uttering something like "squawk squawawk wood." During a test, he kept squawking instead of identifying paper; the questioner noticed that the paper was squarish. Alex correctly identified an alternate, randomly shaped piece, but continued to squawk at the squarish one. We carefully removed all squarish items except the wooden training objects until we could distinguish "three-" and "four-corner" in his speech, and then trained him to identify triangular paper to demonstrate that shape labels could be transferred in the same way as color labels. When we intentionally showed him square paper, he said "paper," but did label it "four-corner paper" after we asked "What shape paper?" On his next six identifications he called it "four-corner paper" four times and "paper" twice.

Novel presentations of other shaped items produced similar results. The first time Alex was given triangular and square pieces of rawhide, he correctly labeled the shape but not the matter. After he chewed the items, however, he correctly identified them with > 80% accuracy.

Contextual/Conceptual Use of Labels

I thus argue that Alex had at this stage acquired contextual/conceptual use of his labels. He did more than associate a particular object and a particular vocalization, but had not yet acquired peri-referential use. He used a symbol—a vocal utterance—to indicate the presence of or seemingly to request not one particular item, but a particular *type* of item (see below). He used vocalizations to transfer information to his trainers about an object, but we could not yet determine if he was commenting on the object or requesting it. He did seem to have some concept of each modifier (e.g., greenness and triangularity) and some limited concept of categories of color and shape, but we did not know the extent to which he could generalize or comprehend the labels he could produce.

Alex's performance was impressive because he was the first bird to demonstrate such abilities, but I was more intrigued by the type of labeling errors he was making than by his success. The reason was that the errors suggested something about how he might be processing information. His most frequent mistakes (and the majority of errors) were of three types: (1) errors of material, in which he identified a relatively novel item that looked or felt similar to a well-known item with the label of that well-known item; (2) initial omissions of color or shape markers; and (3) unclear pronunciations of color markers. The first two types of errors suggested a possible facility for categorization;[8] from the third error, I thought I might infer something about his motivation.

The Beginnings of Categorization?

We assume that animals categorize their world—for example, into food and nonfood; predator and prey—but have little idea of the extent of their abilities and if their divisions are congruent with human divisions. I discuss categorization at length in Chapter 4; for now, the important point is that even at this early stage, Alex's behavior suggested a facility for clumping items. On early tests, for example, rawhide was more often incorrectly identified as "paper" than "key" or "wood," which were his other labels at the time (see Table 3.2). Similarly, half his initial errors on clothes pins ("peg wood") were "wood," and 85.5% of errors on green key were "key." If he were responding randomly, he would have used all labels in his repertoire with equal frequency. Instead, to paraphrase Gertrude Stein, for Alex a key was a key was a key, whatever its shape or color.

We thus tested the extent to which he could identify colors and shapes. This task differed from segmentation, because we were not asking him to *link* color and object (or material) labels to describe novel items; here we wanted to see if he could label attributes of *novel* items. We queried him about items whose colors or shapes, but not materials, he could label. We used red (rose) pens, green plastic paper clips, and variously shaped and colored dog biscuits and buttons. In all cases, he identified correctly, on the first try, colors or shapes (Pepperberg 1981). No formal drills or training preceded such identification, although standard tests, which included some color and shape queries, had been administered earlier on those days. Alex's responses suggested that he had some concept of attributes—what was or was not green, or what was or was not three-cornered—but told us little of his understanding of the labels "color" and "shape." Nevertheless, these data suggested that he might learn the meaning of such categories, and led to the study described in the next chapter.

Labeling Errors as a Possible Indication of Motivation

Trainers sometimes had particular trouble interpreting Alex's color responses (Pepperberg 1981). Although interobserver agreement on his labels by definition had to be at least as good as his test scores, and was occasionally much higher (see the next section), Alex sometimes garbled color labels. Thus during testing he might call a rose or green object "roween," which sounded to some trainers like "rose" and to others like "green." Similarly, he might label a blue item "row-uuw." When asked to reidentify the object, he was correct 95–100% of the time. We never had this difficulty with shape markers. Such mistakes might indicate that Alex saw colors differently from humans—and this possibility was indeed likely[9]—but the nature of the mistakes—most often mixing rose with green, rather than, for example, green and blue—and our use of standardized food colors suggested that spectral separation was not an issue. My hunch was that the problem was motivation. Alex's shape labels were always distinguishable and, furthermore, he occasionally requested shaped objects, whereas he asked for a colored item only if it was the sole one available. He thus appeared uninterested in differently colored items, possibly because they all provided the same reward. Differently shaped objects, in contrast, cracked apart in different ways when he chewed them and provided different tactile experiences when he held them. Although one way to study this possibility might have been to see if associating colors with different flavors improved his accuracy, such a procedure might have further decreased his motivation to label colors of flavors he disliked; learning a label to refuse an item (e.g., "no key") was not actively being trained. Thus, although I could not at the time prove that the problem was motivational, the data again suggested a future experiment: one to determine if Alex could learn to use "want" to request an alternate reward if the object to be identified was undesirable. That experiment is described in Chapter 11.

Pronunciation

Occasional problems in understanding Alex's color labels bring up the question of his overall intelligibility. As noted earlier, all labels eventually became clear enough to be recognized by all trainers at least 80% of the time. At this stage, however, Alex rarely produced a clear label *immediately* after training. Usually, one or two months of work was necessary before he learned to manipulate his vocal tract appropriately to produce a novel sound (Pepperberg 1981). Acquisition of each new label was preceded by a practice period, during which he appeared to experiment with the new sound;[10] for example, initial attempts at "wood" were a series of "whuuuwhuu"s. Although this behavior is distinct from the observed "babbling" stage of juvenile parrots (Nottebohm 1976; we call that stage "speaking Greylish"), the practice period does have some aspects in common with the babbling stage of human infants. I discuss this subject in detail in Chapter 12.

During the practice period, however, all mispronunciations were consistent, and resembled the targeted English word. New trainers, after ~5-8 hours of exposure to Alex, could distinguish different vocalizations with ~80% accuracy; experienced trainers were better than 90% (Patterson and Pepperberg 1994). Interestingly, trainers also agreed on the identity of Alex's incorrect identifications, which is a stronger test because even the person presenting the object could not be cued by the identity of the item in question. Thus, after we omitted labels that both trainers agreed were unintelligible, interobserver agreement on a representative sample of tests, including both correct and incorrect identifications, was 97% (Pepperberg 1981).

Data from audio- and videotaping sessions also helped monitor progress in pronunciation. We could check intra- as well as interobserver reliability by seeing if we interpreted the same vocalization the same way on subsequent hearings. We did not, however, use spectrographic analysis to identify or compare utterances. Unlike bird song, which is fairly stereotypic from one rendition to the next (see, e.g., Greenewalt 1968) and for which spectrographic analysis is widely used, speech production, whether by people or parrots, often differs considerably from one rendition to the next, and the same phrase differs significantly when uttered by different people (see, e.g., Hecker 1971; Todt 1975a,b). Thus we had little reason to expect that comparing sonograms of Alex's vocalizations of any given label on two different occasions would provide any more information than we could obtain by comparing such vocalizations "by ear" (Pepperberg 1981).

Spontaneous Use of Speech

Alex's utterances were not restricted to identifying items, which suggests that he understood that labels could be used to communicate with human trainers. Although data at this point merely provided correlational evidence, the findings again suggested how more convincing data might be obtained; those results are in Chapter 11. At the time, however, my students and I noted instances in which Alex appeared to request objects spontaneously for food or play (Pepperberg 1981). For example, if he saw a trainer take a particularly desirable object (e.g., a cork) from a cabinet, he called for it until the student gave it to him.[11] If the student produced a less desirable item (e.g., a key), Alex did not engage in such behavior. He often requested items not in view. If a trainer presented an item other than one he requested, he generally responded by saying "nuh" (no) and repeating his initial request; occasionally he took the proffered item and threw it back at the trainer or onto the floor. Interestingly, he rarely requested colored objects; as noted earlier, different colors rarely influenced his behavior. If, however, trainers responded to his requests for peg wood (clothes pin) with a chewed-up stump, and he noticed an intact colored one, he repeatedly requested the colored one. In addition, he learned food labels without formal training. Students and I labeled all food items as we gave them to Alex or put them in his dishes; we never, however, modeled use of food labels. He received various fruits, seeds, crackers, and vegetables; he quickly learned to label what appeared to be his favorites, "nut" and "corn."

Alex's Early Achievements

In sum, after approximately two years of training by and interacting with humans, Alex achieved a rudimentary form of communication: contextual/conceptual use of human speech. He could identify, refuse, and apparently request a limited set of objects for play or for food. He also appeared to possess some limited facility for categorizing objects in a manner not unlike that of humans, and could generalize label use to identify objects that differed somewhat from training items. He had also demonstrated a limited capacity to compose multiword phrases from separately acquired labels, which suggested that he might interpret human speech as a collection of discrete units. The data did not prove that he had advanced capacities for information processing, but did suggest new avenues of research. At the very least, my research demonstrated that a Grey parrot was capable of far more than simply the ability to mimic human speech.

4 Does a Parrot

Have Categorical

Concepts?

"Well," said the parrot, turning to me, "I may have started the Doctor learning, but I never could have done even that, if he hadn't first taught me to understand what *I* was saying when I spoke English. You see, many parrots can talk like a person, but very few of them understand what they are saying. They just say it because . . . well, because they fancy it is smart, or because they know they will get crackers given them."

Polynesia, the Grey parrot in *The Voyages of Doctor Doolittle*
(Lofting 1948:29)

In the previous chapter, I proposed that Alex had a rudimentary understanding of categorical concepts. This ability is important, because full understanding of such concepts would demonstrate that he could handle information with some level of abstraction. Thus I needed to determine if he could respond not only to specific properties or patterns of stimuli, like pigeons who respond to positive instances of "tree" (see, e.g., Herrnstein et al. 1976), but also to classes or categories to which these specific properties or patterns belong (see Premack 1978; Savage-Rumbaugh et al. 1980b; Thomas 1980). Could he, for example, go beyond recognizing what is or is not "green" to recognizing the nature of the *relationship* between a green pen and a blade of grass? In the parlance of psychologists, the former ability (recognizing "greenness") is called stimulus generalization; the latter (recognizing the category "color") is called categorical class formation (see Zentall 1996). The differences between these abilities are both subtle and important, and need clarification for understanding the significance of Alex's capacities.

Categorical Class Formation versus Stimulus Generalization

Defining these two abilities is simple, but separating them can be difficult (Pepperberg 1996). Stimulus generalization "implies memorizing a specific reference stimulus or set of stimuli followed by responses to new stimuli based on a failure to discriminate between the reference stimuli and the new stimuli" (Thomas and Lorden 1993:132; note Zentall 1996). Categorical class formation, in contrast, "results in a distribution of affirming responses to exemplars of the concept that is essentially rectilinear" (Thomas and Lorden 1993:133; Zentall 1996), or is the ability "to respond similarly to discriminated stimuli" (Zentall et al. 1986:153). The distinction between stimulus generalization and categorical class formation is clear for classes formed of objects that are physically different, such as arbitrary classes formed by an experimenter—say, squiggles versus straight lines. Subjects may not discriminate among squiggles and still respond to them appropriately. When objects are physically similar, that is, in widely inclusive classes such as "passerine bird," the distinction is less clear, particularly when we ask an animal subject to form the class. In the case of "birds," for example, an animal might not use the same criteria as the experimenter to sort slides into the designated class; responses might be based simply on presence or absence (visibility) of beaks. Slides that omitted the beak might not be sorted correctly. Also, new bird slides (e.g., of a penguin) differing considerably from reference pictures for humans might be indistinguishable to the animal sorter or simply not *be* distinguished. Even if the animal were forced to make an identity/nonidentity judgment, decisions could be based, for example, on size of beaks. A subject in such a study would not have a *human-like* category "bird." Thus the extent to which most animals form categorical classes as defined above is as yet unknown (Pepperberg 1996; Zentall 1996).

One approach to this problem uses symbolic labels (Pepperberg 1983a, 1996). Here, a subject does not simply create a class consisting of a concrete attribute (e.g., red), but instead forms multiple, *hierarchical* classes. Each class, although based on concrete attributes, is nevertheless abstract. A subject learns, for example, to form classes (i.e., separate categories) of color and shape *labels*. Labels relate to physical attributes (e.g., redness, triangularity), but are arbitrary sound patterns ("red," "three-corner") or arbitrary hand or pictorial signals. These labels, which represent distinct classes, are then grouped into higher-order abstract classes whose labels are also made of arbitrary patterns—"color" and "shape."[1] The ability to form these classes or relationships is not elementary, even for humans: Children who learn that one set of objects is "green" and another is "red" may still not understand that there exists an overall concept of "greenness" or "redness" that can be generalized to novel objects (Rice 1980). According to de Villiers and de Villiers (1978), acquiring both the category "color" and the actual labels for colors, for example, means that a subject (1) can distinguish color from other categories or classes such as size, material, or shape; (2) can isolate certain colors as focal points and others as variants of these; (3) understands that each color label is part of a class of labels that are linked under the category label "color"; and (4) can produce each label appropriately. My goal was to see if Alex could learn to respond in such a manner.

Investigating What Alex Already Knew

Given that Alex already responded to particular and novel instances of properties such as redness and squareness, I wanted to learn if he could recognize categorical concepts of color and shape. I initially studied a relatively limited aspect of this problem: Could he learn to associate labels representing the abstract categorical concepts ("color," "shape") with labels representing the various instances of these categories (e.g, "blue," "four-corner")? Specifically, could Alex learn to recognize and respond differently to queries of "What color?" versus "What shape?" when given colored *and* shaped objects (e.g., a blue wooden square)? To be correct, he would have to examine the object, attend to and comprehend the question he was asked, choose the correct response from among several color or shape possibilities, and then encode this response in a vocal label. If correct, Alex would demonstrate some fairly abstract level of information processing, and no bird had yet responded appropriately on such a complex task (Pepperberg 1983a).

By requiring *reclassification,* the task carefully avoided the possibility that Alex would learn to respond in specific ways to specific objects: The same object would have to be categorized with respect to color on one test and with respect to shape on another. Thus Alex was forced to attend to the type of question we asked. And, by requiring flexibility in changing the basis for classification, the task would indicate the presence of "abstract aptitude" (Hayes and Nissen 1956/1971).

Introduction of Categorical Labels

Labels for categorical classes had been introduced, although not formally, when Alex was taught color and shape labels in M/R sessions. The category labels were included casually in appropriate vocal exchanges. Thus our bird heard statements like "OK, Alex, we're gonna work on colors today!" or, after he had correctly identified a triangular piece of uncolored wood, trainers might comment, "Yes, that's right, you know the *shape;* it's *three-corner* wood!" If Alex omitted the relevant marker when asked to identify a colored or shaped object, trainers would ask "What *color* paper?" or "What *shape* hide?" One or two such queries usually elicited the appropriate response (Pepperberg 1981, 1983a).[2]

Such responses, however, could not be taken as evidence that Alex understood that a relationship existed between such categorical queries and the targeted, missing marker. He could have simply interpreted both queries as "say something more."[3] I therefore had to determine what exactly he might have learned.

I used the following protocol to determine if any associative learning had occurred (Pepperberg 1983a). My students and I introduced, at various intervals during play sessions, objects having both color and shape, and queried Alex, "What color?" or "What shape?" We monitored his responses for 6 months, and randomly rewarded him with the object. Because we often used play sessions informally to review his ability to label items, he was accustomed to hearing "What's this?" along with presentation of some familiar item in such a setting. Our procedure thus integrated these new queries and objects into his basic routine without focusing attention on multicategory items: Anything that emphasized the objects or questions might have influenced, in some way, even these very preliminary findings. For the same reason I did not administer a rigorous single-instance test of comprehension (did not test his response to "What color?" or "What shape?" for a first-ever presentation of a colored and shaped item). At this point I was interested simply in three things: the nature and extent of what he might have acquired (no matter how limited), the ramifications of such acquisition in a nonprimate, and if any understanding might be used as a starting point for further instruction.

Preliminary Findings and Their Implications

Although Alex's responses were correct the very first time we showed him an object with multiple attributes (a green wooden triangle) and asked "What color?" and "What shape?", during subsequent presentation of this and other similarly fashioned items his accuracy on categorical queries decreased to about 60% (Pepperberg 1983a). Given that he could choose his responses from two possible materials (wood and rawhide), three possible colors (rose, green, and blue), and two possible shapes (three-corner and four-corner), and that I also considered him to have erred if he responded within the correct category but with an incorrect specific label (e.g., said "green" when the answer to "What color?" was "blue"), the accuracy of his answers was still greater than could be achieved by chance.[4]

Interestingly, Alex's errors were rarely entirely inappropriate and were generally separable into three distinct types (Pepperberg 1983a): (1) Unclear vocalizations—undecipherable sounds, presumably for color or shape markers. Examples were utterances like "roween," noted in Chapter 3. These errors constituted 4% of his total mistakes. (2) Omissions of both markers; we termed these responses "generic." Such errors accounted for 8% of the total. (3) Recitation of both the appropriate color and shape markers; at 12%, these errors were the most frequent. Other errors included: correct category but wrong specific label, 4%; incomplete shape marker (e.g., "corner wood" for "three-corner wood"), 4%; and wrong category, 4%.

Overall, the data suggested that Alex perceived some relationship between the categorical concepts implied by the labels "color" and "shape" and their specific instances. That is, he seemed to understand that "rose" and "three-corner," for example, represented different categories of attributes of a given object. Such data suggested that formal training might impart even greater knowledge.

That type 3 errors were most frequent suggested a particular focus for subsequent training. The current protocol required that, once training had begun on color and shape markers, Alex was not rewarded for responses that omitted the color or shape marker if the object was colored or shaped. Possibly he had inferred the need to provide all relevant data when asked to identify an object. Consequently, he might not have attended to specific questions about color versus shape, but simply responded as if to the common routine, in which he was queried "What's this?" Thus I had to amend his training protocol to emphasize the content of the specific questions being posed. My aim was threefold: first, to provide additional training on the class category concept; second, to separate possible training effects from conceptual knowledge; and third, to introduce the idea that targeted questions were also an integral part of communicatory behavior (Pepperberg 1983a).

Modifications to Our Basic Protocol

Although we continued to use the M/R technique described earlier, we shifted our emphasis to the type of question being asked, and carefully modeled correct responses and the effects of incorrect responses. A typical training interaction is presented in Table 4.1. Note the very

Table 4.1. Excerpt of an M/R training session on color versus shape, 4/30/79. K refers to Kimberley Goodrich (one of the secondary trainers), I to the principal trainer (me), and A to the parrot Alex. This portion of the session lasted approximately 5 minutes. (After Pepperberg 1983a.)

I: (Acting as trainer): Kim, what color? (Holds up green triangular piece of wood.)

K: (Acting as model/rival): Green three-corner wood.

I: (Briefly removes object from sight, turns body slightly away.) No! Listen! I just want to know *color!* (Faces back toward K, re-presents object.) What *color?*

K: Green wood.

I: (Hands over wood): That's right, the color is *green;* green wood.

K: OK, Alex, now you tell me, what shape?

A: No.

K: OK, Irene, *you* tell me what shape.

I: Three-corner wood.

K: That's right, you listened! The shape is three-corner; it's *three-corner* wood. (Hands over wood.)

I: Alex, here's your chance. What color?

A: Wood.

I: That's right, wood; what *color* wood?

A: Green wood.

I: Good parrot! Here you go (hands over wood). The *color* is *green.*

slight differences in emphasis compared to the emphasis in the session described in Table 2.1.

The main change in our protocol involved the criterion used to decide when to initiate testing. The customary criterion of clarity of production could not be used because Alex already clearly produced the targeted color and shape labels. He also responded to questions about objects with only color or shape (e.g., an uncolored wooden square) with 80% accuracy. The criterion for initiating formal testing on categorical concepts was therefore changed to how well training had eliminated type 3 (and, indirectly, type 2) errors. This criterion was independent of Alex's understanding of the relationship between labels representing the abstract class concepts ("color," "shape") and labels representing particular instances of such concepts ("rose," "green," "blue"), because we did not monitor his errors on those relationships. So, before formal tests could begin, Alex had to respond in three successive sessions (outside of formal tests) with a single (but not necessarily categorically appropriate or physically correct) marker 80% of the time when queried "What color?" or "What shape?" for objects having both color and shape. He could have responded at chance levels to these queries and still met the criterion if he responded with a single marker.

We also needed to determine how well Alex would transfer his knowledge to novel instances of color and shape, and to novel objects. Over the course of the study, we obtained colored and shaped keys, and Alex learned additional labels ("yellow," "grey"; "two-corner," "five-corner"). Using test procedures described earlier, we thus could rigorously examine his capacities for categorization.

The Results of Alex's Tests

My students and I inserted 270 queries about the color and shape of familiar and novel objects into more than 170 formal tests; other questions on these tests involved object identification to avoid expectation cuing (see Chapter 2). On these tests, Alex had to respond to objects that embodied five different colors, four different shapes, and three different materials.[5] His accuracy was better than 80% (Table 4.2). For the familiar colors and shapes and both familiar and novel objects, Alex's scores were 84.7% correct on color queries and 83.7% on shape queries; for newer colors and shapes, his accuracy was 92.3% on color queries and 81.3% on shape queries.

These results suggested that Alex had mastered the task I had devised. After viewing an item that could be described with respect to color *and* shape, he could determine from our question which category was being targeted, and then produce, based on this question, the appropriate instance of the category. Alex was also required to add the material or object label to his answer to encourage him to combine labels appropriately (see discussion on segmentation, Chapter 3). Most noteworthy was that he could *reclassify* any object, either immediately or during presentations separated by several days (Pepperberg 1983a).

Comparisons with Studies on Chimpanzees

Before I began my study, Savage-Rumbaugh et al. (1980b) had examined similar abilities in chimpanzees, and a comparison of the two projects will clarify the extent of Alex's accomplishments. The fundamental objectives of both studies were the same: to determine if our respective nonhuman subjects could associate symbolic categorical labels and the symbolic representations of specific instances of these categories. Alex's task, however, was quite different, both in immediate

Table 4.2. Alex's test results for "What color?" and "What shape?" queries. (After Pepperberg 1983a.)

Object	"What color?"		"What shape?"	
	Test scores[a]	Wrong i.d. (no.)	Test scores[a]	Wrong i.d. (no.)
3CBK	6/8	K (2)	7/7	
3CBW	4/5	BCW (1)	8/10	3CH (1), 3W (1)
3CBH	5/6	GH (1)	4/6	3H (1), 3BCW (1)
3CGK	4/4		4/4	
3CGW	5/6	RW (1)	7/9	GW (1), 3W (1)
3CGH	6/9	BH (2),[b] H (1)	4/6	H (2)
3CRK	5/6	K (1)	5/6	3K (1)
3CRW	6/8	U (1),[c] W (1)	5/7	RW (1), CW (1)
3CRH	4/5	UH (1)[c]	4/5	2CH (1)
4CBK	4/5	BPW (1)	6/6	
4CBW	8/10	4CW (1), BCW (1)	10/11	W (1)
4CBH	5/5		4/5	UCH (1)[c]
4CGK	4/4		4/5	GK (1)
4CGW	5/6	4CW (1)	5/6	4CGW (1)
4CGH	6/6		6/6	
4CRK	5/6	4K (1)	4/6	UK (1),[c] CK (1)
4CRW	7/7		4/5	4W (1)
4CRH	5/7	UH (1),[c] H (1)	6/6	
2CBW	2/2		2/3	2CH (1)[d]
2CRW	1/1		1/1	
2CGW	2/3	W (1)	1/2	2W (1)
2CGyW	1/1			
2CGH	1/1			
3CGyW	1/1		2/2	
3CGyK	1/1		1/1	
3CYW	1/1		1/1	
4CYW			1/1	
5CRW			2/2	
5CBW	1/1		1/2	UCW (1)[c]
5CYW	1/1			
5CBH			1/1	

Code: C = corner, R = rose, G = green, B = blue, Y = yellow, Gy = grey, W = wood, K = key, H = hide, PW = peg wood (clothes pin).

a. Test scores are number of correct identifications divided by the total number of presentations.

b. We noted that one of the objects was somewhat bluish looking; after replacing it, these errors ceased.

c. Alex's production of the marker was unintelligible.

d. The error was made on initial presentation.

focus and test protocol. For the chimpanzees, the focus was to learn if they could, after learning to associate symbols for instances of a category with symbols for the category (i.e., "red" with "color"), then properly symbolically categorize novel instances ("green," "blue"). The focus of Alex's task was to learn if he could *decode* symbolic representations (i.e., vocal questions) referring to two categories of abstract attributes ("What color?", "What shape?") when both categories were *simultaneously* relevant. Thus the chimpanzees' task was to encode a standard sorting problem into symbols; Alex's task was to decode symbolic questions.

The distinction between tasks is particularly important for evaluating Alex's test protocol, which required presentation of the same objects on multiple and often successive tests. To determine Alex's abilities with respect to understanding categorical labels, a formal transfer test on novel items (e.g., testing whether Alex would respond "yellow wood" to "What color?" for a novel yellow pentagon after learning the labels "yellow" and "five-corner") would not be particularly informative. Rather, I needed to determine if he could reliably decode, each time, queries about each object: Even if he could classify a blue wooden square as "blue wood" in one instance and "four-corner wood" on another, the critical test was whether he could switch back and forth numerous times. Nevertheless, Alex was given and did indeed pass standard transfer tests. He correctly answered 12 of 15 queries of "What color?" and "What shape?" for first-time ever presentations of objects that varied in newly acquired colors and shapes (e.g., at the first ever presentation of a blue pentagonal piece of rawhide; Pepperberg 1983a). His errors (see Table 4.2) were, for the most part, not directly related to the categorical concepts under consideration, but were primarily what I had termed "generic."

Relation of This Study to Conditional Discrimination Tasks

Alex's task has some relation to standard conditional discrimination tasks, but also differs significantly from such tasks. In the standard conditional task, similar to the one described in Chapter 1, an animal is shown, for example, a red triangle, a red square, and a blue square, and is rewarded for choosing a red item when the objects are backed in white and for choosing a square item when the backing is black (see Thomas 1980). The subject learns something about responding with respect to color versus shape, but is not likely to transfer immediately to novel objects, colors, or shapes. Of course, Alex did indeed learn a form of conditional discrimination: If I hear "A" (one categorical label),

I must respond with an "a" (a label for an instance of that category). And likewise for "B" and "b." Alex, however, could both reliably transfer immediately to novel instances and repeatedly switch between categories without additional training. Thus he had acquired a concept of the relationship between the category "A" and its instances "a" and between "B" and its instances "b."

Moreover, according to Premack (1976), comprehension and use of symbolic *input* adds an additional layer of complexity to the already conditional nature of a productive labeling task. In this view, an animal that can produce labels is already performing a conditional discrimination: Responding "banana" versus "apple" is based on the condition of the presence of one fruit versus the other. Thus an animal that not only can respond "green" in the presence of a green rather than a blue object, but also knows to respond "green" rather than "three-corner" based on the type of question, has demonstrated additional cognitive ability.

Comparing Alex's Understanding of Categorical Classes to That of Other Animals

Alex had indeed learned to form categorical classes. He could comprehend a vocal question, extract the relevant category from the question and from an object that could be classified in multiple ways, and respond with the label for the one, correct instance of this category. Nevertheless, his accomplishments were still fairly rudimentary when compared with those of the nonhuman primates and marine mammals that had already been studied. He could not yet respond to queries such as "How do X and Y differ?" or look at a collection of colored and shaped objects and tell us "What object is green and three-cornered?" And Alex's productive use of the phrases "What color?" and "What shape?", which would be another measure of cognitive ability, was still limited. At the time, he had used the phrase "What color?" only once in a way that could possibly be interpreted as intentional: He queried a student as to the color of his mirror image (Pepperberg 1983a). After hearing replies such as "That's grey; you're a Grey parrot" to six of his queries, he acquired the label "grey" and subsequently used it to identify correctly, on first showings, newly dyed grey objects. Alex had, however, performed at a level that was completely unexpected for a bird,[6] and my challenge was to see what other types of problems he could master.

5 Can a Parrot

Learn the Concept

of Same/Different?

> The poor development in birds of any brain structures clearly
> corresponding to the cerebral cortex of mammals led to the
> assumption among neurologists not only that birds are primarily
> creatures of instinct, but also that they are very little endowed with
> the ability to learn. There is no doubt that this preconceived notion,
> based on a misconceived view of brain mechanisms, hindered the
> development of experimental studies on bird learning. Thus Herrick
> (1924) says: "It is everywhere recognized that birds possess highly
> complex instinctive endowments *and that their intelligence is very
> limited.*" No present-day student familiar with recent work on bird
> behavior would be willing to endorse to-day the words I have put in
> italics. On the contrary, in some contexts it is doubtful whether the
> birds are exceeded in learning ability by any organisms other than
> the highest mammals.
>
> William Thorpe, *Learning and Instinct in Animals* (1964:336)

Despite my studies on Alex's ability to use English speech to label objects and their attributes and to comprehend abstract category labels, questions remained about the ability of birds to use and comprehend abstract symbolic relationships. Specifically, at the time I was examining Alex's capacities, many researchers not only were proposing that species-specific differences existed with respect to comprehension and use of abstraction, but were also arguing in favor of species-based hierarchies (see Premack 1978, 1983). Great apes would be just below humans (Savage-Rumbaugh et al. 1985), other nonhuman primates would follow (see, e.g., Rumbaugh and Pate 1984), and birds (as exemplified by pigeons) barely deserved mention (Premack 1978). These researchers accepted that biological constraints were unlikely to limit an animal's ability to master nonrelational concepts (e.g., to identify classes of objects based on some constellation of properties; Chapter 3), but suggested that data on relational concepts (e.g., darker than, different from) reflected qualitative as well as quantitative species differences. It wasn't that pigeons couldn't do as well as chimpanzees on some tasks, but that pigeons simply couldn't do the tasks at all (Premack 1978). Other researchers argued that demonstrable species differences could be attributed to experimental design rather than differences in underlying cognitive capacities (e.g., Herman 1980; Pepperberg 1981, 1986a; Menzel and Juno 1982, 1985; Kamil 1984; see Macphail 1985).[1] Nevertheless, the suggestion remained that certain tasks existed that just could not be mastered by a bird.

The Rubicon: The Concept of Same/Different

One task in particular, comprehension of same/different, was singled out as requiring abilities not typically attributable to nonprimates, particularly birds (Premack 1978, 1983; Mackintosh, Wilson, and Boakes 1985). Considerable debate existed in the literature concerning exactly what had been proven in animal studies purporting to demonstrate comprehension of same/different (see Edwards, Jagielo, and Zentall 1983; Premack 1983 and commentaries therein). The debate centered on two problems. First, researchers had difficulty determining whether animals tested in standard operant paradigms had shown comprehension of similarity and difference or of something far simpler. Second, some researchers argued that even animals that succeeded on the operant task and thus distinguished similarity from difference still did not understand the *concept* of same/different.

The first problem involves the extent of learning that animals show on match-to-sample and nonmatch-to-sample tasks (such tasks show if an animal can choose something same or different from a given sample). Pigeons initially tested this way merely seemed to learn to respond to the matching alternative, to learn little about the nonmatching alternative, and thus to acquire a concept of *same* but not *different* (see Premack 1978; Zentall et al. 1981; Edwards et al. 1983; Pepperberg 1987a). Related alternative mechanisms could not be ruled out in other studies (e.g., Urcuioli and Nevin 1975; Holmes 1979; Lombardi, Fachinelli, and Delius 1984; Pisacreta, Redwood, and Witt 1984). Pigeons seemed, for example, to have an unlearned predisposition to choose the odd stimulus from a set (Wilson, Mackintosh, and Boakes 1985), and thus had not really learned the *concept* of same versus different.

The second problem dealt with the idea that comprehension of the *concept* of same versus different is more complex than learning to respond to match-to-sample and nonmatch-to-sample or oddity-from-sample. According to Premack (1983), the former requires use of arbitrary symbols to represent *relationships* of sameness and difference between sets of objects *and* the ability to denote the attribute that is different. The latter, in contrast, as tested in the experiments cited above, requires only that a subject show a savings in the number of trials needed to respond to B and B as a match after learning to respond to A and A as a match (and likewise by showing a savings in trials involving C and D after learning to respond appropriately to A and B as nonmatching). Subjects in match-to-sample and nonmatch-to-sample studies might even be responding based on "old" versus "new" or "familiar" versus "unfamiliar" (Premack 1983), that is, on the relative number of times they experience the A sample versus the number of times they see different B samples. A subject that understands same/different, however, not only knows that two nonidentical blue objects are related in the same way as are two nonidentical green objects—in terms of color—but also knows that the blue objects are related to each other in a different way than are two nonidentical square objects, and, moreover, can transfer this understanding to *any* attribute of an item (Premack 1978, 1983). Likewise, such a subject would understand the concept of difference in an equivalent manner.

Although Premack (1978, 1983) argued that such abilities are probably limited to apes and humans, and, because of the requirement of symbolization, to apes that had had some form of language training, when I began my study with Alex, some level of symbolic same/different learning had been shown in pigeons and monkeys. Researchers had trained pigeons to peck distinct symbols for instances of similarity and difference (e.g., Zentall, Hogan, and Edwards 1984). The birds' performances were influenced by the type of stimuli used, but birds did respond symbolically and needed fewer sessions to respond correctly to novel instances of similarity and difference when the task remained the same compared to when it was reversed (e.g., when birds trained on difference were then asked about similarity). Other researchers had shown that monkeys *(Macaca mulatta)* and pigeons could use two-choice symbolic responses to demonstrate same/different concept learning (Santiago and Wright 1984; Wright, Santiago, and Sands 1984; Wright, Santiago, Urcuioli, and Sands 1984; see note 1 for more recent material). Pigeons, although able to respond above chance levels, transferred to novel items poorly compared to monkeys. Such studies, although suggestive, did not demonstrate that birds used symbols for "same" and "different" in a manner fully comparable to that of humans or language-trained chimpanzees, or even appropriately trained monkeys.

Few other studies at the time had investigated concepts of same/different in birds, even though the natural behaviors of individual recognition and vocal dueling and song matching would make categorization based on similarity or difference seem adaptive (see, e.g., Kroodsma 1979; Falls, Krebs, and McGregor 1982; Beecher, Stoddard, and Loesche 1985; Falls 1985; Godard 1991; Godard and Wiley 1995; Stoddard 1996; Naguib and Todt 1998). In laboratory studies, budgerigars seemed to discriminate similarities and differences in calls of canaries *(Serinus canarius);* however, the budgerigars appeared to learn the unique characteristics of individual canary calls (Park and Dooling 1985). Similar results were found in studies on song discrimination with cliff swallows *(Hirundo pyrrhonota)* and barn swallows *(Hirundo rustica;* Stoddard and Beecher, pers. comm., 1986). Related studies demonstrated that great tits *(Parus major;* Shy, McGregor, and Krebs 1986) could recognize test songs as similar to or different from one particular training song (comparable to visual categorization by pigeons; see, e.g., Herrnstein 1984). Song sparrows in a habituation/dishabituation experiment appeared to make fine acoustic distinctions among conspecific syllables in their songs (see, e.g., D. Nelson 1987; Stoddard, Beecher, and Willis 1988), but the data must be interpreted carefully with respect to concepts of same/different because these birds can memorize large numbers of song types (see, e.g., Stoddard et al. 1992), and thus may have responded based on familiarity. Interestingly, studies on field sparrows *(Spizella pusilla)* that compared the relative importance of different sound features in conspecific song recognition and discrimination of conspecific from allospecific song demonstrate just how difficult it may be to design experiments to show that birds understand concepts of same/different in exactly the way Premack would require: Birds differentially weigh information in the various features tested, and thus results could depend on which feature the experimenter chooses (see, e.g., Nelson 1988). Starlings, mockingbirds *(Mimus polyglottos),* and cowbirds, for example, can classify novel series of tones as ascending or descending—that is, as "same" or "different" from ascending or descending reference series—but only for sequences that are within the range of frequencies used in training (Hulse and Cynx 1985). None of these studies, however, showed actual *labeling* of the relation of sameness or difference or, for example, demonstrated transfer to entirely different situations (e.g., calls or songs of different species), nor did any of these tasks demonstrate (to humans) that a bird truly understands what *characteristics* are the same or different (Pepperberg 1987a).[2] Thus when I began my work with Alex, few researchers believed that birds could understand the concept of same/different.

An Experiment Designed to Demonstrate Comprehension of Same/Different in Alex

For the above reasons, Alex's task would have to be functionally equivalent to that used with Premack's chimpanzees (1976, 1983): I would have to ensure (1) that the symbolic concepts tested would be abstract, in contrast to those examined in standard conditional discrimination paradigms (in which alternative explanations might apply); (2) that Alex would be given explicit, equal training on concepts of *both* same and different; (3) that the findings could not be dismissed as the formation of associations specific to the training situation; and (4) that data from initial questions on novel items (first-trial transfer tests) could be examined for their significance (Pepperberg 1987a).

To take into account such constraints and Alex's use of vocal labels, I designed the following task. Alex would be presented with two objects that could differ with respect to three categories: color, shape, or material (e.g., a yellow paper pentagon and a grey wooden pentagon; a green wooden triangle and a blue wooden triangle). He would then be queried "What's same?" or "What's different?" The correct response would be the label of the appropriate *category*, not the specific color, shape, or material marker (e.g., "shape" rather than "five-corner"). Therefore, to be correct, Alex would have to (1) attend to multiple aspects of two different objects; (2) determine, from a vocal question, if the response was to be based on similarity or difference; (3) determine, for the items he was shown, what was "same" or "different" (e.g., were they both blue, or square, or of different material?); and then (4) produce, vocally, the label for this particular category. Thus the task required that Alex perform some form of feature analysis of the two objects: His responses could not be made on the basis of total physical similarity or difference (Premack 1983).

For comparison, remember that most research on same/different in animals (whether on pigeons or monkeys) uses (1) a two-choice design whereby a subject merely indicates if pairs do or do not match; (2) a topographically similar (and thus possibly easier to acquire) response for both answers (e.g., lever pressing or key pecking; see Michael, Whitely, and Hesse 1983); and (3) "same" pairs that are identical in all dimensions and "different" pairs that differ with respect to most, if not all dimensions. Such a task would be equivalent to having Alex view two items and respond merely "same" or "different," the latter response occurring to anything not identical to the original sample. But Alex was required to produce the vocal label of one of three dimensions that was different or same for each pair depending upon the question asked; thus his task was considerably more difficult. Note that even chimpanzees do not respond "different" as reliably for pairs of objects differing on only one of three dimensions as for pairs differing on two dimensions (McClure and Helland 1979).

At the time this experiment began, Alex had a considerable number of possible objects on which he could be trained and later tested. He would thus receive experience adequate for learning the task (see Premack 1976:132; Wright et al. 1990) and still have opportunities to respond to situations different from training. Specifically, Alex already produced vocal (English) labels for five colors (green, rose [red], blue, yellow, grey), several shapes (two-, three-, four-, and five-corner for, respectively, football-shaped, triangular, square, and pentagonal forms), four materials (paper, wood, cork, and hide [rawhide]), and various metallic items (key, chain, grate). He had some comprehension of abstract categories, in that he responded to vocal questions of "What color?" or "What shape?" with the correct label for the targeted attribute (e.g., "green wood" *or* "five-corner hide") for objects that simultaneously incorporated even novel combinations of both variables (green football-shaped pieces of wood, yellow rawhide pentagons; Pepperberg 1983a). During the current experiment, he also acquired labels for orange and purple and six-cornered shapes. Thus labels for *all* these items, as well as labels for foods, locations, and quantity, were *always* available in his repertoire as possible answers. Consequently, Alex was *not* limited merely to choosing between symbols representing "same" or "different," nor to the labels "color," "shape," or "matter," nor to choosing physically between only two objects that were similar to or different from a sample (Pepperberg 1987a).

The design of the task ensured that Alex would have to respond on the basis of same/different rather than on the basis of a simpler alternative. His responses on tests were unlikely to be based on absolute physical properties or on his having learned the answer to a given pair, because the number of possible permutations of question topic, correct response, and combination of objects with various attributes was very large.[3] Moreover, because Alex's response would be a *category* label rather than a specific object or attribute label, tests could also involve totally novel objects whose labels were unknown. Thus, he would have to be able to transfer between like and unlike pairs of colors, like and unlike pairs of shapes, and like and unlike pairs of materials, all of which would vary from training materials; that is, Alex would have to demonstrate transfer among stimulus domains (e.g., colors versus shapes versus materials) as well as among various instances of each domain (see Premack 1976:354–355 for the importance of such transfers in determining that the behavior is not just stimulus generalization). In addition, I planned to determine whether Alex was responding to the content of the questions (i.e., differentially processing "What's same?" versus "What's different?"), and not merely responding to variations in the physical properties of the items (Pepperberg 1987a).

In sum, after training, Alex would be queried to test his concepts of same and different *and* to provide additional evidence for comprehension of categorical concepts. To test same/different, he would be asked, in random order, about pairs of familiar objects not used in training, pairs in which one or both items were totally novel, and probed to determine whether he was indeed processing the content of the questions. Data on categorical concept formation would be obtained as a consequence of same/different testing: To succeed on the same/different task, Alex would have to demonstrate that he could recognize that two novel objects (e.g., pink and brown paper triangles) differed with respect to the category "color" even though the colors were untrained and the specific combinations of color, shape, and material for each item, as well as for the pair, had never before been seen on a test. A correct response would show that he was not responding to specific instances of color, shape, and material, but rather on the basis of the categorical concepts (Pepperberg 1987a).

Training Procedures

My students and I again used the M/R procedure, with two adaptations. The first was that for a correct response, Alex now received *both* objects. The second allowed him to request alternative objects as his reward (see Chapter 2) to motivate him to work with items in which he had little interest. By the time we began this study, he had learned to preface requests with "I want . . ." and to use object labels alone (e.g., "blue key") for identifications (Pepperberg 1988a). Thus during work on same/different, a response of "cork" to a pair of objects was considered an error, whereas "I want cork" was taken as a valid request rather than a mistake. We maintained the intrinsic nature of his rewards, however, by responding to requests for an alternative object only after he had given correct responses to the targeted objects, obtained those objects, and discarded them; that is, he never received a cork directly after telling us, for example, that the color of two objects was different, but only received the alternative reward after taking and dropping those two objects and then asking for the cork.

During training on same/different, Alex observed the following: A trainer held two objects in front of the model, and asked "What's same?" (e.g., for a red wooden triangle and a green rawhide triangle) or "What's different?" (e.g., for a red wooden square and a blue wooden square). Both questions and various object pairs were interspersed in a single session. Training was limited to a subset of familiar items: objects that were red, green, or blue; triangular or square; rawhide or wood—in order to reserve the other items for testing. The model responded with the label for the correct category and was rewarded with the objects (or the right to choose a reward), or erred and was scolded. When the model erred (to demonstrate the consequences of an error), the objects were momentarily removed from view (a "time-out"), then re-presented, and the question was repeated. The roles of trainer and model/rival were frequently reversed (Pepperberg 1987a).

Training occurred in two stages: first on color and shape, and then on material. Because Alex already correctly responded to queries of "What color?" versus "What shape?" for objects that had both color and shape, and occasionally produced those phrases in the presence of various items,[4] initial training contrasted those two categories. The third category label, "matter" (for material), was trained after Alex began to use the labels "color" and "shape" separately from the phrases "What color?" or "What shape?" Training sessions occurred two to four times per week and lasted 5 minutes to 1 hour, depending upon Alex's willingness to pay attention.[5]

In addition, to avoid expectation cuing and maintain Alex's motivation, my students and I concurrently trained and tested him on other tasks. We worked on number concepts (Pepperberg 1987b), additional labels (Pepperberg 1987c), photograph recognition, and object permanence (Pepperberg and Kozak 1986). Maintaining his motivation for same/different trials was more difficult than on some other tasks, because training objects were very familiar, and we were asking him to produce more information for what were essentially the same rewards.

Criterion Prior to Testing

As in the previous study, we had to adjust our criterion for beginning testing (Pepperberg 1983a). Alex already produced "color" and "shape" consistently and clearly; we therefore delayed testing until he could produce the label "mah-mah" (matter). Whether because of difficulties in manipulating his vocal tract or in motivation, Alex could not seem to form the label "matter," but consistently said "mah-mah" from trial to trial.[6]

Testing

The basic testing procedure was described in Chapter 2. During testing, Alex was shown an object or a number of objects, asked "What's this?" "What color?" "What shape?" "How many?" "What's same?" or "What's different?" He then needed to formulate a vocal English response from the 80 labels in his repertoire. Thus even though correct responses to the same/different question were limited to "color," "shape," and "mah-mah," Alex still had to choose from among many possible responses to many other questions and to attend to the *question* to be correct: Given two objects, he could be asked "How many?" "What's different?" or "What's same?"

We tested Alex with three collections of object pairs. The first involved objects that were similar to, but never the same as, those involved in training. The second involved objects that were completely novel. The third involved probes to ensure that he was indeed attending to our questions. I describe each collection in detail to clarify the extent to which Alex's abilities were tested. Examples from various collections and questions are shown in Figure 5.1.

Novelty in the first collection was minimal. Test results thus would not be affected if Alex, like pigeons, was fearful of unfamiliar objects (see, e.g., Zentall and Hogan 1974; Zentall et al. 1981). Items in these tests combined one additional familiar but untrained color, shape, or material, such as blue pentagonal wood: Blue wooden objects had been part of training, but five-cornered ones were not. Such slightly novel objects could now be combined with any other item for queries on same/different. Thus Alex now was tested on variously colored and shaped objects of wood and rawhide, and, later, variously shaped keys. These trials also included later presentations of objects from the second collection—objects that had initially been examined as "novel" items. When these objects were reused, pairings were such that either the question (same versus different) or the appropriate response category (color, shape, or matter) was altered from the previous trial in which the object appeared. Therefore, although individual objects were occasionally used in more than one trial, the *pairings* of objects were always novel, and a particular pair was only re-presented when Alex erred.

The second collection of objects used in transfer test trials were interspersed randomly with those of the first collection. The second

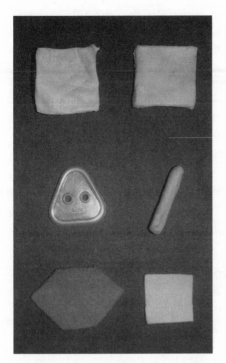

Figure 5.1. Objects used in questions on same/different. Top: In the first set of transfer trials, items were familiar but had not been used in training; in this example, only the materials of the objects were different. Middle: In the second set of transfer trials, items were completely novel; in this example, only the colors of the objects were the same. Bottom: In probe trials, Alex was asked to tell us *one* of the two attributes that were either same or different to see if he was attending to our questions; in this example, both shape and color were different and either response was correct.

collection consisted of object pairs that combined several attributes never used in training or previous tests on same/different (e.g., five-corner white paper), or included at least one totally novel object that incorporated colors, shapes, or materials for which Alex might not even have labels (e.g., pink woolen pompon).[7] Often both objects in the pair were totally unfamiliar to Alex. The items could embody colors or shapes he had probably seen, such as white or round (e.g., on clothing or foods), but could not label; the objects could be items, such as toy cars, with which he at the time had no experience. To avoid a possible fear response (Zentall and Hogan 1974; Zentall et al. 1981) from Alex with respect to novel items, such objects (e.g., toy cars) were shelved and handled in his view for several days prior to testing but never used in training, and the colors, shapes, or materials of these items were never labeled.

Objects in the third collection were not novel, but the focus of the questions differed from that for the other two collections. The reason for this shift was that I was concerned that Alex might not be attending to our queries when formulating his answers, but merely be responding by assessing the physical characteristics of the objects. By looking at the objects, he could have determined the one attribute that was same or different, and simply responded on that basis. Thus, at random intervals, I asked questions for which either of two (of the three) category labels could be the correct response: I showed Alex a yellow wooden triangle and a blue wooden triangle and asked "What's same?" If he was ignoring the content of the question and answering instead on the basis of object attributes and prior training, he would have responded with the one wrong answer, color; if he was answering the question posed, he would have produced one of the two possible correct responses. Having two possible correct answers also provided additional protection against expectation cuing (Pepperberg 1987a).

The Time Needed to Learn the Task

Training Alex on the same/different concept took approximately 9 months, and questions have been raised as to whether the time involved was a consequence of the difficulty of the task or an indication that same/different is not a natural concept for a parrot to acquire. Three reasons other than task difficulty, however, were probably responsible for the amount of time Alex needed to learn the task. An unprovable but plausible suggestion can also be made concerning the relevance of same/different to birds in nature.

One reason for Alex's slowness in learning the task may have involved difficulties in vocal production. Alex must learn to control his syrinx, tongue, and beak (and possibly other parts of his vocal tract) for each novel sound he acquires (see, e.g., Warren, Patterson, and Pepperberg 1996). Thus vocalizations that include novel sounds take considerably longer to learn than those that recombine familiar sounds. Alex, for example, learned "grey" after hearing it in only six sessions; such prompt learning was probably facilitated by prior acquisition of the requisite phonemes "gr" from "green" and "ā" from "pāper" (Pepperberg 1983a). In contrast, the total novelty of the sounds in "mah-mah" might explain the 9 months needed for its acquisition.

Another reason for Alex's slowness might have involved the need to retrain his use of "color" and "shape" and the amount of training time available. For Alex to use "color" and "shape" as labels separate from the phrases "What color?" and "What shape?" (acquired incidentally from trainers in a previous study), he had to learn to separate these phrases into parts and to use familiar utterances in novel ways. Furthermore, concurrent training on number labels proportionately reduced the total time available for training all labels for the same/different task. Focusing on same/different, however, might not have improved Alex's speed of acquisition because of "boredom." Session length depended upon Alex's willingness to attend, and, as noted above, his motivation for this task was not high.

Possibly, too, Alex's slowness in learning the requisite category labels did not reflect his comprehension of the same/different concept. Early in training, he often gave the label of the instance of the category that was same (e.g., "yellow") or one of the two that was different ("three-corner"). Such data, although not proof, suggest early comprehension: Although Alex may merely have been providing general information about the presented objects, the information correlated with the specific question we posed. Comprehension often precedes production in apes (Savage-Rumbaugh et al. 1986) and children (Goldin-Meadow, Seligman, and Gelman 1976; Benedict 1979; Snyder, Bates, and Bretherton 1981); the same may hold true for parrots.

Despite Alex's slowness, we felt that same/different was a concept that should be relevant to birds in nature, and thus readily learned. As noted earlier, some level of same/different knowledge may be relevant to a bird's survival, with respect to individual recognition and song classification, or other aspects of daily life. A bird faced with green and red fruits, for instance, might base its food choice not only on the recollection that an object of the same color (red) and particular shape (oval) indicates ripe and tasty and that green and the same shape indicates unripe and bitter, but also on the recollection that the same color (red) but different shape (round) indicates spoilage and hence the potential for illness. Of course, in these situations, we cannot prove that the bird understands exactly what is same or different, or if the bird can label what is same or different, but can hypothesize only that recognizing sameness and difference may be adaptive.

The Results of Alex's Tests

Alex's test scores demonstrated clear knowledge of same/different (Pepperberg 1987a). He performed at better than chance levels on trials with familiar objects, on trials with novel objects, and on probe trials designed to examine his comprehension of the actual questions. Test details are particularly informative.

For tests involving familiar objects, Alex's score was 99/129 (76.6%) for all trials (first trials plus correction procedures), 69/99 (69.7%) on first trials, $p < .001$, binomial test, chance of 1/3; see Table 5.1. Choice of 1/3 for chance was conservative, because it ignored the possibility

Table 5.1. Alex's responses to "What's same?" or "What's different?" for pairs of objects that were familiar—similar but not identical to those used in training. (After Pepperberg 1987a.)

Question	Correct response (no. times made)	Incorrect response (no. times made)
What's same?	Color, 16	Shape, 2; matter, 2
	Shape, 16	Color, 4; matter, 2
	Matter, 16	Color, 3; shape, 2
What's different?	Color, 17	Shape, 2; matter, 2
	Shape, 16	Color, 3; matter, 4
	Matter, 18	Color, 3; shape, 1

that Alex could have produced many vocalizations besides "color," "shape," and "matter"—he could have given labels for relevant colors, shapes, or materials, as well as any other item in his repertoire. In all cases, however, his first single vocalization was a category label. Other phrases he produced encoded requests for other items or actions (e.g., "Want X") and were not errors in identifications. Similarly, his score for just those pairs made of objects that were no longer novel (additional presentations of items previously presented as "novel"), but that contained a color, shape, or material he couldn't label (e.g., plastic) was 13/17 (76.5%) for all trials ($p = .0003$), 10/13 (76.9%) for first trials ($p = .0014$).

Of particular importance were Alex's scores on trials involving novel objects. Such trials are considered formal transfer tests because they examine how well he transfers his knowledge of same/different to new situations. On these tests, Alex's score was 96/113 (85%) on all trials, 79/96 (82.3%) on first trials, $p < .0001$ with chance again at 1/3 (Table 5.2). A list of some object pairs in the order of their presentation and Alex's responses are in Table 5.3. Item pairs and queries are listed in the order they were presented, and Alex's responses, except for requests for other items ("I want X"), are reported in the order they were produced. His scores for pairs containing one versus two totally novel objects (respectively 86% and 83% for first trials, $p < .0001$) differed little. Based on a test for differences in proportions (.05 confidence level), his scores were not significantly better for questions involving novel items versus familiar ones for all trials, and were only just significantly better when examined for first trial performance only. At first, I was perplexed by Alex's accuracy, because subjects usually perform less well on transfer tests; I was concerned that he was being cued in some manner. Close analysis of our testing situation revealed no cuing,

Table 5.2. Alex's responses to "What's same?" or "What's different?" for pairs of objects significantly different from those used in training, including objects made of colors, shapes, and materials for which he might not have labels. (After Pepperberg 1987a.)

Question	Correct response (no. times made)	Incorrect response (no. times made)
What's same?	Color, 17	Shape, 1; matter, 2
	Shape, 16	Color, 3; matter, 1
	Matter, 17	Color, 3
What's different?	Color, 15	Shape, 1
	Shape, 16	Color, 1
	Matter, 15	Color, 3; shape, 2

but rather an interesting phenomenon. Remember that the objects themselves were Alex's primary reward, and that his incentive to pay close attention to both objects and his responses was greater when these rewards were new and potentially interesting to chew apart, to try to eat, or to use for preening. In other words, we had given him the equivalent of Christmas in July, and he seemed to take advantage of the situation (Pepperberg 1987a).

Finally, results of the probe tests demonstrate that Alex was indeed processing the questions, as well as responding to the physical properties of the objects. Data are presented in Table 5.4. His score was 55/61 (90.2%) on all trials, $p < .0001$, chance of 2/3 (note that two possible answers were correct); 49/55 (89.1%) on first trial performance, $p < .0001$.

The Implications of Alex's Performance

Alex, if not all parrots, could master the concept of same/different according to Premack's criteria (1978, 1983). His performance can be examined with respect to three issues: (1) How did he compare to chimpanzees? (2) Did he demonstrate abilities beyond those shown for pigeons? (3) Was his performance affected by his capacity to communicate with humans via English speech?

Alex's test conditions, although not identical to those of Premack (1976, 1983), were at least as rigorous as those of the initial chimpanzee study. Alex's task involved symbolic comprehension of "same" and "different" in contrast to the production task commonly used with language-trained chimpanzees; that is, Alex responded *to* questions of "What's same?" and "What's different?" denoting exactly what attribute

Table 5.3. Some sample data for Alex's results for trials involving the first presentation of objects that were never part of the same/different training or that were entirely novel. C refers to color, S refers to shape, and M refers to material; "1" designates Alex's first response, "2" his second response. (After Pepperberg 1987a.)

Object pair	Question		Responses		
	Different	Same	C	S	M
5CWhP-3CWhP	x			1	
©RH-4CYH		x			1
4CGPlay-Doh-©GPls		x	1		
YWoolenBall-PkWoolenBall	x		1		
Peg Wood-©OWBead		x			1
MtlCup-3CRK		x			1
CurtainHook-©GyK	x			1	
©YPls-Washer		x	1		
Mostaciolli-Shell	x			1	
4CPkPlay-Doh-4CPpPlay-Doh	x		2	1	
YMtlTruck-3CPpK		x			1
RWBead-PpWBead	x		1		
RWBead-WhWCube		x			1
BCstClay-BCstP	x				1
Banerry-4CRK		x	2		1
RWCube-GWCube	x		1		
©OK-2COPlsBead		x	1		
3CPkPlay-Doh-4CPkPlay-Doh	x			1	
BPlsBox-5CBPlay-Doh		x	1		
6CRPls-4CRPlsK	x			1	
OMtlBox-BPlsBox		x	1	2	
PpCube-©PpBead	x			1	
GyPlsPen-YWPencil		x		1	
GMtlCar-GCube		x	1		
GSpool-BrSpool	x		1		
RPlsHeart-RPlsRing	x			1	

was same or different, whereas chimpanzees responded merely with the labels "same" or "different." We did not test, and thus still do not know, whether Alex can learn to respond with respect to analogic "relations between relations"—whether he can understand questions such as "Is the relation between A1 and A2 (same/different) as that for B1 and B2?" (Premack 1983). This latter task is the strongest test of the concept of same/different.

Alex's task was considerably more complex than that given pigeons. He was shown object pairs—including objects never before tested—

Table 5.3.   (continued)

Object pair	Question		Responses		
	Different	Same	C	S	M
Penny–Dime	x		1		
Die–GWCube		x	1	2	
GChalk–WhChalk	x		1		
RPlsCup–YPlsBarrel		x			1
GyMtlBox–GyWCube	x				1
GTape–YTape	x		1		
BrPlsCube–BrWCube	x				1
YPlsBox–BPlsScrew		x	1		2
GyLeatherShoe–GyLeatherGlove	x			1	
PkMtlNeedle–YPencil		x		1	
RPlsRing–YPlsTube		x			1
RMtlSpangle–BMtSpangle	x		1		
Almond–2CGW	x		1		
GPlsBead–GWBead	x				1

Code: R = rose, G = green, Y = yellow, O = orange, B = blue, Gy = grey, Pp = purple, Pk = pink, Wh = white, Br = brown, C = corner, © = round, Pls = plastic, H = rawhide, W = wood, K = key, P = paper, Mtl = metal, Cst = crescent.

simultaneously, rather than successively, so that, unlike animals that are trained on successive match (or oddity) to sample, he was unlikely to respond on the basis of "old" versus "new" or "familiar" versus "unfamiliar" (Premack 1983:127). Likewise, Alex had to respond symbolically on the basis of both (1) the instances of the categorical classes that the objects represented and (2) the symbolic relationship (the questions involving "same" or "different") posed by the experimenter—and not on the basis of whether two items basically looked alike or were objects he had seen before. Alex's task was therefore unlikely to be interpreted as a forced-choice conditional discrimination of the type generally used to study nonhuman conceptual abilities. Alex clearly had progressed beyond match-to-sample or oddity-from-sample, comprehended the vocal symbols "same" and "different," and demonstrated a more abstract concept of category than in our previous study (Pepperberg 1983a). His responses were not stimulus-specific, because he performed as well on novel objects as on familiar ones, including those he could not label. Thus he appeared to learn the task in terms of the concepts of "color," "shape," and "material" (Pepperberg 1987a).

Table 5.4. Alex's responses to probes designed to determine if he was responding on the basis of the experimenters' questions or to the physical variations in the objects. (After Pepperberg 1987a.)

Question	Correct response (no. times made)	Incorrect response (no. times made)
What's same?	Color, 10	Matter, 1
	Shape, 11	Matter, 2
	Matter, 11	
What's different?	Color, 8	Color 1
	Shape, 7	Shape, 2
	Matter, 8	

Arguably, Alex did learn one type of conditional discrimination: Given two objects, he responded to "What's same?" and "What's different?" with categorical labels, whereas he responded to "How many?" with "two X." Nevertheless, comprehension and use of such relatively complex conditional constraints on the already inherently conditional nature of the labeling task (see the discussion in Chapter 4) probably reveals additional cognitive capacities (Premack 1976:140–142; 1983:129), rather than detracts from Alex's demonstrated abilities (Pepperberg 1987a,b).

Although many researchers have discussed the relationship between cognitive abilities and training on language-like tasks for animals (e.g., Premack 1983 and commentaries therein; also Pepperberg and Kozak 1986; Pepperberg 1987a,b),[8] we cannot determine whether Alex's demonstrated capabilities are the result of his extensive training on object and category labels. Evidence from children suggests that labeling is probably not a prerequisite for comprehending the same/different concept: Children in grades K-2 could more easily state whether pairs of objects were "same" or "different" than label the individual items (Reese 1972). The children's understanding of same/different may, however, have been limited, because they were not asked to label which attributes were same or different. Also, although object discrimination may be enhanced if different objects are associated with different responses (see, e.g., Reese 1972) or rewards (Brodigan and Peterson 1976; Saunders and Sailor 1979; Peterson 1984; also Edwards et al. 1983), the studies cited here did not necessarily involve labeling because the connections between the objects and the responses or rewards were not always referential (see Chapter 2). These studies also generally used a small number of items and repeated presentation of these items. Alex, however, performed equally well for a wide variety of object pairs, even when the colors, shapes, or materials could not be labeled. Most likely, Alex's previous training merely provided him with experience in forming relationships (e.g., between object and referent; see Marschark 1983).

Alternatively, we can view Alex's success on the same/different task as a measure of his communicative competence in what, for him, is an *exceptional* (nonspecies-specific) communication code—that is, as a measure of the degree to which the exceptional code (English vocalizations) has acquired functional significance (Pepperberg 1985, 1986a,b). Communicative competence in birds is often judged by their ability to defend territory and mate successfully (for a recent review, see Kroodsma and Miller 1996). In humans, communicative competence is often considered an indication of cognitive abilities, although the data are clearest with respect to comprehension rather than production (Bates 1993). Obviously, the standard avian index cannot be used to determine the communicative competence of a bird that has acquired some semblance of a human code. Rather, the bird must be tested in human terms. A valid index might be the degree of abstraction of the code in question: whether the subject has not only learned to associate an object or its attributes with particular labels, but also acquired some abstract concept of what these labels represent (Premack 1983). Comprehension of categorical concepts and the concept of same/different suggests that Alex has some level of communicative competence in the human code, although clearly one not nearly as complex as that attained by humans or even other language-trained species (Pepperberg and Kozak 1986; Pepperberg 1987a,b).

In sum, Alex's data suggest that symbolic comprehension of same/different is no longer the exclusive domain of primates. These data are, however, unlikely to determine exactly where the Grey parrot may fit on any relative "intelligence scale" (see, e.g., Thomas 1980), because the training and testing methods are not identical to most paradigms used for nonhumans. Nevertheless, even if tasks indeed exist that can be learned only by primates or maybe even only by humans and great apes, Alex's data show that competence with respect to same/different is not such a task.

6 Can a Parrot Respond
to the Absence
of Information?

> I sometimes amuse myself by imagining an intelligent visitor from
> another planet arriving on this earth just before the differentiation
> of the human stock—say somewhere about one million years ago.
> If such a visitor had been asked by an all-seeing Creator which
> group of animals he supposed would the most easily be able to
> achieve a true language, I feel little doubt that he would have said
> unhesitatingly, "Why, of course, the birds." And, indeed, if we now
> look at the birds together with all the mammals other than man,
> we have little hesitation in saying that the birds are by far the most
> advanced both in their control of their vocalisations and by the
> way in which they can adapt them collectively and individually
> to function as a most powerful communication system.
>
> William Thorpe, *Animal and Human Nature* (1974:129)

An intriguing cognitive ability generally studied in young children
rather than in animals is that of expressing negation, because negation
in its most complex sense is a very advanced concept, and one that is
assumed to require some level of language-like ability (see Litowitz
1993). Children develop use of negation in stages (see, e.g., Bellugi
1967; Bloom 1970; Pea 1980; Goodwin 1985). The first stage involves
rejection: refusing objects or actions, either gesturally or vocally. The
next stage involves comments on absence (nonexistence) or on some
malfunction. The third stage is characterized by comments about items
that are dropped (maybe to the floor) or that have disappeared. The
final stage involves commenting on the veracity of a statement.[1]

Few of these stages have been examined in detail in animals, probably because the earliest stage seems too simple to warrant study and
the more advanced stages too difficult for animals to master. The first
stage can be observed in almost all animal species and, at the most
basic level, be defined as a tropism: instinctual movement away from

a negative stimulus. A slightly more complex level within this first stage also seems relatively common: We often, for example, observe and interpret signals such as the warning bark of a guard dog as an indication that we should desist in whatever action we are performing, and we note the harsh calls of a jay when we intrude upon its territory during a walk in the woods. Thus these behavior patterns are rarely considered interesting from a cognitive standpoint.[2] As for the next, more intriguing stages, parallels between humans and other animals are unclear, and little research has been devoted to examining various animals' understanding of, or capacity to express, negation or nonexistence.

Negation's second stage is particularly interesting. The ability to understand and comment upon nonexistence, or even the slightly more basic notion of absence, although seemingly simple, denotes a relatively advanced stage in cognitive and linguistic development (Brown 1973). An organism reacts to absence only after acquiring a corpus of knowledge about the expected *presence* of events, objects, or other information in its environment—only when a discrepancy exists between the expected and the actual state of affairs (see, e.g., Skinner 1957; de Villiers and de Villiers 1979; Hearst 1984).[3] Thus on a foray to the refrigerator in the late evening you respond to the absence of chocolate cake only if you *expect* the leftover piece from dinner to be present. At this level, researchers have tested many animal species using a Piagetian paradigm (object permanence; see Chapter 10), and some animals do react to the disappearance or nonexistence of items they expect to be present (see, e.g., Pepperberg, Willner, and Gravitz 1997). Bloom (1970), however, suggests that not only comprehension but also verbal production of terms relating to nonexistence is necessary before an organism can be considered to have acquired the *concept* of nonexistence.

Experimental demonstration of the concept of nonexistence thus can be difficult, even in humans. For example, humans perform better on tasks that require searching for the presence rather than absence of certain features, or making judgments based on affirmation and occurrence rather than on negation and nonoccurrence, even if the tasks are purportedly of equal difficulty (Hearst 1984). The problem does not involve comprehension: Even when subjects clearly comprehend the nature of a task concerning nonexistence, performance may not reflect their level of comprehension. This performance/comprehension disparity can be avoided by requiring positive actions in response to nonoccurrence so that subjects can recognize the salience of "absence" or other aspects of negation (Fazio, Sherman, and Herr 1982;

Hearst 1984). Sometimes subjects are helped by receiving a different reward for indicating presence instead of absence (Hearst 1984).

Demonstrating that animals have a concept of nonexistence is even more difficult, although some behavior patterns in nature suggest that birds may be appropriate subjects for the study of this concept. Ethological studies, although not specifically designed to examine nonexistence, show that some birds naturally exhibit positive behaviors that correlate with the absence of certain signals. Some species react to the absence of signs of territorial defense (e.g., song) from conspecific neighbors with obvious acts of territorial invasion (great tits, Krebs 1977; red-winged blackbirds, *Agelaius phoeniceus,* Peek 1972; Smith 1979), and some male warblers change the proportion of song types in their repertoire after loss of their mates (Adelaide's warblers, *Dendroica adelaidae,* Staicer 1996; yellow warblers, *Dendroica petechia,* Spector 1991, 1992; chestnut-sided warblers, *Dendroica pensylvanica,* Kroodsma et al. 1989; golden-winged warblers, *Vermivora chrysoptera,* Highsmith 1989). Each of these reactions, however, corresponds to a single situation, and these reactions may result as much from removal of a factor that was inhibiting their expression (e.g., a neighbor ready to do battle) as from recognition of the absence of a particular stimulus. Such behavior, although suggestive, can thus be considered only indirect evidence for a concept of nonexistence.

Most psychological experiments with animals and even humans also provide only indirect evidence for concepts of nonexistence or absence (reviews in Hearst 1984; Herman and Forestell 1985). Subjects in such studies generally performed "feature negative" discriminations (looked for differences between two samples and reacted positively when one lacked something, such as a small bright dot) and thus probably based their responses on something other than nonexistence (Pepperberg 1988b). These subjects may have responded on the basis of whether they had previously experienced an object (e.g., as part of a test of memory), or whether two items were identical or somewhat dissimilar ("not same")—that is, they used a match-to-sample or nonmatch-to-sample strategy. Moreover, subjects were not required to generalize their responses beyond their very specific training paradigms. In some cases, subjects may merely have learned, over repeated trials, where to look for a difference (e.g., a subject could have learned to check the center of a green square for a white dot; see Sokolov 1963; Hearst 1984). Even if tasks required subjects to respond to the presence or absence of a particular stimulus, little learning took place unless this *single,* arbitrary stimulus was extremely salient. Thus humans often did not learn the discrimination if the critical factor was part of a compound stimulus such as "smoke" emanating from a chimney in a cartoon (Hearst 1984; note Jenkins and Sainsbury 1969, 1970).

For these reasons, human comprehension of nonexistence is often studied descriptively rather than experimentally. Researchers, for example, observe how use of terms for nonexistence and absence emerges as part of the development of the syntax, semantics, and pragmatics of negation (Bellugi 1967; Bloom 1970). Thus animals that have learned human-based communication codes (codes based on rules presumed to underlie human systems; see Premack 1976; Gardner and Gardner 1978; von Glasersfeld 1977) are usually used to compare animal and human negation abilities. At the time I began studying the concept of nonexistence with Alex, some research had indeed suggested that animals were capable of commenting upon the nonexistence of expected items.

Studies with Animals Trained in Interspecies Communication

Marine mammals and chimpanzees that were taught interspecies communication have demonstrated some understanding of nonexistence. A dolphin (Herman and Forestell 1985) and a California sea lion (*Zalophus californianus,* Schusterman and Krieger 1984) learned to respond to questions about the existence of objects in their environment. The dolphin pressed a paddle placed to its left (the "no" paddle) to designate that an object was absent, or a different style of paddle placed to its right (the "yes" paddle) if the object was present. The sea lion "balked" (did not respond) when asked to perform an action (e.g., a flipper touch) on an absent object; it engaged in a visual search but then refused to leave its station.[4] Gardner and Gardner (1978) reported that their chimpanzees used American Sign Language to comment upon the absence of a familiar object at a customary location, and Rumbaugh and Gill (1977) reported similar behavior in a chimpanzee trained to communicate through a computer.

Although laboratory studies on pigeons did not demonstrate comparable understanding of nonexistence (Hearst 1984), some studies suggest that pigeons can use absence in certain ways to accelerate learning (Zentall and Hogan 1978),[5] and my data suggested that Alex might be a viable avian subject for such experiments. He already reacted to an object's absence by saying "nuh" to refuse an object offered in place of one he had requested (Pepperberg 1981, 1987c). He also routinely used "nuh" to reject an object unacceptable for other reasons, such as a toy so well chewed as to be of little interest or possibly unrecognizable. Also, in a Piagetian object permanence study, he reacted strongly to the absence of a specific object he expected to find hidden under one of three plastic cups: He produced a clear "yip" (a sound correlated with being startled), and on subsequent trials in which an object was similarly hidden, knocked over all cups simultaneously with a single motion of his head; previously he had simply turned over the appropriate covering object (Pepperberg and Kozak 1986; see Chapter 10). Such data provide little direct evidence for a general concept of nonexistence, but the results did indicate an awareness that something was not present and that additional study would be interesting.

Two other reasons existed for subjecting Alex to a formal study on nonoccurrence. First, he had demonstrated other conceptual abilities at levels comparable to those of nonhuman primates (labeling, Chapter 3; comprehension of categories, Chapter 4; comprehension of same/different, Chapter 5). Could he perform at a similar level on this task? Second, his training on same/different would provide a practical paradigm for the experiment. Because he already responded to "What's same?" or "What's different?" for pairs of even totally novel objects on the basis of categorical concepts (rather than specific instances) of color, shape, and material, a logical next step was to test whether he could generalize this ability to report on the *absence* of similarity or difference between objects (Pepperberg 1988b).

A Task Designed to Examine Alex's Response to the Absence of Information

Two constraints existed for Alex's task. First, it had to be functionally equivalent to those used with chimpanzees and marine mammals to provide comparative data. Second, the design had to be consistent with Alex's prior learning, including his use of vocal signals.

A task using the same/different paradigm would satisfy these constraints. Given that Alex answered "What's same?" and "What's different?" with a category label ("color," "shape," "mah-mah") rather than a specific object or attribute label, I could extend this paradigm to include queries to which he could reply "none" when information about abstract categories was *not* present—when two objects were completely different or identical. Because he responded correctly on the same/different task to totally novel items, transfer tests on "none" would be straightforward. Such a protocol would ensure the following: First, the symbolic concepts tested would be more abstract than those examined in standard "conditional discrimination paradigms" (Chapter 5 reviews the difference between responses based on match-to-sample or oddity-from-sample versus those based on same/different). Second, correct responses could not be based on remembering an object or its location (e.g., if an item had been previously presented). Third, correct responses would demonstrate comprehension of arbitrary symbols representing attributes of real-world objects. Fourth, the findings could not be dismissed as stimulus-specific associations; to succeed, Alex would have to generalize the concept of absence rather than learn rote responses to particular sets of objects (i.e., first trial test results on novel objects would show if he could transfer his knowledge to objects other than training items). Finally, the performance/comprehension disparity noted in other studies would be irrelevant because Alex would make a positive response to designate absence (Pepperberg 1988b).

In sum, training and testing protocols would replicate those of the same/different study. Alex would be shown two objects that could be identical or differ with respect to one, some, or all three properties of color, shape, or material (e.g., two round rose metal keys; a yellow rawhide pentagon and a grey wooden pentagon; a green wooden triangle and a blue wooden triangle; a yellow rawhide pentagon and a grey wooden square). He would then be queried "What's different?" or "What's same?" The correct response would be "none" or the label of the appropriate *category*, not the specific color, shape, or material marker that represented the correct response (e.g., "color," *not* "yellow"). To be correct, Alex, as before, had to (1) attend to multiple aspects of two different objects; (2) determine, from a vocal question, if the response was to be based on "sameness" or "difference"; (3) determine, by looking at the items, what, if anything, was "same" or "different" (e.g., were they both blue, or triangular, or made of wood?), and then (4) produce, vocally, the label for the appropriate category or the response "none" (Pepperberg 1988b).

The complexity of Alex's task is best appreciated by contrasting it with procedures used in most research on nonoccurrence in animals. Other projects commonly use (1) a two-choice procedure whereby subjects merely indicate whether a *single, learned* stimulus or part of a stimulus (e.g., a green dot) is or is not present or if some correspondence exists between sets of objects; and (2) a topographically similar (and thus possibly easier to acquire) response for both answers (e.g., lever pressing or key pecking; see Michael, Whitely, and Hesse 1983), or a single response versus withholding that response. Subjects in these protocols, which do not require a fine-grained analysis of the same/different paradigm, need not identify the basis upon which they form their response (Pepperberg 1987c). Alex, however, could not answer correctly without processing all information on the several similarities or differences: Pairwise trials of total similarity or difference were to be intermixed with those for which he must respond with the vocal label of *one* out of three dimensions that was same or different depending upon the question asked. Alex's task was also challenging because most animals are less accurate when responding on the basis of whether a *single* attribute is present or absent than when greater degrees of contrast exist (see, e.g., McClure and Helland 1979; note Jenkins and Sainsbury 1969, 1970).

Training

Training, using the M/R technique, began in September 1986 (after the same/different study was completed) and lasted through November 1986; sessions were of similar length and frequency as those on same/different. Training to produce and comprehend "none" involved a small set of objects initially used in the same/different study—items that were red, green, or blue; triangular or square; rawhide or wood. So that Alex would learn "none" in the appropriate context, sessions included questions for which "color," "shape," or "matter" ("mah-mah") were also correct. Training ended and testing began based on the clarity of Alex's production of "none."

Testing

The same general test procedures were followed as in previous studies, and followed the pattern of tests on same/different with respect to questions involving familiar and novel objects (see, e.g., Pepperberg 1987a). Testing began in December 1986 and continued through December 1987. My students and I began with object pairs that were familiar to Alex from other studies (e.g., Pepperberg and Kozak 1986) but that were not used in training any aspect of the same/different task (Pepperberg 1988b). These items combined all additional colors, shapes, and materials available in the laboratory, and included new combinations of objects that had been paired as "novel" items for transfer tests in the earlier same/different study. Thus although individual objects had been used in previous questions on same/different, pairings were always novel. These trials were interspersed randomly with (1) transfer trials that, beginning at the end of May 1987, used totally novel items never before presented on a test (these were items of colors, shapes, and materials that Alex often could not even label, such as a miniature pink rubber flamingo;[6] (2) test trials on unrelated subjects such as vocal comprehension and two-dimensional representations; and (3) training on numerical concepts (Pepperberg 1988b).

Alex was thus shown an item or a number of items, asked questions on any of several topics ("What's this?" "What color is the metal key?" "What shape is the wood?" "How many?" "What's same?" "What's different?"), and had to formulate a vocal English reply from the more than 80 possible responses in his repertoire. Thus, even though the range of correct responses to queries of "What's same?" or "What's different?" was limited to 4 choices ("color," "shape," "matter," or "none"), in any one trial he had to choose from among numerous valid responses to other questions such as "How many?" or "What color?" to be correct (Pepperberg 1988b). Shape and color queries were used for a separate comprehension task (Chapter 8), or as second questions when he responded to "What's this?" by identifying correctly only the material of a colored or shaped object. (Such "generic" answers were considered errors; Pepperberg 1981.) "How many?" was used during concurrent training on numerical concepts (Pepperberg 1987b).

Questions on same/different were asked one to four times per week, and neither Alex nor his trainers could predict which questions on what topic would appear on a given day. No transfer trials were administered during July–August, and few during September of 1987, because of trainer vacations and a laboratory relocation. Some previous trials were repeated during this period to avoid a break in routine, but results of such trials were not included in the data analysis (Pepperberg 1988b).

Alex's Acquisition of "None"

Alex first produced "none" in his trainers' presence after about 7 weeks of modeling. Four more weeks of training were necessary before "none" was reliably recognized aurally by all trainers. Based on transcriptions of taped samples, interobserver agreement for Alex's utterances of "none" was 88.9–92%. Remember that the length of time needed to train him to produce "none" reliably did not necessarily correspond to the time needed for comprehension of the task for which it was being trained (Pepperberg 1988b).

Interestingly, even though Alex had previously acquired the label "one," training to add the "n" took longer than expected.[7] "One" had been trained in the context of numerical work (with respect to "How many?"); conceivably our use of a similar sound pattern in a different context confused Alex and caused him to learn more slowly. Such an explanation is merely speculative, although subsequent studies on other Grey parrots (Chapter 14) seem to support this view.

Results of Tests with Familiar and Novel Objects

Alex's score for questions on objects similar, but not identical, to training items was 76/94 (80.9%) on first trials ($p < .0001$, binomial test with chance at 1/4). Chance at 1/4 was conservative, because it ignored the possibility that Alex could utter anything in his repertoire other than "color," "shape," "mah-mah" (matter), or "none." On all trials (first trials plus correction trials) his score was 94/112 (83.9%). Table 6.1 gives a breakdown of his performance. In all cases, his first single utterance was one of these four labels (Pepperberg 1988b). Other responses encoded requests for other objects or activities (e.g., "I want X!"), and were not actual errors. His score for queries for just those pairs of objects for which the answer was "none" was 20/24 (83.4%) for first trials ($p < .00001$) and 24/28 (85.7%) for all trials.

Alex's score for questions of same/different for novel objects or object pairs was 66/91 (72.5%) on first trials ($p < .0001$ with chance again at 1/4), and 91/116 (78.4%) for all trials (Pepperberg 1988b); see Table 6.2. His score for questions for just those pairs for which the answer was "none" was 18/23 (78.3%) for first trials ($p < .0001$); 23/28 (82.1%) for all trials. As before, Alex's first single utterance in response to our queries was one of the four appropriate labels, although requests for other items or changes of location were also prevalent (Pepperberg 1988b).

The data suggest that Alex shows some comprehension of the concept of absence. His test conditions, although not identical to those for mammalian subjects (see Premack 1976; Gardner and Gardner 1978; Schusterman and Krieger 1984; Herman and Forestell 1985), were structured to provide evidence for comparable abilities. Alex did respond on the basis of the absence of similarity and difference in a manner that was neither a test of memory nor a form of match-to-sample or oddity-from-sample. He responded not to specific instances or sets of objects, or from a repertoire with a limited number of responses, but rather provided a description of attributes, if any, that were shared by two objects. He conveyed his knowledge, by means of arbitrary symbols (English labels), about either the presence or absence of specific categories these objects had in common. This ability extended immediately to items not used in training and to totally novel objects whose attributes often could not be labeled. The actual level

Table 6.1. Alex's responses to queries of "What's same?" or "What's different?" to pairs of objects similar but not identical to those used in training, including training for response to absence. Results are for all trials (first trials plus correction trials). (After Pepperberg 1988b.)

Question	Correct response (no. times made)	Incorrect response (no. times made)
What's same?	Color, 11	None, 1
	Shape, 10	Matter, 1; none, 2
	Matter, 14	Color, 2
	None, 12	Shape, 1; matter, 1
What's different?	Color, 13	Shape, 1; matter, 1; none, 1
	Shape, 12	Color, 2
	Matter, 10	Shape, 1; none, 2
	None, 12	Color,[a] 1; shape, 1

a. The shades of blue of one pair of triangular metal keys were slightly different in that one was somewhat more greenish than the other. This difference might have been more pronounced to Alex than to his trainers, because his peaks of color sensitivity are unlikely to be identical to those of humans. Parrot eyes, unlike those of many other birds, have few oil droplets, and the supposition is that parrots' color sensitivity, although not identical to that of humans, is more similar to that of humans than to that of other birds (Walls 1963). Parrots do, however, see in the ultraviolet, which could affect blue perception (see Bowmaker 1986).

of Alex's abilities might best be appreciated by examining the precautions taken to ensure the abstract nature of the task (Pepperberg 1988b).

Precautions to Ensure That Alex Was Using Abstract Processing

These results might be viewed as primarily providing information about Alex's concepts of "same" and "different" and additional evidence for his comprehension of categorical concepts, but the data also provide evidence for comprehension of absence, that is, recognition that a pair of novel objects may or may not have something in common. Alex's scores suggest that he responded not to specific instances of color, shape, and material (particular sets of items or classes of stimuli), but rather on the basis of abstract concepts, even when the particular attributes of each object were unfamiliar and the specific combinations of color, shape, and material for each item, as well as for the pair, had never before been seen on a test.

Table 6.2. Alex's responses to queries of "What's same?" or "What's different?" to pairs of objects completely different from those used in previous tests, including objects made of colors, shapes, and materials that he could not label. These results are for all trials (first trials plus correction trials). (After Pepperberg 1988b.)

Question	Correct response (no. times made)	Incorrect response (no. times made)
What's same?	Color, 11	Matter, 2; none, 1
	Shape, 11	Color, 1; matter, 2; none, 1
	Matter, 11	Shape, 1; none, 2
	None, 12	Color, 1; shape, 1
What's different?	Color, 11	Shape, 1; matter, 1; none, 1
	Shape, 12	Color, 2; matter, 2
	Matter, 12	Color, 1; shape, 2
	None, 11	Color, 1; shape,[a] 1; matter, 1

a. One of the objects was a perfect cube and the other was a rectangular solid; we expected Alex to respond on the basis of the number of corners, but he responded on the basis of the overall shape.

This study was specifically designed so that Alex could not respond on the basis of matching (recognizing total similarity or identity) or nonmatching (recognizing difference in any or all attributes) of object pairs (Pepperberg 1988b). Because queries on presence or absence of similarity or difference were presented as an extension of the previous study on same/different, the majority of questions were about pairs of objects that still had a single attribute that was same or different— queries for which "color," "shape," or "mah-mah" ("matter") were appropriate answers. Thus completely identical or dissimilar pairs made up only a small proportion of the total possible combinations, and Alex, to maintain his high accuracy, had to attend to the questions and do a feature analysis of the object pairs before responding (see Premack 1983 for further discussion). Specifically, Alex was not trained simply to respond "same" or "not-same (different)" (see Premack 1976). "None" was the correct response only in cases of complete identity or complete difference; partial similarity (or partial difference) required categorical responses.

Furthermore, because probe trials in the same/different study had shown that Alex could successfully report what was different if more than one attribute was different or what was same if more than one attribute was same (e.g., could accurately answer "What's same?" when shown a pink plastic square and a brown plastic square), similar probes could now be given that still required feature analysis and provided a safeguard against responses based on total similarity or difference. Alex could thus be asked "What's same?" for identical objects and thus be required to answer "color," "shape," or "mah-mah" rather than "none." I used two such trials to learn if Alex would perform the task. My students and I did not administer a complete set of such probes because of the large number of trials needed to establish statistical significance with a chance value of 3/4. Nevertheless, for each trial Alex did respond with the label of an appropriate category rather than "none," suggesting that he was indeed both processing our questions and performing some form of feature analysis (Pepperberg 1988b).

The study was also carefully designed to ensure that Alex's responses were not based on memory, recognition of specific stimuli, or some form of generalization rather than on the notion of absence. Because Alex's responses were "none" or a category label rather than a specific object or attribute label, any possible set of objects could be used in testing, including items with which he had never interacted or could not label. Even objects used in other studies could be combined for testing in many different ways. Moreover, he could be queried about labels for seven colors (green, rose [red], blue, orange, yellow, grey, and purple), five shapes (two-, three-, four-, five-, and six-cornered objects), four materials (paper, wood, cork, and hide [rawhide]), various metallic items (key, chain, grate, and truck [toy car]), and be asked to combine these labels to identify about 80 different items. He could also respond to queries about color, shape, and material for objects that incorporated novel combinations of these attributes.[8] He was thus unlikely to respond during testing on "none" on the basis of absolute physical properties of objects and could not learn the answer to a given pair, because the number of possible permutations of question topic, correct response, and combination of objects was extremely large. Moreover, as noted above, labels for all objects, materials, colors, and shapes, as well as for various foods, locations, and quantity, were always in Alex's repertoire as possible answers; yet his first responses, even when incorrect, were always category labels or "none." Such data suggest that he understood the questions and concepts being targeted. Consequently, Alex was not, like animal subjects in other studies (e.g., Premack 1983), limited merely to choosing between physical (or even vocal) symbols representing "same" or "different" (or "present" versus "absent," Herman and Forestell 1985), or even among categorical labels and "none." Nor was he limited to choosing physically between items that were similar to or different from a sample. His responses thus had to be based on abstract categories rather than memory or recognition of specific stimuli (Pepperberg 1987a, 1988b).

Furthermore, Alex had to transfer his knowledge across stimulus domains as well as among instances of each domain; that is, he had to transfer among like and unlike pairs of colors, shapes, *and* materials, all of which varied from the training items. Premack (1976) emphasizes the need for such transfer to show that an animal really understands same/different (and consequently, absence of sameness or difference): Premack claims that transfers within domains simply demonstrate "stimulus generalization"; that is, that an animal trained to respond to yellow-yellow as "same" will probably respond similarly to orange-orange, not because it understands the concept of "same," but because the stimuli are very similar. He suggests that the strongest possible test is to present that subject with trials involving uncolored shapes, because success would demonstrate a capacity not explainable by stimulus generalization.

Having Alex respond "none" also avoided problems associated with the performance/comprehension distinction (Pepperberg 1988b). As noted earlier, subjects required to withhold a response to designate nonoccurrence often performed relatively poorly, whereas subjects required to make a positive response to nonoccurrence performed at a level demonstrating their actual knowledge about nonexistence (see, e.g., Hearst 1984). Requiring Alex to respond "none" in the absence of any similarity *or* difference thus demonstrated his comprehension of absence of such information.

Finally, my students and I were extremely careful to ensure that Alex's responses to test questions were not based on the presence or absence of novelty (e.g., "same" as on a previous trial, Pepperberg 1988b). This precaution was important because, by the time Alex was ready for transfer tests, we sometimes had difficulty finding objects that were physically safe for him to handle or chew and that differed simultaneously in all dimensions from those used in previous transfer tests on same/different. Thus for a few object pairs, only one or two of the three categories were novel; for instance, car-shaped rubber erasers provided only novel material. We therefore counterbalanced questions on such trials so that, for example, the familiar material was not always "same" or the unfamiliar color was not always what was "different." Without such precautions, Alex could have responded on the basis of familiarity or unfamiliarity (Pepperberg 1988b). Premack

(1983) devotes considerable discussion to the problem posed by familiarity/unfamiliarity in studies on concept formation in animals.

Remaining Cautions about Alex's Competence with Respect to Comprehending Absence

Alex, of course, may not have acquired a broad concept of absence or nonexistence. He may simply have learned to respond appropriately within the context of the task, or by first recognizing identity/nonidentity and then by performing a further discrimination. Although neither of these possibilities can be completely eliminated, the data suggest that they are not likely.

Let's first look at the likelihood that Alex's understanding was limited to the same/different task. On a single transfer trial in which he was asked to respond to presence or absence of particular objects, as opposed to their attributes (see Herman and Forestell 1985), the result was equivocal (Pepperberg 1988b). The trial was part of a simultaneous investigation of vocal comprehension (see Chapter 8). In that study Alex viewed trays of various collections of differently shaped and colored objects of different materials and was asked questions such as "What color is the metallic key?" or "What object is three-cornered?" (Pepperberg 1990a). He could answer correctly only by processing the information in the question and eliminating irrelevant objects from consideration. For the first question above, for example, the tray might include paper, metal, and plastic keys, a toy car, and a metallic nail file; Alex would have to focus on the concept of color and ignore other concepts and all objects other than the key (see Essock, Gill, and Rumbaugh 1977; Granier-Deferre and Kodratoff 1986). We adapted this task to test generalization of "absence" by asking "What's purple?" in the absence of anything purple. Instead of responding "none," Alex said, "Want grape." Although the likelihood of his requesting that particular item was small (e.g., 1/11 even if we limit "chance" to foods rather than to his entire repertoire), the request might not have been correlated with the absence of a purple item because requests for grapes as well as other foods occur frequently. Only additional study to learn if Alex would respond to the absence of yellow items with "Want banana" or "Want corn," or absence of orange items with "Want citrus" or "Want carrot" (i.e., requests for items that come in only a single color) could test such a correlation. Data from a subsequent study (Pepperberg and Brezinsky 1991), in which Alex was shown two objects and asked to respond to questions of "What color bigger/smaller?" did suggest that he might extend his understanding of absence to a different task; those data will be discussed later (Chapter 9).

One might argue that Alex responded "none" to questions of "What's different?" for identical pairs not because he understood the concept of nonexistence, or because a feature analysis provided no basis for difference, but rather because he quickly recognized the special case of identity. In some studies, a short latency to respond suggests such a mechanism (e.g., Dooling et al. 1987). The implication is that the subject first decides "identical" versus "not identical," and, if the latter, *then* decides what is "not identical"; thus responses to identity are quicker than responses to other situations. Alex's response latency, however, would not provide relevant information, because his readiness to respond to a set of objects is correlated with his level of interest in obtaining these items rather than with any other factor (Pepperberg 1987b,c). Three other pieces of information, however, suggest that he did not view questions on difference as a special case: First, he was no more accurate on these questions than others involving same/different; second, he responded appropriately when asked to report on an attribute that was same for an identical pair; and, third, if queried "How many?" for an identical pair, he consistently responded "two X" (see Pepperberg 1987b). The last two points specifically demonstrate that he was not responding to the qualities of objects, but to the questions we posed. In any case, questions on difference for identical pairs could not be omitted because they were needed as a contrast to "What's same?" for totally dissimilar pairs—that is, Alex could not be allowed to regard "none" as a response solely to "What's same?" (Pepperberg 1988b).

The Implications of Alex's Data

The data suggest that Alex has at least limited use of the concept of nonoccurrence or absence (see, e.g., Bloom 1970), and that his use is directly comparable to that of other animals that have undergone similar training (see, e.g., Rumbaugh and Gill 1977; Gardner and Gardner 1978). Whether Bloom's criteria concerning verbal production are necessary for determining an animal's understanding of absence is unclear, given that marine mammals respond to absence appropriately, even without verbal production, at the same level as that of a child—a sea lion indicated nonexistence by "failing" to respond to an absent object (Schusterman and Krieger 1984); a dolphin touched either of two paddles to indicate presence or absence (Herman and Forestell 1985). Alex's response level may have been more sophisticated, but we do not know if his comprehension was greater than that of the marine mammals. All we can state with respect to his overt "language-like" training is that a two-way communication code allows us most efficiently to examine such concepts as same/different and absence, and, in particular, enables us to perform cross-species comparisons in as direct a manner as possible (Pepperberg 1986a).

7 To What Extent Can a Parrot Understand and Use Numerical Concepts?

> Since mathematics is a product of culture, an individual's understanding will depend largely on local cultural practices, and especially on educational practices. Cultures vary widely in their use of mathematics and in the ways in which their members experience mathematics in formal and informal settings. For example, some cultures appear to use mathematical concepts scarcely at all, and their languages contain no words for the numbers, though there is no systematic research published on numerical abilities in these cultures . . . Research on children suggests that there is a set of core mathematical concepts we all possess, independent of our cultural experiences. It is still not clear whether these concepts depend on a specific component of our genetic endowment, which we perhaps share with other species, or whether we construct them using intelligent inference from more general ideas.
>
> Brian Butterworth, Editorial in *Mathematical Cognition* (1995:1)

Studying numerical competence in animals is particularly challenging. Even for humans, researchers disagree on what constitutes various stages of numerical competence (see Gelman and Gallistel 1986; Davis and Pérusse 1988; Gallistel 1988; Fuson 1988; Siegler 1991; Sophian 1995) and which are the most complex, advanced stages (see Fuson 1988, 1995; Gallistel 1988; Frydman 1995; Starkey and Cooper 1995). Furthermore, controversy surrounds the role of language in numerical competence, a particularly important point for animal studies. Some scientists (e.g., Lenneberg 1971) argue that development of language and development of number skills depend upon the same underlying cognitive capacities (see Dehaene 1992); others believe that labeling merely separates the less and more advanced numerical skills (e.g., Gallistel and Gelman 1992). Moreover, although data show that for some brain-damaged individuals, "numerical abilities are severely impaired when other cognitive abilities are intact, and that they can be spared when memory, language, and reasoning are all severely compromised" (Butterworth 1995:1), other data refute the "notion of a 'center for calculation' or a single brain area where numerical knowledge would be concentrated" (Dehaene and Cohen 1995:115). This dichotomy may be resolved by findings suggesting that basic number meaning (knowledge of quantities and their relations) is probably encoded in a particular brain area (intraparietal cortex, with correlates across species), but that other numerical abilities require coordination among many areas (possibly lacking in nonhumans; Dehaene, Dehaene-Lambertz, and Cohen 1998). Such data render animal studies even more daunting.

Before I began number work with Alex, many researchers had investigated nonhuman numerical capacities (review in Rilling 1993; for additional references, see Koehler 1950; Ferster 1964; Thomas and Chase 1980; Davis and Memmot 1982; Seibt 1982). Few had, however, examined animals that had also been trained with human-based communication codes (e.g., Dooley and Gill 1977; Matsuzawa 1985; Matsuzawa et al. 1986). I was particularly interested in learning how a parrot with considerable experience in vocally labeling and categorizing objects, colors, and particularly shapes (see Chapters 3, 4, 5) might fare on numerical tasks. Such a study would not only yield insights into Alex's processing abilities, but, like my other projects, also provide cross-species comparisons with "language"-trained primates. Given that previously studied Grey parrots had demonstrated sensitivity to quantity and basic concepts of numerosity and numerousness (see, e.g., Koehler 1943, 1950; Braun 1952; Lögler 1959), I thought that Alex would be a good subject. Detailed understanding of the issues surrounding number studies are, however, necessary to appreciate both the complexity of the task I and my students were undertaking and Alex's eventual level of competence; I therefore review various numerical tasks used with animals, describe how Alex was trained, and try to evaluate how his "language" abilities might have affected our results.

What Constitutes Numerical Competence?

Humans of normal intelligence generally recognize and label different quantities of objects; this skill uses number as a descriptor or categorical marker (Gallistel 1993). This skill can be further broken down into various levels and differs, according to many researchers (e.g., Fuson 1988; von Glasersfeld 1993), from "counting," in which a subject (1) produces a standard sequence of number tags; (2) applies a unique number tag to each item to be counted; (3) remembers what already has been counted; and (4) knows that the last number tag used tells how many objects are there (Fuson and Hall 1983). Another enumeration process, "subitizing," exists that also differs from counting and supposedly uses preattentive mechanisms in a fast, effortless, perceptual apprehension of quantity ≤ 4 (Taves 1941; Kaufman et al. 1949; Stevens 1951; Klahr 1973; Atkinson, Campbell, and Francis 1976; Aoki 1977; Oyama, Kikuchi, and Ichihara 1981; Wolters, van Kempen, and Wijlhuizen 1987; Dehaene and Cohen 1994). Not surprisingly, researchers argue as to the relative complexity of subitizing and counting and the validity of the "≤ 4" value to categorize subitizing (e.g., Davis and Pérusse 1988; Gallistel and Gelman 1992; Starkey and Cooper 1995). Probably the best review (though now slightly outdated) of these arguments is in Davis and Pérusse (1988) and the commentaries following their article; I thus provide only a summary. I start by describing forms of numerical competence that involve categorization because that was the subject of Alex's first set of experiments; the second set of experiments began to tackle the subitizing versus counting problem.

Number as a Descriptive Category

A subject, human or animal, who uses number as a descriptor or category need not specifically "label" a group of items as "three" or "six" to demonstrate recognition of quantity (see Pepperberg 1987b). Most animal studies prior to the 1970s avoided the issue of symbolic labeling by having subjects (1) choose a particular set of items from among several competing arrays, presumably based on their recognition of quantity; (2) choose between observable arrays or their symbolic representation with respect to "more" versus "less"; (3) perform match-to-sample problems for various quantities and denote quantities in separate, simultaneously presented arrays as "same" or "different"; or (4) respond with a specific behavior to a particular quantity of sequential events. When I began number work with Alex, only Ferster (1964; Ferster and Hammer 1966) and Matsuzawa (1985; Matsuzawa et al. 1986) had demonstrated some level of symbolic labeling in chimpanzees (for subsequent studies, see Boysen 1993). Differences among the above tasks, and why they often failed to demonstrate number comprehension, are best clarified through examples.

In the first task, subjects such as canaries (Pastore 1961), rhesus monkeys (*Macaca mulata*, Hicks 1956), and a raccoon (*Procyon lotor,* Davis 1984) ignored color, shape, size, brightness, pattern, and so forth, to select one quantity from simultaneously presented arrays of other quantities. If the designated quantity was three, subjects supposedly responded based on "threeness." Animals in these studies, however, did not respond to different quantities based on different cues presented at random, responded to only one quantity at a time, and could not (or were not given tests to) transfer their responses from simultaneous to sequential presentations. A conservative interpretation of the results is that these animals recognized the particular aggregation as a single perceptual unit that provided reward (Mandler and Shebo 1982; von Glasersfeld 1982; Pepperberg 1987b).

In the second task, subjects such as chimpanzees (Hayes and Nissen 1956/1971; Dooley and Gill 1977), squirrel monkeys (*Saimiri sciureus*, Thomas, Fowlkes, and Vickery 1980; Thomas 1992), rhesus monkeys (Lin and Gong 1989), pigeons (Honig and Stewart 1989; Honig and Matheson 1995; Emmerton, Lohmann, and Niemann 1997), and rats (e.g., Church and Meck 1984) probably used prenumerical abilities to choose between arrays of "more" versus "less"; such choices did not require recognizing a specific quantity (note Koehler 1950). Even when animals symbolically indicated "more" versus "less" and objects varied in size, brightness, and so forth, the task still did not require labeling specific quantities (Pepperberg 1987b).[1]

In the third task, animals succeeded to different extents on simultaneous match-to-sample problems, indicating if two arrays were identical in quantity. Grey parrots, ravens *(Corvus corax),* and jackdaws *(Corvus monedula)* succeeded for quantities up to eight; pigeons reached five or six and chickens managed two or three (Koehler 1943, 1950; Braun 1952; Lögler 1959). Chimpanzees initially reached only four (Hayes and Nissen 1956/1971; Woodruff, Premack, and Kennel 1978; Woodruff and Premack 1981), but in subsequent studies succeeded for quantities up to eight (Boysen 1993). How this task relates to labeling of quantity or counting is, however, unclear: Koehler called the match-to-sample task "thinking in un-named numbers" or "nonnumerical counting," because distinguishing same versus different requires only a 1:1 correspondence and not necessarily understanding or perceiving actual quantity (Gelman and Gallistel 1978; Fuson and Hall 1983). Also, the mechanisms involved in match-to-sample versus labeling are probably different: At least in humans, separate brain areas apparently mediate these two tasks (Geschwind 1979).

In the fourth task, animals produce a particular action in response to a certain number of events (review in Pepperberg 1987b). Thus a rat responds to auditory sequences of three but not two or four (Davis and Albert 1986). According to Seibt (1982), however, these data indicate that animals merely form a 1:1 correspondence with some internal pattern such that external events are judged as to whether they match this pattern. Koehler's birds may have performed a more sophisticated version of this task: They learned to open boxes randomly containing zero, one, or two baits until they obtained a fixed number (e.g., four). The number of boxes to be opened to obtain the precise number of baits varied across trials, and the number being sought depended upon independent visual cues: Black box lids denoted two baits, green lids three, and so forth. Koehler claimed that his birds simultaneously learned four different problems of this kind. He did not state, however, whether different colored lids were presented randomly in a single series, and thus whether colors indeed "represented" particular quantities.

The question, then, was whether Alex, like Matsuzawa's (1985) chimpanzee, could go beyond these tasks and use number as a categorical label. Specifically, could Alex learn to use vocal numerical labels to distinguish various numerical arrays and to denote an attribute of the collection, the same way he labeled its color? If successful, could he generalize (transfer) his behavior to any novel collection, including novel items placed in random patterns? And, finally, could he respond to a collection solely in terms of its quantity, that is, respond with equal accuracy to mixtures of different objects (e.g., balls and blocks)? If so, he would demonstrate numerical competence comparable to that of chimpanzees and young children (Pepperberg 1987b).

Training and Testing Alex to Label Arrays of Familiar Objects

We used M/R training and the testing procedures of previous studies, with one adaptation because Alex had already used the number labels "three" and "four" with respect to shapes: I had deliberately chosen the phrases "three-corner" and "four-corner" for triangular and square objects to introduce a preliminary concept of labeling quantity (Pepperberg 1983a). By September 1979, Alex could accurately identify (~80%) "three-corner" and "four-corner" items made of wood, paper, or rawhide; occasionally, however, he called a wooden triangle "three wood" or a paper square "four paper" (Pepperberg 1981). These occurrences were usually treated as simple errors. Rarely, a questioner repeated Alex's response along with an appropriate set of objects— e.g., held up three pieces of wood and said, "Here's three wood"— then re-presented the triangular item with the query, "Now, what's *this?*" Alex produced the correct shape response after one or two such interactions. When we began training on numbers, however, we used these instances to model appropriate responses to collections of items (Pepperberg 1987b). Alex quickly learned to respond correctly to queries of either "What's this?" or "How many?" We built on these abilities to train him via M/R sessions on "two" and "five" with respect to object arrays rather than shaped items. We used two different sets of objects (two keys or wooden sticks; five pieces of paper or wood) to train each quantity so that Alex would not associate the label with one particular set, but instead focus on the sole aspect—that of quantity—that the collections (the five pieces of wood or paper) had in common.[2] Using a limited number of sets simplified testing Alex's ability to generalize to sets of familiar objects that had not been used in training (e.g., corks); unfamiliarity of test items (and possible fear responses to novel items) would not then bias his results (Zentall and Hogan 1974; Zentall et al. 1981). Training on "five" began only after testing began on "two"; training on "six" began 2½ years after Alex acquired "five," when a hissing noise entered his repertoire; sibilants had been problematic until then (Pepperberg 1987b).

Training Alex on number tasks was challenging. He quickly lost interest during sessions of entirely numerical queries on only two different sets of objects (e.g., keys and wood). Although he would quantify a collection when he found he could not obtain any object until he provided the label for quantity as well as material (e.g., "five wood," not just "wood"), acquisition of several identical objects did not provide enough incentive for him to continue the task beyond two or three trials. He then ceased to work, began to preen, or interrupted with successive requests for other items (see Pepperberg 1983b). This problem was solved by intermingling training (and later testing) on quantity with training on labels for other novel objects or colors (Pepperberg 1987b).

During training and initial testing, we always presented objects by hand (Pepperberg 1987b). This procedure did not rule out the possibility that Alex recognized a particular pattern rather than a particular quantity, but at this stage, we were not interested in the actual mechanism he used to identify collections. We simply wanted to see if he could categorize with respect to something related to quantity as well as color, shape, and matter. However, by presenting collections in the same manner as all other items used in training and testing, we avoided undue emphasis on quantity. Note that Matsuzawa (1985) also presented objects by hand in his numerosity study on the chimpanzee.

Tests were carefully designed (see Pepperberg 1987b). Alex was never tested on exactly the same items used in training. Even for familiar items (corks or wood), test queries always involved pieces of somewhat different sizes or shapes than those used in training. Moreover, to be correct, Alex had to include both number *and* object label, that is, respond "three cork" or "five wood." We required him to identify material so that we could maintain, for later experiments (see below), his awareness of *what* was being quantified and collect information on his ability to combine contextually appropriate labels. We were particularly interested in whether he could generalize to sets of objects that, although themselves familiar, had never been observed as a collection. We did not require use of plural "s" because Alex still had difficulty producing final sibilants; "six," for example, is even now pronounced as "sih." Finally, we tested novel shapes, to learn if he could transfer his knowledge to identify the number of corners on football-shaped ("two-corner"), pentagonal, or hexagonal objects.

Identification tests showed that Alex used abstract labels to distinguish sets of small quantities, even if his mechanism was some form of pattern recognition. His overall score was 78.9% (Table 7.1). Interestingly, over half his errors (~60% of the 20% wrong answers) were "generic"—correct identification of an object ("key"), but omission of the quantity label. Although not a misidentification, his response did not provide all the requested information and was therefore considered wrong; Alex was correct 95% of the time after an additional query of "How many?"[3] Furthermore, because criteria for correct responses included object labels (e.g., "key"), Alex was considered wrong if he correctly identified quantity but mislabeled the item. His scores would be better had I considered only the "numerical" part of his response or omitted generic errors (81.7% and 91.7%, respectively). For his first attempts at identifying novel shaped objects (two-, five-, and six-cornered wood, paper, and rawhide), he erred only once, labeling a pentagon as "wood" (Pepperberg 1987b).

Although, as noted above, Alex might not have used a numerical discrimination to arrive at the correct answers, he was probably not responding to perceptual cues of mass, surface area, odor, or brightness (see Swenson 1970; Pepperberg 1987b). Paper items used in testing were randomly sized strips of white, unlined index cards and "wood" was represented by wooden craft sticks (1 × 11.4 mm) and plant stakes (1.6 × 14.7 mm), both of which, like clothes pins and corks, had often been altered in size and shape by Alex's prior chewing. A cork in a given trial could be half or twice the size of corks in previous trials, depending upon what was available in the stockroom. Only metal keys were constant in size, and performance with these objects was no better than with others; note, however, that colors and shapes of keys varied among trials. Alex was never tested on foods, so odor

Table 7.1. Alex's use of number labels to categorize arrays of objects. U represents an unrecognizable vocalization. (After Pepperberg 1987b.)

Object	Quantity	Score	Error (no. of times)
Paper	2	9/10	Paper (1)
	3	9/11	Paper (2)
	4	9/11	U paper (1), paper (1)
	5	7/11	Paper (2), 4 paper (1), U paper (1)
	6	3/3	
Key	2	10/13	Key (3)
	3	9/14	U key (1), key (3), 4 key (1)
	4	9/11	Key (2)
	5	6/7	Key (1)
	6	3/4	Key (1)
Wood	2	11/12	Wood (1)
	3	7/11	Peg wood (1), 3 peg wood (1), wood (1)
	4	9/11	U wood (1), 3 wood (1)
	5	7/10	3 wood (1), 4 hide (1), 4 wood (1)
	6	3/4	Peg wood (1)
Peg wood	2	9/10	Peg wood (1)
	3	8/9	4 wood (1)
	4	9/12	Peg wood (2), 3 (1)
	5	6/8	4 peg wood (2)
	6	3/4	(6 wood, peg wood) as one phrase
Cork	2	11/16	Cork (5)
	3	9/10	Cork (1)
	4	10/12	3 cork (1), 4 wood (1)
	5	5/5	
	6	2/3	Cork (1)

could not be a factor, and objects he identified precluded factors such as brightness. For shaped objects, size and surface area were not cues, because areas of different batches of wood, paper, and rawhide shapes differed up to 10%, and I deliberately intermingled objects from different batches.

Of particular importance, however, was not Alex's overall score, but his accuracy the first time he had to pair number and object labels for familiar objects presented in *novel collections* (e.g., first presentation of two cork after training on two key and two wood), and his ability to label novel shapes, based (presumably) on their number of corners (e.g., his first response to two-corner wood after seeing only groupings of two pieces of wood). On these first trials he made no errors on material labels (except for once saying "wood, peg wood" in response to one item), and, with the exception of some generic responses, transferred immediately from training to testing items.

Alex's accuracy, however, may not be as impressive as it seems for two reasons. The first involves the format of the test questions; the second involves the arrangement of test objects. Both reasons must be considered, although each can be discounted to some extent.

Alex's task may have been simplified because novel collections included only familiar items and were introduced on a number-by-number basis (Pepperberg 1987b). Although tests involved many different topics and thus only one or two number queries, a novel collection always represented the most recently acquired numerical label. Alex may have used his most recently acquired *numerical* label on seeing a novel collection, obtained a reward, and then memorized the 25 different sets—even though components of each set could vary with respect to size and shape. Similar criticism can be leveled at his transfer from quantities of objects to shaped objects: Novel objects may have been associated with the most recently acquired numerical label.[4] Numerical labels, however, were not the only "novelty" because he was concurrently learning additional color and object labels. Another argument in favor of Alex's competence is that Matsuzawa (1985; Matsuzawa et al. 1986) also examined generalization on a number-by-number basis, but his chimpanzee, unlike Alex, did not immediately generalize to novel collections of familiar items. Nevertheless, we needed to perform more rigorous transfer tests that, in contrast to the work of Matsuzawa, involved collections of entirely novel *objects*.

The question of pattern recognition is also important: Did Alex respond to perceptual units rather than number? Although I knew that Alex's initial tasks would not differentiate between counting and pattern recognition, I did attempt to lessen possible effects of pattern recognition (see Pepperberg 1987b). I ensured that trainers and examiners varied orientation and angular separation when presenting various items; furthermore, no two humans held objects in precisely the same manner (see Matsuzawa 1985 for similar hand-held presentations to chimpanzees). Nevertheless, the number of ways to present two corks by hand was limited. Alex's immediate transfer between quantities of objects and shaped items—e.g., between five wood and five-corner hide, or two key and two-corner paper—suggested that any pattern recognition was not based, at least initially, on a single perceptual unit; but once a pentagonal shape had been related to five, and a "football" to two, subsequent shape recognition could have been by form. Thus Alex's data did not preclude the possibility that identifications were based on similarity either to regular polygons or to approximately linear patterns. Mandler and Shebo (1982), for example, found that "canonical patterning" (e.g., use of a triangular array for three, a square for four) simplified recognition of quantity for numerosities greater than three (see von Glasersfeld 1982). Clearly, I needed to test Alex on novel arrangements of objects to preclude pattern recognition and learn whether he was sensitive to quantity.

Experiments Involving Novel Objects and Arrays

Alex was thus trained to identify collections arrayed linearly on a felt-covered metal tray. Use of the tray would allow us eventually to present all types of novel objects and random patterns. Training with the tray also included nonnumerical sessions so that the tray itself was not a cue for numerical responses (Pepperberg 1987b).

Forcing Alex to respond to essentially a repetition of tasks on which he had been trained and tested for several years presented a challenge. His interest in the tray itself, its cover, and the noise he could make by drumming his beak on its surface initially impeded progress. Other disruptive behavior patterns could occur. He might refuse (by saying "nuh") to identify even the material of the now very familiar objects, walk along the perch away from a trainer (it was considerably harder to make him attend to objects on a tray than to a collection of items waved in front of his face), or request objects not in sight ("I want X") and successive changes of location ("Wanna go Y"). Fortunately, the task also required that we introduce quantities of possibly interesting novel objects whose labels were unknown (washers, thimbles, wing-nuts, etc.). As in other studies, we placed these items in Alex's view for several days before their use to avoid a possible fear response (Zentall and Hogan 1974; Zentall et al. 1981).

Before we could test Alex on novel arrays, he also had to learn to produce only numerical labels upon presentation of sets of items. This protocol was another significant departure from the standard task. We did not take data on Alex's transfer to the first sets of new objects: Too many simultaneous changes in procedure existed for transfer tests to be meaningful. We worked with Alex on both new and old objects on the tray until he regained his ~80% accuracy (Pepperberg 1987b).

I then purchased a second set of novel objects (e.g., small bottles, raisins, wooden spools, erasers, candied almonds).[5] I allowed Alex to see, but not interact with, these items for a few days. I also used wooden beads from the "same/different" study (Chapter 5). I wanted to test the effect of introducing novel objects separately from the effect of introducing novel patterns; I thus placed objects in linear arrays, but varied spacing between objects so that array length could not be a cue (Pepperberg 1987b). I presented two different quantities of each item to see if Alex would transfer between different quantities of the same object: I wanted to learn whether, once having seen and labeled two novel items, he might confound the label "two" with the objects themselves. This procedure meant that he was not questioned on all quantities for each of these novel items.

Alex succeeded on this task. For first presentations of novel collections, he scored 15/20 (75%); overall, he succeeded in 20/25 trials, or 80% (i.e., he was always correct on his second response if not his first). Only three errors were on quantity; in two other cases he used familiar labels to identify objects he couldn't label (e.g., "key" for toy cars). These results suggested that Alex could proficiently identify quantities of entirely new objects that varied considerably in size and shape from original training items: He had not simply learned to identify sets of familiar objects. Because all but one set of transfer items was entirely novel and accuracy was based on first trial data (Gardner and Gardner 1978), our tests were significantly more stringent than those used in the study on labeling of quantity by chimpanzees (Matsuzawa 1985; Matsuzawa et al. 1986).

To determine whether Alex would be equally accurate if objects were presented in random arrays, I queried him about sets that, except for the first trial, were formed by tossing objects onto the surface of the tray (Figure 7.1). Arrangements were altered only if an object was occluded by placement of other objects. In the first trial, six candies were aligned in two rows of three because a student misunderstood the instructions; however, Alex had never before encountered this pattern. I used a variety of novel and familiar objects, and, as before, two different quantities for each item.

Alex also succeeded on this task. For first trials on novel presentations, he scored 15/20 (75%); for all presentations, 20/26 (77%). This experiment still did not provide specific information about mechanisms he might have used to arrive at the answers, but did eliminate simple pattern recognition for most questions. The two pencils, for example, landed in an approximate "X" pattern; the two short cylinders of pasta fell out in a jagged but linear array. The pattern for three thimbles, however, appeared somewhat triangular, but orientation of three wooden craft sticks looked "random" to the human trainer. Such problems did not occur for larger collections.[6]

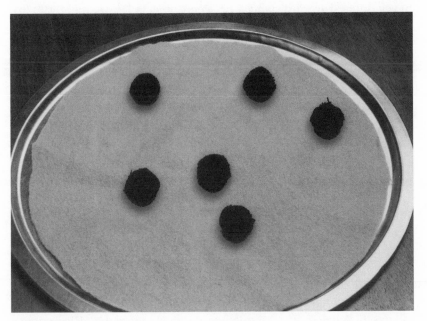

Figure 7.1. Random array of woolen pompons, used to test whether Alex was basing his responses on pattern recognition.

Initial Experiments with Heterogeneous Arrays

Another procedure to determine whether Alex understood quantity as a category, or used another mechanism to label the number of items in an array, was to see if he could label quantity for a targeted subset within a heterogeneous array. The overall pattern of the array would then be irrelevant; to be correct, he would have to observe a multi-object array, decode our question to determine which items were targeted, and then produce vocally the appropriate label for their quantity (Pepperberg 1987b). The task was feasible because Alex could already decode different queries (on color *or* shape) about a single object (Chapter 4); he could probably transfer this ability to an array.

I therefore tested Alex on collections made of two sets of objects. He could label all objects except nails, which were combined with jacks to see how he would respond to items he could not label. Given that he might ignore the "unlabelable," thus simplifying the task, I included only one such set. I aligned items on the tray in two ways: Items in the two sets were either placed side by side in separate groups ("contiguous array") or randomly intermingled (Figure 7.2), but identical objects were never clustered—that is, distances between groups never differed from distances between objects (Pepperberg 1987b). Alex was shown the collections and queried "How many X?" where X was the label for one set of objects (e.g., X = key).

Note that Alex had never before been presented with anything resembling a heterogeneous set except for an accidental grouping of differently shaped and colored keys. Specifically, he had never been given a *collection* that represented a choice of objects about which to comment or act. Thus his responses were of great interest (Pepperberg 1987b).

Alex succeeded (Table 7.2). In 7/10 cases, he produced the appropriate numerical label for the targeted items ($p = .0035$, chance of 1/4, because only 4 number labels were involved). He thus had some ability to look at a heterogeneous array, target a subset based on the question posed, and correctly label quantity even when sets of objects were interspersed.

Alex's responses to the first heterogeneous set were particularly intriguing; I thus provide a transcript of his exchanges with trainers. Although he might simply have erred when asked about the number of keys, he labeled the total number of objects presented. His response was notable because humans label the total number of objects in a heterogeneous set *if* they have been taught to label homogeneous sets exclusively (see, e.g., Siegel 1982; Greeno, Riley, and Gelman 1984). Children (aged 3–4 yrs) initially trained to quantify only homogeneous sets, however, did not spontaneously label the total number of items for collections containing two sets of objects, but quantified the larger set (Gast 1957). Premack (1976) found similar limitations with chimpanzees. Gelman (1980) and Gelman and Gallistel (1986), however, found that children as young as 2½ years have no problem with heterogeneous sets, and suggest that earlier findings were a consequence of the training procedures (cf. Siegel 1982). Gast (1957) and Klahr and Wallace (1973) suggest that children first need to understand how all members of a heterogeneous set fit into one category (e.g., "toy animals") before they can judge the numerosity of a set containing two groups of items (e.g., toy cats and dogs; see von Glasersfeld 1981; Greeno et al. 1984). At the time, I had no evidence that Alex had categorical concepts of "food" or "toy," although he understood categorical classes of color and shape (Pepperberg 1983a) and had begun to comprehend a class based on material. He had had no training on the labels "food" or "toy," though he heard students make comments such as "We're almost out of food," "You need fresh food," or "Let's work with some toys!" I therefore decided to see if Alex could label the overall number of objects in a heterogeneous set. I developed the

Figure 7.2. Intermingled set of objects, used to test how Alex would respond to a heterogeneous array.

following protocol so that I would not first have to train labels such as "food" or "toy" (Pepperberg 1987b).

I presented different quantities of two different subsets of items; one subset had objects Alex could not label (e.g., "washers"), and he was asked about the quantity of this group. As far as he was concerned, questions could mean either "How many (specific) X?" or "How many (total) X + Y?" If he responded to the unlabelable items, he would have excluded a familiar label to identify the quantity of items he couldn't label; I felt such an event was unlikely. If he viewed the set as a "collection," he would label the total quantity. We presented only five trials, because more trials might have clued him into what we were asking; specifically, differences in our vocal queries (e.g., "washer," "thimble," "bobbin," "hearts," "plastics") might have cued him that we were referring to sets of unusual objects rather than the collection as a whole. We nevertheless repeatedly asked him to identify the novel subset, even if he initially correctly identified the total collection; we wanted to see if he could indeed understand that the novel label referred to novel items (Pepperberg 1987b).

Despite the small number of trials and his training on homogeneous sets, Alex's initial responses on the first four groups referred to the

Table 7.2. Data on trials concerning heterogeneous groupings of objects. Alex is asked to produce the label for the number in the targeted subset. A complete transcript of his responses to the first collection is included. C refers to a contiguous grouping of objects in a subset; I to a randomly interspersed array. (After Pepperberg 1987b.)

Object set	Array	Targeted subset	Response	Correct
2 wood, 3 key	C	Keys	Wood	No
	C	Wood	2	Yes
	C	Keys	Key	No
	C	Keys	2 wood	No
	C	Keys	5	No
	C	Keys	3	Yes
2 rock, 3 pennies	C	Rock	2	Yes
2 cork, 3 key	I	Key	3	Yes
2 wood, 3 clips	I	Wood	2	Yes
2 nails, 4 jacks	C	Jacks	4	Yes
2 rose wool, 4 yellow wool	C	Yellow wool	4	Yes
2 cork, 4 pennies	I	Cork	2	Yes
2 key, 3 rock	I	Rock	Rock	No
	I	Rock	3	Yes
2 wool, 5 wood[a]	C	Wood	Wool	No
	C	Wood	5	Yes

a. This test was performed as a probe, to see if Alex would err if the total number of objects in the array were more than he could label.

total number of objects in the heterogeneous collection. When we then said something like "That's the number of *toys;* how many *thimbles?*" he responded with the subset number on the second query. By the fifth and final trial, he produced the quantity for the unfamiliar label (Pepperberg 1987b).

The Implications of Alex's Apparent Use of Number Labels as a Category

Alex could thus label different cardinal sets of objects, even when items were novel and were displayed in random configurations or in mixed sets (Pepperberg 1987b). Yet the results do not elucidate the mechanisms he used to respond correctly. The data did not, for example, tell if Alex could count. Animals that label quantity may not comprehend number labels as a standard ordered set, and any of several noncounting mechanisms may account for their behavior. We could claim only that Alex used number as a distinctive feature in an

Table 7.2. (continued)

Collection 1 transcript: A = Alex, T = student trainer, I = Irene
Pepperberg. Alex is shown 2 plant stakes and 3 keys on a tray;
I does not know what is on the tray.

T: How many key?
A: Wood.
I: Are there any wood?
T: Yes.
I: Try that.
T: OK, tell me, how many wood?
A: Two.
I: Two?
T: Yes. (Alex is given wood; he chews it and a new stake is placed on the tray.)
T: Now, how many key?
A: Key.
T: That's right, key. How many?
A: Two wood.
T: There are two wood, but you tell me, how many key?
A: Five.
I: Five?
T: OK, Alex, that's the number of toys; you tell me, how many key? (To our knowledge, Alex does not understand the label "toy"; the trainer simply responds the best she could.)
A: Three.
I: Three?
T: Good boy! Here's a key!

identification, much like color, which, of course, was the ability we had set out to test. Alex could have subitized—used a perceptual mechanism (Kaufman et al. 1949). As noted earlier, however, considerable discussion exists as to whether subitizing precedes counting or vice versa (Klahr and Wallace 1973, 1976; Mandler and Shebo 1982; Fuson and Hall 1983; Gelman and Gallistel 1986), and I will discuss this controversy in detail later. Alex may have used a form of pattern recognition for the smaller quantities, but some researchers propose a connection between subitizing and pattern recognition that may confound the two processes (von Glasersfeld 1982). Mandler and Shebo (1982) and Maertens, Jones, and Waite (1977) suggest that routine use of canonical patterns (e.g., a rectangular array of dots for four) facilitates recognition of quantities of four or more; for small quantities, subitizing may involve *learning* to use canonical patterns (a doublet for two, a triplet or triangle for three; Mandler and Shebo 1982; Trick, Enns, and Brodeur 1996). Alex's data on labeling quantity for novel and randomly arrayed sets and for subsets of heterogeneous collections suggest that he does not use canonical pattern recognition exclusively; however, a more sophisticated form of pattern recognition is possible, specifically, a "clumping" or "chunking" procedure (Jevons 1871; von Glasersfeld 1982; review in Mandler and Shebo 1982). For example, when humans were shown five or six wooden pencils in hand-held linear arrays, queried "How many?" without time constraints for response, and asked "How did you know?" only one reported counting; others reported visually "clumping" arrays into groups of twos, threes, and fours, and then "recognizing" combinations as familiar (Pepperberg 1987b). This strategy could work equally well for random displays of small quantities (von Glasersfeld 1982). Whatever Alex's strategy, he ignored object identity, shape, color, and so forth, to respond with an abstract number label. When I completed Alex's experiments, such data had not been reported for other animals. The next question, of course, was whether Alex had a more complex sense of number.

Subitizing versus Counting and Alex's Competence

Whether different processes are used to enumerate small versus large quantities and the relative complexity of such processes must be known to compare numerical competence across species and to use number studies to compare cognitive abilities across species. Thus the subitizing versus counting debate must be addressed before I describe further studies with Alex. As noted earlier, full discussion of this controversy is beyond the scope of this chapter. I agree with researchers (Cole and Scribner 1974; Davis and Pérusse 1988; Starkey and Cooper 1995) who maintain, despite arguments to the contrary (e.g., by Beckmann 1924; Mandler and Shebo 1982; Gelman and Gallistel 1986), that subitizing is the simpler process. One way to address the subitizing/counting debate is to argue that subitizing is really two processes that develop at different times: to assume that one process, precounting subitizing, develops in children (and possibly animals) at an extremely early age and does not require subjects to handle abstract concepts (Pepperberg 1994a), and that another process, postcounting subitizing, develops with practice, handles larger quantities (i.e., 4–6 items), and for those quantities probably involves clumping or chunking (Trick et al. 1996). The cognitive overlap of subitizing and counting proposed by Gallistel and Gelman (1991, 1992) seems undermined by data showing that the overlap is a form of subitizing (Davis and Pérusse 1988;

Starkey and Cooper 1995). Unfortunately, most studies in which animals select or label quantity used collections within subitizing range (≤ 6; e.g., Davis 1984; Matsuzawa 1985; Matsuzawa et al. 1986; Pepperberg 1987b; Boysen and Berntson 1989). So far, few nonhumans have met the criteria described at the beginning of this chapter for human counting. Some data on nonhumans suggest a capacity for counting (e.g., in rats, Capaldi and Miller 1988; Capaldi 1993; cf. Davis and Albert 1986; Davis 1993), but only chimpanzees show abilities similar to those of young children. One chimpanzee demonstrated ordinality and labeled, with cards depicting Arabic numerals, sums of two arrays separated in time and space (Boysen and Berntson 1989, 1990; Boysen 1993).[7]

Questions nevertheless still exist as to (1) whether processes used by animals to recognize quantities ≤ 8 involve anything more than perception (see, e.g., Mandler and Shebo 1982); (2) whether animals use, for larger amounts, enumeration processes in which humans ultimately obtain proficiency; and (3) whether overall cognitive capacity can be judged by numerical competence if perceptual mechanisms are used to recognize small collections. Clearly, animals must be tested for competence in the more complex counting procedure (Pepperberg 1994a). Might Alex show such capacities? If, as some researchers suggest, capacities necessary for numerical competence are used for other complex cognitive tasks (Gallistel 1993), the answer should be affirmative.

Choosing a Task to Demonstrate Alex's Numerical Competence

Demonstrating Alex's number sense is particularly challenging, because birds are extremely good at recognizing sequential auditory patterns (Seibt 1982), and, unlike some species (e.g., rats, Davis and Albert 1987), might transfer this ability to visual patterns. As noted earlier, homogeneous sets of external events (and, if transfer occurs, visual patterns) may be judged by whether they match or "fill up" specific perceptual patterns, even for quantities that humans might count (Seibt 1982; Meck and Church 1983). Some examples for passerine birds suggest such mechanisms exist. Crows *(Corvus brachyrhynchos)* probably use various numbers of caws with different temporal patterning to identify individuals in flocks (Thompson 1968, 1969); each crow thus judges several different acoustic patterns to some degree based on quantity. Other birds apparently recognize particular sets of repetitions of different neighbors' vocalizations so as to respond appropriately (carduelid finches and their hybrids, Güttinger 1979; European blackbirds, *Turdus merula,* Wolfgramm and Todt 1982; wood pewees, *Contopus virens,* Smith 1988; eastern kingbirds, *Tyrannus tyrannus,* Smith and Smith 1992). In the laboratory, great tits, after learning to respond differentially to songs with large or small numbers of notes, respond to songs with intermediate numbers of notes on a relative basis (Weary 1989). Such behavior is not unlike human determination, without necessarily counting, of how many fa-la-la's to use in a familiar Christmas carol; birds, however, seem particularly able to memorize many such perceptual sets (Pepperberg 1994a). This behavior is termed "serial subitizing" (Burns 1988:581; note Davis and Pérusse 1988). Given the possibility of avian auditory-to-visual transfer, the sequential addition process used for the chimpanzee Sheba (Boysen and Berntson 1990) might not be valid for testing Alex.

An inferential test to distinguish perceptual recognition from counting was, however, suggested by a paradigm for humans based on visual processing mechanisms involving "distractors" (Trick and Pylyshyn 1989, 1991:36). Humans enumerate items under two conditions: (1) white horizontal *or* green vertical lines among green horizontals, or (2) white vertical lines among green vertical *and* white horizontals. In both cases, nontargeted items are the distractors. Subitizing occurred for one through three only in the first condition. Subitizing thus apparently fails when identification of visual items requires simultaneous processing of several different types of information, that is, when subjects must distinguish among objects defined by a *collection* of features (e.g., color *and* shape). Such findings are consistent with Glanville and Dallenback's (1929) suggestion that the number of items that can be apprehended simultaneously decreases as the amount of information to be perceived about them increases. Given that, by the time I began this study, Alex could use a conjunctive condition to identify a single object within a collection (e.g., a red key within a collection of colored keys and other red items; see Chapter 8)[8] and could label the quantity of a subset of items in a heterogeneous group, I thought he could be given tests similar to those used for humans. Specifically, he could be asked to label the quantity of a subset of items defined by a conjunction of properties. Success would demonstrate that his competence, if not necessarily his strategy, was equivalent to that of humans (Pepperberg 1994a).

I thus queried Alex about the quantity of one of four subsets of items that varied with respect to 2 color and 2 object categories; items in each subset shared either color or form with those of the other subsets (e.g., red cups, blue cups, red keys, blue keys). Given Alex's extensive repertoire, I could form collections using 7 different colors and 12 different types of objects; many object categories furthermore contained items of different materials (e.g., paper *or* plastic cups), thus increasing the number of different collections. In each trial, Alex had to produce vocally the label for the quantity of a subset defined by both color and object category (e.g., answer "How many blue key?"; sample data in Table 7.3 and Figure 7.3). Each question thus required processing two types of information: (1) that about the topic under study (e.g., subset quantity versus attributes of items in these arrays), and (2) that designating the subset targeted by the search (by conjoining labels for color and object category, e.g., "blue and key," not "blue and truck"). To respond correctly, he had to process each type of information without error, then recognize and encode as a vocal label (e.g., "two") information about quantity (Pepperberg 1994a). Some or all of this behavior probably occurred as separate steps, each step adding to the complexity of the task (Premack 1983).

Alex's task was more complicated than the human one because he had one more set of distractors (see Trick and Pylyshyn 1991): He had to label the appropriate quantity from among four rather than three subsets of items that shared either color or object labels. I added one distractor to counter the possibility that the concreteness of objects in Alex's collections might have simplified his task. Whereas Alex had objects, such as colored keys and nails, humans had colored lines of different orientations (Pepperberg 1994a).

Alex's task was also more complex than those given most animals. As in all my studies, the protocol differed from ones used with other animals in two ways (Pepperberg 1994a). First, Alex had to use the vocal mode; second, each trial was presented intermittently during training and testing of unrelated topics also under study (e.g., middleness, phoneme labeling). Alex thus had to choose responses from his entire repertoire (> 90 vocalizations, including food and location labels) and from several topics involving various items and questions in each session. This design not only increased task complexity, but also prevented several forms of cuing (Premack 1976:132; Chapter 2).

To compare Alex with humans, the range of numbers tested had to begin at "one," but Alex's ability to label single items had not been tested. "One," however, was already in his repertoire (Pepperberg, Brese, and Harris 1991). I thus used accuracy in labeling single objects as the criterion to begin testing this new task. Using standard laboratory protocols (Chapter 2), my students and I queried Alex from November 1989 to January 1990 about various single objects ("How many?"). He reached criterion, 24 initially correct responses in 30 consecutive trials (80%), after 35 sessions.

Table 7.3.   Representative trials for a "confounded" number set. For these
trials, wood refers to wooden tongue depressors, wool to woolen pompons,
rock to Play-Doh tubes, nail to nails partially covered with plastic mollies.
Singular labels were used for all questions to avoid cuing "one." (After
Pepperberg 1994a.)

Trial	Objects set on tray	Question	Response
1	1 orange chalk, 2 orange wood, 4 purple wood, 5 purple chalk	How many purple wood?	4
2	1 yellow block, 2 grey block, 4 yellow wool, 6 grey wool	How many yellow block?	1
3	1 rose wood, 2 blue nail, 3 blue wood, 5 rose nail	How many rose nail?	5
4	2 grey truck, 3 grey key, 4 orange key, 5 orange truck	How many grey key?	2, 3
5	1 blue box, 3 green box, 4 blue cup, 6 green cup	How many green cup?	6
6	1 purple rock, 2 green rock, 3 purple plastic key, 4 green plastic key	How many green rock?	2

Testing Alex on Heterogeneous "Confounded" Sets

Before beginning the tests my students and I made two adaptations to
standard testing procedures (Pepperberg 1994a). First, trials began by
allowing Alex to touch his tongue to each object in the collection. This
step enabled him to distinguish certain materials that appeared visually
similar (e.g., Play-Doh and plastic). A trainer then scattered objects
randomly, by heterogeneous handfuls, onto the surface of the tray.
Arrangements were altered only if an object was obscured by other
items or if subsets were obviously clumped. Each object was generally
< 2 in. from the nearest other item. Alex saw four different quantities,
such as sets of red and blue blocks and balls, on the same felt-covered
tray previously used in number work and other studies (e.g., object
permanence; Pepperberg and Kozak 1986). To ensure further that the
tray was not associated solely with number tasks, we used the tray for
intermittent queries about colors, shapes, or materials of specific items
or collections, and for training phoneme recognition and middleness
(Pepperberg, in prep.).

I used two different values of chance to learn whether Alex's results
were statistically significant. First, I based chance (1/4) on the number

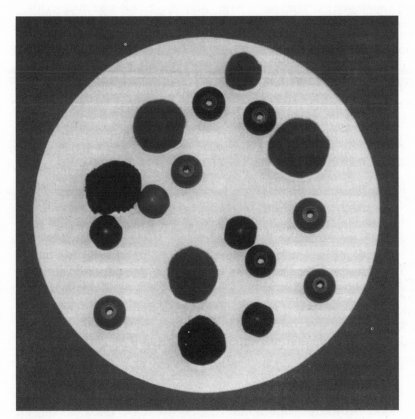

Figure 7.3. Sample of a "confounded" number set: a complicated heterogeneous array used to see how close Alex could come to achieving the results shown by humans in similar tasks.

of labels relevant to the task, as if I thought that Alex might randomly guess the quantity of one presented subset. Second, I based chance (1/3) not on a completely random choice of subsets, but on one of the targeted categories of subsets, as if I thought Alex might, for example, guess the answer to "How many green truck?" after targeting only green items. In that case, he could respond to the number of (1) total green objects, (2) green trucks, or (3) items in the other green subset. All calculations are conservative in that they assume 100% comprehension of labels identifying the subset; that is, they ignored the probability of Alex's misunderstanding a question and correctly identifying the quantity of a wrong subset. Given that Alex's accuracy was ~80% on somewhat more complicated comprehension tasks (Pepperberg 1990d, 1992a), the probability of such a misidentification was small but not nonexistent. Chance could also be based on 1/6, as if I thought that, when hearing "How many . . .?" Alex randomly guessed after limiting his choice to numbers (one through six). Alex could also have uttered any label in his repertoire, but calculations assume he would always ($p = 1$) attend and respond correctly to the "How many?" part of the question.

Alex's Results on the Heterogeneous "Confounded" Number Task and Analysis of His Errors

Alex clearly succeeded on the task. For first trials, overall accuracy was 45/54 (83.3%). For all values of chance, $p \ll .01$ on a binomial test. Table 7.4 shows data separately for each number tested, and lists the errors. Scores varied from 9/9 (100%) on "four" to 6/9 (66.6%) on "three."

How Alex's scores vary according to number is important. According to Gallistel and Gelman (1992), a subitizing mechanism can be defined by data that conform at least qualitatively to Weber's law: The greater the reference numerosity, the more imprecisely will a subject distinguish between it and nearby numerosities. Thus if Alex used a perceptual strategy, rather than some type of counting, we would see more errors for larger numbers (see Mandler and Shebo 1982). His errors, however, varied randomly with respect to the number of items to be identified. The error pattern nevertheless had to be tested formally for randomness. Because the pattern of such data cannot be tested for significance by a normal linear regression, I performed a regression instead on severity of the error. Testing whether the severity of Alex's errors (that is, the distance between a correct answer and his response) increased as the number of items increased would provide both a measure of his accuracy and evidence for either a subitizing or a counting-like mechanism.[9] A sequential canonical analysis (Gorsuch and Figueredo 1991), in which direction of the error was partialed out (because the direction of error had to go toward smaller numbers as the quantities increased), showed that the severity of Alex's errors was not significantly correlated with the increase in the number of items to be identified (Pepperberg 1994a).[10]

Interestingly, after examining Alex's responses in detail, I found that some errors might involve his misunderstanding of questions rather than numerical ability (Pepperberg 1994a). As in other comprehension tasks (e.g., Pepperberg 1992a), errors could arise from four sources unrelated to number: (1) confusion of labels that sound alike; (2) misunderstanding of the label that directs the search; (3) problems involving perceptual boundaries, e.g., differences in avian and human

Table 7.4. Results and errors on Alex's heterogeneous "confounded" number task. All probabilities are based on chance values of 1/4 and 1/3; all probabilities can be multiplied by a factor to take into account the probability that Alex was able to identify the targeted subset correctly. (After Pepperberg 1994a.)

Quantity	Score	Probability		Wrong label	Error type[a]
		1/4	1/3		
1	7/9	.0012	.0073	2, 3	C, O
2	8/9	.0001	.0009	4	R
3	6/9	.0087	.0341	2, 2, 4	O, O, O
4	9/9	<.0001	.0001		
5	7/9	.0012	.0073	2, 3	C-O, O
6	8/9	.0001	.0009	4	C

a. Error types were C = color (Alex's response was consistent with the number of items in the subset that corresponded to the targeted objects but not the targeted color); O = object (Alex's response was consistent with the number of items in the subset that corresponded to the targeted color but not the targeted object); C-O = color and object (Alex's response was consistent with the number of items in a subset, but a subset that did not correspond to either the targeted color or the targeted object); R = Alex's answer was random, that is, did not correspond to the number of items of any presented subset.

color perception; or (4) failure to understand that information from *two* categories must be used to determine the search objects ("conjunction" of information; see Chapter 8). Interestingly, only one error was "random" (i.e., did not correspond to any presented quantity).

In another study (Chapter 8), almost half of Alex's errors involved either comprehension or production of labels that, at least to humans, sound similar (e.g., *rock* and *block*). I tried to eliminate such confounds in the present task, but a few trials combined *block* and *truck*. Such labels match only in their final consonant cluster, but Alex twice gave the number of trucks, rather than blocks, in the presented array. On one of these trials, however, he also erred with respect to color.

Determining whether Alex misunderstood one or both labels that directed the search is not easy because such errors cannot be distinguished from numerical errors (Pepperberg 1994a). In Chapter 8, however, we will see that he scored ~80% on tests requiring comprehension of attribute labels; scores on the present task were comparable. Of course, he may have misinterpreted the defining labels, then correctly quantified the incorrectly targeted subset. Indeed, eight of nine errors were the correct number for an alternative subset.

Perceptual errors are also difficult to track. Few of Alex's nine errors, however, could be attributed to perceptual confounds (Pepperberg 1994a). I avoided collections combining objects such as rocks (Play-Doh) and rawhide, which are difficult even for humans to distinguish by sight. A different perceptual problem is that psittacine and human visual systems have different color boundaries (Bowmaker 1986; Bennet and Cuthill 1994; Bowmaker et al. 1994, 1996), and as noted previously, Alex may confuse orange with rose or yellow, and purple with blue or rose. I thus tried to avoid such combinations. In the two trials in which Alex responded with the number of a subset identified by the wrong color, he labeled the number of green, rather than yellow, wool and yellow, rather than purple, papers.

In all cases, Alex used the conjunctive condition (Pepperberg 1994a). He never quantified a subset defined by only one attribute (e.g., provided the number of all the trucks). In eight of nine errors, he responded with the quantity of a nontargeted subset that was still defined by a conjunction of color and object label. A sequential canonical analysis (Gorsuch and Figueredo 1991) tested whether the number of errors on the attributes of color and object identity were correlated with numerical errors and whether the number of color and object errors increased as the number of objects to label increased. The test found no significant correlation (Pepperberg 1994a).

Latency to Respond: A Way to Differentiate Subitizing from Counting?

Many studies use response latencies to distinguish subitizing from counting (Averback 1963; Klahr and Wallace 1976; Trick et al. 1996). The rationale is that subitizing requires less time than counting (e.g., Mandler and Shebo 1982). Response latency is not, however, a viable criterion for Alex, because, as noted earlier, his readiness to respond to any item is correlated only with his interest in acquisition. In this task, as in earlier number tasks, familiarity of test items generally correlated with his lack of interest. He often requested a nut or a cork after completing a trial and ignored the targeted items. Response latencies therefore could neither be meaningfully measured nor compared with those of other subjects, human or animal (Pepperberg 1994a).

The Implication of Alex's Data

At least three possible interpretations exist for the data. Alex might have been (1) subitizing because his capacity for perceptually recognizing quantity greatly exceeds that of humans; (2) counting according to conditions described by Trick and Pylyshyn (1989) and Gallistel and Gelman (1992); or (3) using some combination of these processes. Arguments can be made for each interpretation.

Alex might have been subitizing, particularly if avian numerical perceptual capacities are superior to those of humans in the visual mode (Pepperberg 1994a). He might, for example, perceptually segregate the set to be enumerated from all other items, then judge whether the segregated set "matches" some internal pattern (Dehaene and Changeux 1993; Starkey and Cooper 1995). Because targeted items are scattered among ≥ 9 distractors, this strategy seems unlikely, but I cannot dismiss the existence of extended perceptual capacities. Avian numerical perception probably surpasses that of humans in the auditory, sequential mode (Thompson 1968, 1969; Wolfgramm and Todt 1982), and this ability might be transferred to the simultaneous, visual mode (see, e.g., Seibt 1982). To ensure that Alex was beyond the subitizing range, I would have to test him on visual simultaneous quantities larger than those perceived auditorially and sequentially in nature. Data on how Grey parrots process natural vocalizations are, however, nonexistent.

If Alex's perceptual capacities are no better than those of humans, his data are consistent with counting rather than subitizing (Pepperberg 1994a). If he did subitize, accuracy for smaller numbers would be greater than for larger numbers and he would be further from the correct number as quantities increased (see, e.g., Gallistel and Gelman 1992; Starkey and Cooper 1995). But his accuracy did not vary and no significant correlation existed between numerosity and severity of errors. Moreover, humans cannot subitize when asked, like Alex, to quantify a subset of items distinguished from other subsets by a conjunction of qualifiers (Trick and Pylyshyn 1989). Such data may, however, tell us little about Alex, because comparable performance across species cannot ensure use of comparable mechanisms, and procedures that control for alternative explanations of behavior in one species may not provide controls in others (Pepperberg 1994a).

Alex might, of course, have combined mechanisms. He had no time constraints that could separate counting from subitizing. Possibly, he used a perceptual process to overcome the distractor effect to subitize quantities ≤ 3 and then switched to a more accurate counting strategy for ≥ 4. He did demonstrate a small rise in number (but not in severity) of errors at 3 (Pepperberg 1994a). Such a combination of strategies would be intriguing: Alex would, then, have shown numerical competence not unlike that of humans at the higher quantities.

Competence, however, is not sufficient for claiming that Alex counts. Whatever strategies he used in the heterogeneous "confounded" number task, he cannot technically be said to "count" unless he understands that number labels represent an ordered set of tags that can be paired with each object (Fuson 1988). Studies to demonstrate explicitly Alex's (and other juvenile parrots') sense of ordinality are under way (Pepperberg, in prep.). These experiments are complicated because simple ordered vocal production (i.e., "one, two, three . . .") would, especially in a parrot, indicate only rote memorization (see, e.g., Dehaene 1992). And demonstrating 1:1 pointing is difficult: Parrot claws and beaks are less suitable than primate digits for pointing (Pepperberg 1988a), and training Alex to produce specific actions, such as head movements toward targeted objects, would increase the likelihood of his use of rhythmic patterns or sequential subitizing (Burns 1988).

Thus, despite intriguing data, I cannot claim that Alex counts. Determining if he counts in the human sense, however, may not be useful: The ability can always be redefined to exclude whatever animals cannot do (see Fouts 1978; Davis and Pérusse 1988; Pepperberg 1988c, 1990a). Nevertheless, the research is a step toward determining the extent of Alex's cognitive competence and whether he might eventually acquire more complex capacities, such as labeling of larger quantities and understanding of addition and subtraction. Of particular interest, too, is examining whether Alex is "different" from other animals in that he uses vocal numerical labels to designate quantity. Remember, researchers such as Gallistel and Gelman (1992) are particularly concerned with preverbal versus postverbal processes. Does Alex's labeling affect his competence?

The Effect of Labeling Capacities

As noted at the beginning of this chapter, arguments exist both for and against relating language and numerical capacities. Dehaene and Cohen (1995), for example, describe models that involve different categories of mental representation for different types of number manipulation, but believe labeling may not be necessary for simple recognition of quantity. Other researchers, however, argue that the ability to label quantity may separate counting from subitizing. Starkey and Cooper (1995) believe counting develops from subitizing when labeling begins: A child learning to label objects also learns to associate number labels with subitized quantities; the child thus learns the significance of each number name, and begins to associate other number names with larger quantities. Similarly, the number label for the subitized numerosity becomes correlated with the final number label used in counting; such processes eventually lead to true counting. Hurford (1987) and Dehaene (1992) also suggest a close correlation between labeling and number skills, in that numerical cognition is "a layered modular architecture, the preverbal representation of approximate numerical magnitudes supporting the progressive emergence of language-dependent abilities such as verbal counting" (Dehaene 1992:35). Interestingly, children with Specific Language Impairment syndrome, who appear normal except for delayed language development (see, e.g., Leonard 1989), have difficulty with numerical tasks such as rote counting to 50, but perform normally when allowed to use actual objects to count (Fazio 1996). Conceivably, Alex's extensive labeling training did enable him to exhibit behavior that could not occur in a "preverbal" bird (note Premack 1983).

Clearly, Alex's labeling may cause us to view his data differently from that of animals unable to label. But of utmost importance is that although Alex appropriately uses abstract vocalizations at a representational (and referential) level, his behavior is not, in most respects, isomorphic with human language (Pepperberg 1994a). His ability to label quantity simply allows him to be tested in a manner similar to that used with humans, and labeling may help him form a more precise representation of quantities in a subset. Whether labeling carries him to Gallistel and Gelman's "verbal" level of counting is unclear. Rice (1980), for example, believes that linguistic input cannot teach a nonlinguistic concept for which a subject is unready. Thus although Alex uses "five" to label a specific number of objects, the label may still represent a perceived attribute of the subset, like color, rather than be the result of a counting process. That the presence of distractors prohibits use of such a perceptual strategy in humans (Trick and Pylyshyn 1989) may not be important because birds and humans may have different perceptual capacities. Most probably, Alex's ability to separate out, visually collect, and enumerate items in the targeted subset is a consequence of his years of task-oriented training; his ability to produce a vocal label for the result merely provides an efficient testing mechanism.

What Alex Has Demonstrated

Whether or not Alex can indeed count is still under study. He can, however, vocally label the quantity of a specific subset of objects in a complex heterogeneous array (Pepperberg 1994a). The subset is uniquely defined by the conjunction of one color and one object category and the array is designed to provide maximal confounds, or distractors, as in comparable work with humans (Trick and Pylyshyn 1989). These data suggest that a nonhuman, nonprimate, nonmammalian subject has, whatever the mechanism, an overt level of numerical competence equal to that of humans on the present task, and the "take home-message" is that we should not underestimate the numerical capacities of nonhumans, particularly nonmammals.

8 How Can We Be Sure That Alex Understands the Labels in His Repertoire?

> Through careful analyses of the circumstances under which a chimpanzee was able to use a particular symbol, Savage-Rumbaugh and her associates showed that none of the previously mentioned functions of a word can be taken for granted. Learning to use a symbol to request some object in one context does not imply competence to request that object in another context or the knowledge that the symbol in question could function as a name for identifying an object. Savage-Rumbaugh also showed that the competence to request an object with a particular symbol does not imply comprehension of that symbol when another organism uses it to make the same kind of request—whether that organism is the teacher or another chimpanzee.
>
> Herbert Terrace, *Ape Language: From Conditioned Response to Symbol* (1986:xvii)

Previous chapters show that some of Alex's communicative and conceptual abilities—labeling, categorization, and abstract concepts of same/different and absence—compare favorably to those of great apes and marine mammals. I achieved these comparisons with training and testing procedures that relied on interspecies communication derived from a human code. To obtain additional comparative data, particularly on comprehension, I used this communication code to study "recursive" competence: the sequential, hierarchical processing of multiple forms of information.

Because such studies test both language-like and cognitive behavior in nonhumans, three issues must be clarified so that the results can be appropriately evaluated. First, such studies rely on the animals' *comprehension* of the human-based code that has been taught, and serious questions have arisen as to whether animals comprehend the codes as well as they seem to produce the codes. Thus results of these studies can be seen as a strong test of comprehension. Second, such studies must specify exactly what is considered evidence for cognitive behavior, because as many definitions exist for "cognition" as there are researchers in the field (Pepperberg 1998). Third, as a consequence of the first two points, these studies again raise questions about relationships between language-like training and cognition. I thus briefly address these issues before discussing how I tested Alex's competence on this difficult task.

Competence in Comprehension

The level at which an organism comprehends a communication code need not correspond to its competence in production. A child's first wordlike utterances may simply be imitative, and gain meaning only after an adult responds in a "predictable and interesting way" (Tomasello 1996:283; note Bates 1979)—production may precede comprehension. More often, however, comprehension precedes production (Goldin-Meadow, Seligman, and Gelman 1976; Benedict 1979; Snyder, Bates, and Bretherton 1981), and researchers who initially studied chimpanzees' acquisition of human-based codes assumed that their animals acted similarly (Savage-Rumbaugh 1986). Although apes *can* acquire competence in comprehension and production in the same way and order as do children (Savage-Rumbaugh et al. 1986), researchers studying interspecies communication have learned that an animal's ability to comprehend as well as produce a human-based code crucially depends upon what training techniques have been used (Savage-Rumbaugh, Brakke, and Hutchins 1992; Pepperberg 1997; Williams, Brakke, and Savage-Rumbaugh 1997). For particular training paradigms, respective levels of an animal's competence in comprehension and production could be reversed. Data suggest that some chimpanzees trained to *produce* specific symbols in appropriate circumstances may have learned little about what the symbols really meant (see Savage-Rumbaugh et al. 1993). Those apes merely learned to associate particular actions with particular circumstances. Even an animal's ability to use a symbol both to label *and* to request depends upon its training: Lana, a chimpanzee trained to use computer symbols predominantly as requests (Rumbaugh 1977), required retraining to use the same symbols to label specific items (Savage-Rumbaugh 1986). Thus, I needed to show that Alex, whose training differed greatly from that of these apes, had comparable levels of comprehension and production if I wanted to use his human-based code to demonstrate advanced cognitive skills. Because Alex always received the items he labeled as his initial reward, I also had to show he could distinguish identification from requesting, but that topic is reserved for Chapter 11.[1] Here I concentrate on comprehension and cognition.

What Is Cognitive Behavior?

Much debate about complex cognitive capacities in animals arises because researchers have yet to agree on criteria for defining "cognitive behavior" or for ordering animal capacities with respect to complexity (see, e.g., Macphail 1987; Kamil 1988; Pepperberg 1990d, 1998; Zentall 1993). Even researchers who infer these capacities from positive results on tasks such as those discussed in previous chapters and who propose relative hierarchies for intelligence may not specifically define "cognition" (e.g., Thomas 1980; cf. Premack 1986). The difficulties of devising an appropriately general definition of cognition have been discussed elsewhere (e.g., in Kamil 1988). For specific experiments, however, we can make *operational* definitions that characterize both cognitive behavior and tasks that facilitate cross-species comparison of such behavior (Pepperberg 1990a).

Generally, cognition is viewed as an organism's ability to make a decision by evaluating, or processing, current information based on some representation of prior experience (see, e.g., Kamil 1984). The supposition is that the organism does not react mindlessly to environmental stimuli, but rather processes the stimuli and chooses to react in certain ways. As noted earlier, a foraging bird, seeing green and red fruits, for instance, might recall that red indicates ripe and tasty and green indicates unripe and bitter, and thus choose red. For the study described in this chapter, however, cognitive behavior involves an additional capacity or criterion. I require the decision-making process also to include *the capacity to choose,* from among *various possible sets of rules* that have been acquired or taught, *the set that appropriately governs the current processing of given data.* Such a capacity enables the foraging bird discussed above to recognize conditions under which selecting green fruit might be wise, such as when red indicates spoilage. Under this criterion, organisms limited to examining information according to a single set of rules (e.g., in a matching procedure) do not have the chance to demonstrate complex cognitive processing. Cognition thus defined also often requires an ability to transfer a skill learned in one domain to another (Rozin 1976). My criterion is more an operational guideline than a definition of cognition, but facilitates our drawing distinctions and comparisons between levels of complexity in problem-solving abilities (Pepperberg 1990a).

Most tasks that examine animal cognition do not require an animal to evaluate the rules governing the processing of data as well as the actual data. Recursive tasks have been proposed to test this form of cognition because they require sequential processing, and thus force the animal (or human) subject to process rules as well as data. Let's first look at the qualities of a recursive task, then why it requires processing rules *and* data (Granier-Deferre and Kodratoff 1986).

In a recursive task, a subject is presented with several different objects and one of several possible questions or commands concerning the attributes of these items (Pepperberg 1990a). Task complexity is based on the number of items present, the number of different queries or commands, and the complexity of the individual queries or commands. Each question contains multiple parts, the combination of which uniquely specifies the targeted object and the action to be performed. A question's complexity is determined by its context and the number of its parts (e.g., the number of attributes used to specify the target and the number of actions from which to choose). A subject must divide the query into these parts and (recursively) use its understanding of each part to answer correctly. A subject thus demonstrates competence by reporting on only a single aspect (e.g., color, shape, or material) of, or performing one of several possible actions (fetching, touching) on, one item among many differently colored and shaped items of various materials. A relatively complex question is "What color is the item that is circular and rawhide?" To reply correctly, the respondent must exclude all circular items of other material, all rawhide items of other shapes, and all categories other than color, and must discern and then encode information about the appropriate instance of color as some form of label (e.g., a vocalization, a computer symbol) after the object is isolated. A query can be simplified in two ways: The number of features defining a targeted item can be decreased,[2] so that a subject is asked, "What color is the rawhide item?" Alternatively, a subject can be told, "Touch that which is circular and rawhide," so that no additional information must be processed after a target is found. The task nevertheless retains its recursive nature, because sequential processing is still needed to respond (e.g., in the latter case, an animal must first choose between "bringing" and "touching").

A recursive task fulfills the requirements I define as necessary for cognitive processing because it requires the subject both to evaluate information based on representational knowledge *and* to choose the set of rules upon which to base its response (Pepperberg 1990a): The subject must comprehend each element that *represents* several possible actions (e.g., "fetch," "touch") or object attributes (e.g., "shape," "color") in a transient question (e.g., a query consisting of vocalizations or hand signals) in order to choose among a set of possible responses. The subject must then use its understanding of additional symbols to determine the set of information to which it will selectively attend (e.g., "circular" and "rawhide" items) in the context of any number of different situations (i.e., for any collection of items; Granier-Deferre and Kodratoff 1986). Finally, the subject must determine its course of action and encode its response into an appropriate physical motion or verbal representation of an object or attribute (Pepperberg 1990a). The problems posed by the tasks cannot be solved by responding to a single set of criteria (e.g., match-to-sample based on color). Nor can the queries be answered by performing an action involving a relatively simple 1:1 correlation ("Pick up X," when options are either only picking up or only X). The tasks do not involve responding to single questions (e.g., "What's this?") for even a large number of different items, or chaining two independent responses to different objects ("Do X to A *and* Y to B"). This last factor is important because chaining, although requiring considerable memory, does not demand hierarchical processing (for a detailed discussion of this point, see Premack 1986 and material later in this chapter). And although recursive tasks are related to conditional tasks ("*If* tray is green, *then* do match-to-sample; *if* tray is black, *then* do nonmatch-to-sample"), the number of concepts involved and the number of response options in recursive tasks are considerably greater than in conditional tasks (see below; Thomas 1980, 1996).

Possible Relationships between Language-Like Training and Cognitive Abilities

In humans, the study of cognitive behavior is closely tied to study of language development (see, e.g., Bloom, Lifter, and Broughton 1985; Kuczaj, Borys, and Jones 1989; note Menyuk, Liebergott, and Schultz 1995; Nelson 1996), and several researchers suggest that the study of animal cognition may be affected by animals' lack of human language (Premack 1983; Terrace 1985; Roitblat 1987; Gallup 1989; note Greenfield 1991). My stance on this matter, discussed in previous chapters and various articles (e.g., Pepperberg 1986, 1990b,c; Pepperberg and Funk 1990), bears repeating. I argue that language training (or its lack) probably affects only the *ease* with which animals can learn and be tested on certain concepts, not *whether* learning occurs. Two-way communication may enable a researcher to teach a concept that an animal might not easily otherwise acquire, but such learning is unlikely if the animal lacks the cognitive capacity to acquire the concept. Once a concept is acquired, interspecies communication codes, because of their common basis in human language, then provide the most direct means of comparing cognitive ability across species (note Bradshaw 1993).

Nevertheless, as I have mentioned earlier, success on a cognitive task is often used to judge a subject's communicative competence. Successful solution of the questions or commands posed in recursive tasks, for example, requires an animal to comprehend each of the different symbols and concepts in the questions or commands (Pepperberg 1990a). Researchers (e.g., Granier-Deferre and Kodratoff 1986; Premack 1986) thus argue that success on complex recursive tasks can *define* a relative level of comprehension (see Bradshaw 1993). But comprehension of a language-*like* code, although clearly indicating some complex processing ability, does not necessarily imply comprehension of the full complement of elements that constitute human language (Kuczaj and Kirkpatrick 1993). This point is important because of the endless debate about the extent to which animals can acquire human language. To my mind, the relationship between research on human language development and studies on interspecies communication is unclear (see Herman 1988, 1989; Schusterman and Gisiner 1988, 1989). Many animal studies use, both to instruct and to test their subjects, codes based on human language, and consequently demonstrate certain comparable cross-species capacities for abstract, symbolic use of human-based communication codes (Pepperberg 1986). Although of considerable interest to researchers in numerous fields (e.g., Greenfield and Savage-Rumbaugh 1984, 1991), such language-like behavior is not necessarily equivalent to human language, and direct cross-species comparisons about competence in a human-based code are neither always possible nor useful. Full competence implies production as well as comprehension, but marine mammal studies, for example, predominantly involve comprehension; the one project with a human-dolphin computer interface is still in its infancy (Reiss and McCowan 1993). At the cost of belaboring a point, I reemphasize that valuable comparative work can be done if we focus on cognitive, rather than linguistic, abilities, and if an animal's language-like behavior is viewed as an investigative tool (Pepperberg 1986, 1990a).

All this now said, researchers successfully use interspecies communication and recursive tasks to demonstrate the competence of chimpanzees and marine mammals with respect to cognition and communication (e.g., Essock, Gill, and Rumbaugh 1977; Herman 1987;

Schusterman and Gisiner 1988; Savage-Rumbaugh et al. 1993). Note, too, that recursive tasks, in the form of relative clause comprehension, are often used to study children's cognitive and communicative competence (see, e.g., Crain 1991).[3] No study, however, had evaluated the competence of a nonprimate, nonmammalian subject relative to these other subjects. I thus designed a study to provide such data on Alex.

Alex's Previous Competence on Comprehension

Alex already displayed the prerequisite for recursion—comprehension of his communication code (see previous chapters). He not only vocally labeled objects and their attributes, but also understood how the labels fit into categories and the meaning of category labels: Thus he responded differentially to "What color?" *or* "What shape?" for items simultaneously incorporating both attributes. Experiments on same/different also relied on his ability to comprehend and respond to multiple questions about a set of objects: Given two items that were identical or varied in some or all of their attributes, he responded differentially to "What's same?" *or* "What's different?" with an appropriate category label or "none." And to ensure that he attended to such queries, he was given probes for which two possible correct answers existed (e.g., "What's same?" for red and blue round plastic keys). If his answers were based on prior training and the physical aspects of objects rather than on question content, he would have determined the *single* attribute that was same or different and responded with the one wrong answer. Also, Alex's tests included queries on all possible topics; he thus always had to comprehend the question topic to determine from which set of responses his answers should come (e.g., for two keys, he could be asked not only about same/different, but also "How many?"; Pepperberg 1987a,b). Finally, my study on requesting showed that he comprehended his own utterances. When presented with an item other than one he had requested, he would refuse the alternative object (say "nuh") and often repeat his initial request (Chapter 11).

These data, however, did not demonstrate the *extent* of Alex's skills. Tasks contrasted two or, at most, three questions for any given set of items, although tests always contained queries on many different topics. A recursive task, related to those given chimpanzees (Essock et al. 1977) and marine mammals (Herman 1987; Schusterman and Gisiner 1988), would involve more questions and a larger set of objects. I did not replicate exactly the marine mammal studies, that is, require Alex to perform physical actions in response to a verbal command or question. He had been trained to work in the vocal mode, and that mode was well suited for recursion. My goal was to see if his ability—to ascertain the subset of information being targeted in a given situation—could transfer to more complex questions and situations. If successful, he would demonstrate levels of comprehension and cognitive processing comparable to those of great apes and marine mammals.

The Task Given to Alex

To compare Alex to other animals, his task had to be functionally equivalent to theirs; however, my protocol differed in two important ways from those used previously. First, Alex used the vocal mode. Second, unlike some studies (see below), each of his trials occurred intermittently during training and testing of other unrelated topics (e.g., number competency, photograph recognition). He was thus exposed to a wide variety of possible items and queries during each session (see below, also Premack 1976), and was required to respond from a repertoire of over 80 vocalizations, including labels for foods, locations, and quantity. Furthermore, because this protocol intermixed recursive trials and other tasks, "proactive interference" errors were less likely. Proactive interference is the inappropriate transfer of information from previous experience to current behavior. An example is how the memory of being rewarded for choosing red versus green in one trial on a match-to-sample problem can interfere with appropriate choice of green on a subsequent trial. When recursive trials are grouped, animals may make similar errors (Schusterman, Gisiner, and Hanggi 1988); mixing types of trials, however, prevents the problem.

Alex's task involved the following steps. On each trial, he was asked one of four possible vocal questions ("What color is object-X?" "What shape is object-Y?" "What object is color-A?" or "What object is shape-B?") about a collection of 7 objects. Each collection differed for each trial; objects were chosen from among 100 items of various combinations of shapes, colors, and materials. Every question contained three pieces of information (Pepperberg 1990a): (1) information about the topic under study (e.g., the attribute of 1 item in a collection versus the quantity of the collection); (2) information that designated (in the form of the label for a particular instance of a category, e.g., "green" and not "rawhide") the 1 object of 7 that was the specific target of the search; and (3) information that designated (in the form of a *different* category label, e.g., shape, not color or material) the particular category from which a response should be chosen. To respond correctly, Alex had to process each piece of information correctly, then recognize and encode as a vocal label (e.g., "blue") information about the appropriate instance of the targeted category (Pepperberg 1990a). Some or all of these actions would probably occur as separate steps in the overall process, with each step adding to the complexity of the task (Premack 1983).

How Alex Was Trained on the Recursive Task

Although optimally Alex would have been tested without any training whatsoever, pragmatically most animal subjects' fear of novelty decreases their ability to respond to a new task involving new contexts and new questions (Zentall and Hogan 1974; Zentall et al. 1981). My students and I therefore conducted three probe trials on familiar objects at widespread intervals to determine the feasibility of the study, and Alex's first response was correct on each trial (Pepperberg 1990a). We nevertheless were concerned that he might not respond appropriately during a long series of trials; thus we began training on the task beginning in October 1986. We used the standard M/R technique.

We needed to ensure that context did not cue the nature of the task. During training, items of different colors, shapes, and materials were thus placed on the same felt-covered tray we used for numerical studies (Chapter 7) and studies on object permanence (Chapter 10). The tray therefore would not be associated solely with recursion. We also took the following precautions (Pepperberg 1990a): We varied the number of items on the tray and continued to ask questions of "How many?" as well as those about the color and shape of specific objects and the material of particular colored or shaped items. We also included training on more complicated tasks (e.g., "What color is the four-corner metal key?") for subsequent studies (Pepperberg 1992a). We recorded which objects were included in each training trial so that we would never use the same exact collection during testing.

Alex's behavior during training differed from that during probe trials. His accuracy declined considerably. This decrease appeared, however, to have little to do with understanding the task: Alex began to (1) label and grab favored items before we asked questions, or (2) respond to many queries with either the label "green" (while pulling at the green felt tray liner) or "tray" (while biting at the tray). My students and I thus had to determine the bases for such behavior and develop appropriate strategies to counter Alex's actions so that training could proceed (Pepperberg 1990a).

We surmised that Alex's inattentiveness and consequent inaccuracy was caused by the lack of novelty of the rewards. We used novel combinations of items in each trial, but each object had to be familiar because Alex had to be able to label all attributes.[4] As noted previously, when familiar objects are repeatedly used as rewards, Alex ceases to work, begins to preen, or interrupts with requests for other items ("I want X") or changes of location ("Wanna go Y") (see, e.g., Pepperberg 1988a). In contrast, remember that during the same/different experiment (Chapter 6), Alex's attention span increased and accuracy improved on transfer trials, which by definition incorporated novel (i.e., conceivably more interesting) items.

To counter what thus appeared to be "boredom," we emphasized certain training aspects and adjusted others (Pepperberg 1990a). First, we continued to intersperse trials with other training and testing tasks. Second, we emphasized that a correct response would allow Alex to request a favored item or activity. Third, we syntactically altered queries within tests to provide additional novelty: We alternated "What color is the key?" and "Wood, what color?" Altering question forms, thus forcing Alex to attend more closely, might also have lessened what psychologists call "retroactive interference"—information in the second label or category in a query interfering with processing that in the first label or category. Retroactive interference effects occurred in some comprehension studies on marine mammals (e.g., Schusterman and Gisiner 1988; Schusterman et al. 1988).

Criterion Prior to Testing

Our criterion for beginning testing differed from the criteria in some previous studies. The earlier criterion based on the clarity of Alex's speech was inappropriate because the present study used his current repertoire, and clarity was not an issue. We decided instead to initiate testing when we eliminated his interruptive behaviors. We thus required that three consecutive sessions did not include irrelevant responses of "tray," "green," or grabbing of preferred items before testing began.

How Alex Was Tested on the Recursive Task

Tests including a single comprehension question were held one to four times per week from February through December 1987. The basic procedure was as described in Chapter 2. We adapted the procedure slightly, however, to correspond to the needs of the recursive task.

We repeated some protocols of the number study. Comprehension trials began by allowing Alex to touch each item in the collection with his tongue so that materials that appeared similar could be more easily distinguished. These objects were then scattered onto the surface of the tray. Only if an item was obscured by another object was the arrangement altered in any way. Each object was generally spaced $\leq 2''$ from any other item (Figure 8.1).

So that Alex could not tell which question was to be asked from the composition of items on the tray, all trials contained seven objects, including shape trials for which only five labelable choices existed (Pepperberg 1990a). Thus although Alex could not label circles, we often included circular objects so that he could not distinguish shape-related questions from color-related questions by the number of objects on the tray. Furthermore, on all trials some items varied in both color and shape even if only one of these categories was targeted. Chance, however, was calculated on the number of possible answers, not the number of objects (see below).

Alex's Competence on the Recursive Task

Alex responded with considerable accuracy. Table 8.1 provides sample queries. Test scores, as usual, were calculated for both first and all trials. Alex was correct on 39 of 48 of his first trials (81.3%). For all trials, he scored 48/57 (84.2%). Scores for both types of color queries were 75.0% (first trial) and 80.0% (all trials); for shape queries, 87.5% (first trial) and 88.9% (all trials). On binomial tests, $p < 0.0001$ for first trials, with chance of either 1/5 (shape queries) or 1/7 (color queries). I also calculated scores for individual queries ($p \leq 0.0001$; Table 8.2). Chance values of 1/5 and 1/7 were conservative in that they were based on the number of possible responses to a targeted category (rather than to all categories) and ignored the possibility that Alex could have produced *any* label in his repertoire. I used 1/5 for shape trials even though there were seven items, because Alex had only five possible shape responses (see Herman et al. 1984 and Schusterman and Gisiner 1988 for discussions about appropriate choices for chance).

Alex's responses not only were well above chance, but also demonstrated that he indeed processed information about the type of questions we asked: For 47/48 questions, his first single utterance was a label from the appropriate category, even if it was an incorrect instance of that category (Pepperberg 1990a). Other responses were either repetitions of parts of our queries (e.g., "What color?") or encoded requests for other objects or actions (e.g., "Want X"), and were not identification errors (Pepperberg 1987b,c).

Figure 8.1. Object array used to see if Alex could perform a recursive task: Could he provide the label for one attribute of an object defined by another attribute? He had to understand all the terms in a question such as "What color is the key?" to respond correctly.

Of particular interest are Alex's types of errors, because they tell us not only what he does or does not comprehend, but also something about how he may process information in questions, about objects, and in his responses. Errors could arise primarily from three sources (Pepperberg 1990a): (1) confusion of labels that sound alike; (2) misunderstanding of the label that directed the search; and (3) correct selection of the targeted item, but mislabeling of the targeted attribute.

Of Alex's nine errors, four probably involved comprehension or production of labels that, at least to humans, sound very similar. On one question, Alex gave the color of the rock when asked the color of the block. Interestingly, his initial production of "box" was "bock," which may explain why he once responded with "rock" rather than "box" and on a different trial with "box" rather than "rock" when asked, "What object is blue?" And, although "grey" and "green" may not sound similar in standard American English, Alex sometimes produces "gree" for "green" such that it might be mistaken for "grey"; by our criteria, such a response caused one error (see, e.g., Pepperberg 1981). Thus about half Alex's errors—and < 9% of his total responses—could be

Table 8.1. Representative sample of sets of objects, questions, and Alex's responses in trials on the recursive task. (After Pepperberg 1990a.)

Trial[a]	Objects in set[b]	Question	Responses
1	Purple key, yellow wood, green hide, blue paper, orange clothes pin, grey box, red truck	What object is grey?	Box
14	Purple paper, yellow clothes pin, green box, blue truck, orange key, grey wood, red hide	What color is truck?	Blue
20	Circular plastic, 2-corner nut, triangular wood, square paper, pentagonal key, hexagonal hide, barrel	What object is 2-corner?	Nut
27	Circular rock, 2-corner wood, triangular plastic, square wool, pentagonal paper, hexagonal key, chalk	What shape is wool?	4-corner

a. Numbers refer to the actual trial numbers in the experiment.

b. Colored items were of various shapes, and shaped items were of various colors, to ensure that Alex could not determine the question of the trial on the basis of the types of objects he was given.

attributed to possible perceptual problems (both for trainers and for Alex), rather than lack of cognitive ability. Such confusion is significant and not unexpected, and suggests that Alex interprets acoustic cues in an orderly manner, possibly with respect to the formants and speaking rates of his trainers. These findings are similar to those seen for humans (Chapter 16; Miller 1987; Lieberman and Blumstein 1988).

Alex's errors that arose from the second and third sources could not easily be distinguished (Pepperberg 1990a). Thus, if we asked, "What color is the wood?" and Alex said the incorrect color, he could either have understood and appropriately chosen "wood" but mislabeled its color, or correctly identified the color of the wrong object. Previous data suggest that he erred on the attribute: Attribute labeling accuracy averages ~80% (Pepperberg 1987c, 1990c), and he thus had to be close to perfect on question comprehension to achieve his current scores.

The Implications of Alex's Abilities

I designed this study to compare Alex's abilities with those of great apes and marine mammals. His level of cognitive and comprehension

Table 8.2. Alex's results for each type of question in trials on the recursive task. The scores are the number of first trials of the 12 presented for which he was correct. (After Pepperberg 1990a.)

Question	Test score	Errors Type	No.
What object is designated color?	10/12	Wrong object label	2
What color is designated object?	8/12	Wrong color label	4
What object is designated shape?	12/12		
What shape is designated object?	9/12	Wrong shape label	2
		Correct color label	1

ability can best be appreciated by comparing his task with those used for these other species. Protocols can be compared with respect to task complexity, the variety of questions posed, the variety of items, the animals' familiarity with items used in the tasks, the overall diversity of the task, and the type of cognitive ability tested.

Some animal studies lacked complexity; unlike Alex's task, tasks in these studies tested comprehension of only one label per trial (Pepperberg 1990a). Subjects needed only to indicate the item among several to which this one label referred (e.g., "Point to X"; Premack 1976; Patterson 1978; Savage-Rumbaugh and Rumbaugh 1978). Other tasks such as "? color-of X?" ("What color is object X?") could potentially demonstrate comprehension of several categories and labels, but often did not (Premack 1976): To test a subject adequately, the task must use an "X" that can vary on each trial, and require responses to be chosen from a repertoire that includes attributes other than color. Instead, "X" typically was an item of invariant color, such as an orange, and subjects' response choices were often limited to two color symbols. In contrast, Alex's trials each tested comprehension of several categories and labels. For 48 collections of variously colored and shaped objects, he had to answer queries such as "What color is key?"—which tested comprehension of the object label ("key"), the targeted category label ("color"), and also the label that specified the object's color (e.g., "green"). And Alex had four different types of such questions.

The number of different possible types of questions a recursive task entails is crucial for determining an animal's capacities. In some studies, researchers serially ask subjects one type of question about, or ask them to perform a single action on, a given set of objects in a repetitive set of trials (Premack 1976; Essock et al. 1977; Savage-Rumbaugh et al.

1983; see Herman 1987). These animals may have used context cues to simplify the task: They may have realized, after a few trials, that responses could be limited to one attribute (e.g., color), and that they need not attend to, or process, the meaning of other elements in the query (Pepperberg 1990a). Thus, beyond the first few trials, a simpler (though still interesting) form of comprehension would have been examined rather than the more advanced (recursive) form the study was designed to test. Even the great ape given two different queries (e.g., about color or material) could demonstrate only limited comprehension because trials repeatedly used only a small set of items and trials on the same question were often clustered (Essock et al. 1977).[5]

Similarly, comprehension tasks in some studies appear complex, but subjects may, after a few trials, have derived and used one rule to simplify the tasks (Pepperberg 1990a). The questions combined several different action and object labels, but involved one theme, for example, "Do X (action) to Y (object)." Such repetition over all trials may have enabled subjects to use the rule "action-to-object." These tasks did demonstrate simultaneous comprehension of novel combinations of action and object labels (Schusterman and Krieger 1984, 1986; Herman 1986, 1987; Herman, Richards, and Wolz 1984; Schusterman and Gisiner 1988), but these trials provided only limited tests of complex comprehension. Tasks given some chimpanzees required comprehension of numerous simultaneous correspondences among symbols, English utterances, photographs, and objects (see, e.g., Savage-Rumbaugh et al. 1986), but still lacked generality: Subjects had only a limited number of possible responses (~3), and tasks involved only one rule, "What corresponds to X?"

Alex, however, was given different types of queries in any given session: Trials involved one of four questions that were intermixed not only with each other, but also with questions on other topics (e.g., same/different, absence, number concepts, photograph recognition). Similarly, some studies with dolphins (e.g., Herman 1987) and sea lions (Schusterman and Gisiner 1988) combined recursive and "relational trials" (of the type "Perform action X to relate objects Y and Z"; e.g., "Take Frisbee to surfboard"), *and* trials concerning the presence or absence of objects ("Is object X here?"; Herman and Forestell 1985). These marine mammals thus could not anticipate the type of trial and, to succeed, had to comprehend all elements in the question or command.

Chimpanzees in some studies may not have demonstrated their real competence because they could interact with only a few objects. These animals were trained *and* tested in a series of trials that used the same limited number of objects (e.g., six items that could each have six different colors; Essock et al. 1977). Thus, by the time the final set of trials occurred, each object had been presented, individually and in small groups, in numerous single topic tests (e.g., series of queries on color or material). Such a paradigm tests an animal's competence in the given task, but transfer tests are then needed to examine the ability to generalize to novel situations and thus show that the concept, rather than a specific set of problems, has been learned (Pepperberg 1990a).[6] I was careful in Alex's study to ensure that the number of possible items from which collections were formed was over 100 and that each combination for each trial was novel—that he had never seen that particular collection of items, and that no item in a collection had ever before been the target of a search for the information being requested. Thus the extent of his competence could more easily be determined.

Overall, Alex's task resembled that of the marine mammals more than that of the apes in the diversity of experience (Pepperberg 1990a). Diversity for marine mammals, however, did not derive from the number of physical objects used to form collections, which was smaller than Alex's. In fact, dolphins and sea lions had significant experience with each of their objects (e.g., Frisbees, pipes). Rather, diversity in marine mammal studies derived from the addition of spatial and hue modifiers, the number of different possible physical actions that could be directed toward these objects, and the number of possible topic questions that could be asked (see, e.g., Herman 1987).

Alex's task, however, differed from those given marine mammals in one important way: The relational questions/commands given marine mammals specifically tested whether they could respond *on the basis of word order*, but Alex's task did not: His task emphasized semantic comprehension, whereas the marine mammal work also emphasized comprehension of, for example, "Take pipe to Frisbee" versus "Take Frisbee to pipe." Because word order was intentionally removed as a cue for Alex (see above), his comprehension tests lacked that complexity, and even now I do not know if he can learn to respond on the basis of word order. The trade-off, however, was that Alex could not use word order to determine the meaning of the questions to be answered (Pepperberg 1990a).

In sum, Alex succeeded on a task requiring cognitive, representational skills comparable to those of great apes and marine mammals, and thus showed that such skills can be learned by a nonmammal. The extent of Alex's abilities, however, was still undetermined. Could he manage a more difficult task? Specifically, could he respond to queries combining modifier and object labels? Such tasks involve conjunction as well as recursion, and thus demonstrate more advanced levels of information processing (Granier-Deferre and Kodratoff 1986). Conjunctive tasks had already been given squirrel monkeys (*Saimiri sciureus*, Burdyn and Thomas 1984), dolphins (Herman 1987), and sea lions (Schusterman and Gisiner 1988); thus I began experiments with Alex.

More about the Conjunctive Task and Its Use in Evaluating Animal Intelligence

Conjunctive tasks, as defined by Thomas (1980), evaluate a complex form of intelligence: use of a logical operation ("and") to connect any number of specific concepts when many such concepts are present. A subject must respond to "red and square," *or* "blue, wool, and triangular," and so forth, on any given trial. A subject thus not only evaluates stimuli with respect to several different concepts, but also realizes that evaluation is based on the *process* of combining and recombining concepts (Pepperberg 1992a): Even a subject that performs the search as a parallel rather than a sequential task (Langley and Riley 1988; Woods, Alain, and Ogawa 1998) must form a combined representation of the two (or more) relevant stimuli. Moreover, because conjunctive tasks require data to be stored about multiple concepts, the tasks also test the capacity for processing various *amounts* of information.[7] How might Alex fare on such a task?

Conjunctive tasks are rarely used to study animal intelligence, possibly because their power to assess cognitive ability is unappreciated and because designing a truly conjunctive task is difficult. Specifically, researchers often mistakenly apply the conjunctive label to three other types of tasks. These other tasks may or may not be simpler than, but are distinctly different from and test different abilities than, tasks involving conjunction (see Thomas 1980).

Often, tasks requiring a subject to respond to a subset of *specific* attributes (or even signs for these attributes) are misidentified as conjunctive (Thomas 1980, 1991). Subjects learn to respond to one configuration, for example, "A and B, not A and C"—blue and round, not blue and square (see Wells and Deffenbacher 1967; Rescorla 1981)—then are tested on the same configuration. Such specific responses do not demonstrate sensitivity to, or understanding of, the abstract (and flexible) combinatorial concept basic to conjunction, because subjects cannot show whether they can transfer the *combinatorial* concept to another configuration (Thomas 1980, 1996).

Conjunctive tasks may also be confounded with those that involve decoding information about compound stimuli. Conjunctive tasks test how a subject processes questions about various combinations of several abstract categories. In contrast, tasks involving compound stimuli generally examine how combining two or more specific attributes affects accuracy in discriminating among objects (see Langley and Riley 1988). Decoding tasks test whether it is more difficult for a subject to learn, for example, to choose red when samples are a red circle and a green square rather than red and green circles—that is, whether shape information interferes with processing color information.[8]

Finally, some studies claiming to test conjunction test instead a conditional discrimination. The basis for choosing a triangle if test items are on a red tray and a square if the tray is green may be "for reward, choose triangle *if* red" as well as "for reward, respond to triangle *and* red" (Thomas 1991). In a conjunctive task redness is *not* the criterion for choosing triangle (Pepperberg 1992a); rather triangle and red are considered equally. Conditional tasks thus involve sequential processing to test the relationship between one of generally two initial conditions (e.g., red versus green) and the appropriate choice of a subsequent action. Conjunctive tasks instead test how *(re)combinations* of any number of initial conditions (which can be processed in any order) affect choice. Thus several instances of redness without triangularity and of triangularity without redness, in addition to the red triangle, must be presented, so that the *conjunction* of red and triangular directs a subject's response; furthermore, the task must test whether a subject can respond equally well to other combinations, such as green and square.

These problems may be avoided by adding conjunction to a recursive task. Understanding how combining conjunction and recursion achieves this objective is not simple. I have already discussed processing required by a recursive task; now I describe the effect of adding conjunction and how that addition tests more advanced capacities in our animals.

Adding Conjunction to Recursion

Three effects must be considered when adding conjunction to recursion (Pepperberg 1992a): (1) how conjunction affects the *recursive* complexity of the task; (2) how recursion ensures that the task is conjunctive; and (3) how adding conjunction affects *overall* task complexity.

As noted above, conjunction need not alter a task's *recursive* complexity. A recursive task without conjunction defines a target by *one* category (e.g., "What color is the item *that is paper?*"). Conjunction adds features (e.g., "What color is the item *that is paper and blue?*"). A subject could, however, search for "paper" and "blue" simultaneously, that is, use parallel rather than sequential processing. Parallel processing does not require an additional recursive step, and recursive complexity need not increase when conjunction is added.

Combining conjunction and recursion lessens the possibility of designing a task that fails to test conjunction. First, because trials in a recursive task can involve series of unique collections of different objects (e.g., subsets of a large collection of items that each vary with respect to color, shape, and object label) and one of several different possible commands concerning the attributes of these objects, the subject cannot respond to any single, specific subset of attributes or to one type of object (Granier-Deferre and Kodratoff 1986; Pepperberg 1990a). Second, a conjunctive, recursive task is not about decoding individual attributes of a complex object (e.g., something red and square) but rather is about responding to (i.e., processing information about) *successive* combinations of concepts or attributes. On one trial, for example, a subject must label the shape of an item defined by color and material and on the next trial label the material of an item defined by shape and color. Third, the structure of a recursive task lessens the likelihood of its being solved as a conditional discrimination. Assume, for example, that a subject is given a tray of items, including red nonmetallic and nonred metallic objects, and answers a question about the shape of a red metallic object by the apparently conditional process "If I find several red objects, then I'll choose the metallic one" (or "If I find several metal objects, then I'll choose the red one"). The subject has not, however, used a conditional process, for the condition red is not the *exclusive* reason for choosing something metallic. Neither the condition red nor metal alone is sufficient for a correct response; rather, the subject must combine ("conjoin") instances of color and material. In sum, potential problems inherent in testing for conjunctive understanding are considerably lessened when tests are administered as recursive tasks.

Finally, the *overall* difficulty of a recursive task does increase when conjunction is added. A subject must not only process information needed to solve the standard recursive task but also understand how different (re)combinations of categories affect its search for the targeted object. Moreover, a subject must store additional information about abstract categories to complete the search; that is, additional memory is required.

The Conjunctive, Recursive Task and How It Was Presented to Alex

Training was identical to that for recursion (Pepperberg 1990a, 1992a). One conjunctive question was added to one training session each week starting in November 1986. Training ended in February 1987 when my students and I began testing the recursive task without conjunction.

Each training and testing trial followed the protocol of the recursive task, except for the added complexity. Alex was given different 7-item sets, each chosen from 100 objects of various combinations of shapes, colors, and materials, and was asked to label the specific instance of one category of an item uniquely defined by instances of *two* other categories: "What color is (item designated by shape-X and material-Y)?" "What shape is (item designated by color-Z and material-Y)?" "What item is (designated by color-Z and shape-X)?" (Figure 8.2; Table 8.3). Each question contained three types of information: (1) information about the topic under study (e.g., the attribute of one item in a collection versus the quantity of the collection); (2) information that designated (by conjoining labels for *particular instances* of two categories, e.g., "paper and blue" versus "rose and five-corner") the 1 object of 7 that was the specific target of the search; and (3) information that designated (in the form of a different category label, e.g., material, rather than shape or color) the category from which a response should be chosen. (This last piece of information could, of course, be deduced from other information; because the question involved the only category for which specific information was lacking.) To respond correctly, Alex had to process each type of information correctly, then recognize and encode as a vocal label (e.g., "key") information about the appropriate instance of the targeted category.

Alex's disruptive behavior was even more pronounced on this task than on the pure recursive task. We could obtain data only by stopping testing for long periods so that the task recovered some novelty. Testing thus ceased from May 1988 to May 1989, and from August 1989 to May 1990. Testing also ceased between September 1990 and February 1991 for two different reasons: First, Alex contracted aspergillosis in September 1990; second, the lab was moved to Arizona in late 1990.

The Results of Alex's Tests on the Conjunctive, Recursive Task

On first trials, Alex scored 26/34 (76.5%; Table 8.4). His results did not differ significantly from recursive trials based either on a test for dif-

Figure 8.2. Object array used to see if Alex could perform a recursive, conjunctive task: Could he provide the label for one attribute of an object defined by two other attributes? He had to understand all the terms in a question such as "What color is the three-corner wood?" to respond correctly.

ferences in proportions or on a chi-square test $[X^2 (1) = 0.2709]$. Thus adding conjunction did not affect accuracy. Because he was already familiar with the recursive task and conjunction did not increase the complexity of the recursive part of the task, scores probably reflect competence on the conjunctive condition.

I calculated statistical significance using two different chance criteria. First, I based chance on the assumption that Alex could have randomly selected an item to label. Second, I assumed that he could have selected an item on the basis of one but not both targeted properties,

Table 8.3. Representative sample of sets of objects, questions, and Alex's responses for the conjunctive, recursive task. (After Pepperberg 1992a.)

Trial[a]	Objects in set[b]	Question	Responses
1	2-corner rose hide, 3-corner purple rock, 3-corner orange wood, 3-corner blue paper, 3-corner yellow hide, 4-corner green hide, 5-corner grey hide	What color is 3-corner hide?	Yellow
11	Circular green rock, 2-corner green wood, 3-corner blue paper, 4-corner green paper, 5-corner orange paper, 6-corner green wool, yellow paper cup	What shape is the green paper?	Corner, 4-corner[c]
24	2-corner purple wood, 3-corner purple key, 5-corner green paper, 5-corner purple rock, 5-corner orange wood, 5-corner yellow hide, purple chalk	What object is 5-corner purple?	Rock

a. Numbers refer to the actual trial numbers in the experiment.

b. Hide is rawhide, rock is a shaped piece of Play-Doh, and wool is a pompon or a shaped piece of felt.

c. Alex occasionally initially responds to questions about shape with the vocalization "corner" followed without pause by the numbered label, such as "2-corner." Because this pattern includes no pause and thus no need for intervention by the trainer, such a response is not considered an error.

that is, as though he did not use conjunction. I thus evaluated his performance at two levels of complexity (Pepperberg 1992a).

To compare Alex's results to values based on a random choice of item to label (i.e., to test whether he responded to objects and queries), his data must be evaluated separately for each topic. Remember, chance for each type of question had to be based on the number of possible responses to the targeted category rather than all possible categories (Herman et al. 1984; Schusterman and Gisiner 1988; Pepperberg 1990a). Thus, for color and material queries, I used chance as 1/7, because seven possible responses existed—seven colors or seven objects on the tray. I used chance as 1/5 for shape trials even though Alex saw seven objects, because the number of shapes he could label was five. Chance values were conservative because they ignored the possibility that Alex could have produced *any* label in his repertoire. On binomial tests for color queries, $p < 0.0003$; for shape and material queries, $p < 0.0001$.

Chance could also be calculated based not on random object choice, but on choice based on one category in the query (e.g., responses to

Table 8.4. Alex's results for each type of question on the conjunctive, recursive task. Scores are the number of first trials for which he was correct. (After Pepperberg 1992a.)

Question	Test score	Errors	
		Type	No.
What object is [item designated by color and shape label]?	9/11	Wrong object	2
What shape is [item designated by color and object label]?	10/12	Wrong shape	2
What color is [item designated by shape and object label]?	7/11	Alternate item cited	1
		Wrong color	3

a subset defined by color *or* shape). On each trial, four items embodied one targeted category, four embodied the other, and only one item exemplified both. If Alex had chosen randomly after targeting one categorical subset, chance would be 1/4 for color and material trials; his actual p values, respectively, were 0.0064 and 0.0001. On five shape trials, two of four items in one subset could not be labeled, so chance was 1/2; on such trials Alex's score was 5/5 and $p = 0.03$. For the other seven shape trials, one of four objects in each subset was unlabelable, giving chance as 1/3; his score for these trials was 5/7, $p = 0.04$ (Pepperberg 1992a).

The number and type of Alex's errors for each question are intriguing. In eight trials in which he initially erred (Table 8.4), he responded correctly on the second presentation. In each of two additional trials (one on color, one on material), however, he did not respond correctly on the second try. He produced instead *each* of the wrong possible answers from the appropriate category, *repeated* each wrong answer, then grabbed the tray liner and tossed all objects to the floor. The probability (binomial test) of his producing, by chance, each wrong answer *in turn* is 0.0061; the probability of his repeating that behavior is even smaller ($\ll 0.0001$). His consistent omission of the correct answer suggests that his errors were not caused by a failure to understand the task; remember that his disruptive behavior patterns (including grabbing the tray liner) had caused numerous pauses in testing. These two trials were considered mistrials and are not included in the data.

Other data show he understood which category was targeted (Pepperberg 1992a). As in the recursive study, Alex's first single utterance was generally from the appropriate category, even if it was an incorrect instance of that category. With the exception of his response to one question (in which he labeled an item not present), all other initial responses were either repetitions of parts of our questions (e.g., "What color?") or encoded requests for information, other items, or actions (e.g., "You tell me, what shape?" "I want X"), and were not scored as errors. His use of a label from an appropriate category did not necessarily imply that he merely responded to the initial part of the query: Remember, we often varied the order of phrases in a question; he was as likely to hear "What shape is the green wood?" as "Green wood; what shape?"

Examining Alex's errors shows not only what he does and does not comprehend, but also something about the cognitive processes he may use in order to respond. Errors could arise from five sources (Pepperberg 1992a): (1) confusion of labels that sound alike; (2) misunderstanding of a label that directs the search; (3) problems with respect to perceptual boundaries, e.g., differences in avian and human color perception; (4) inability to comprehend conjunction, that is, not being able to understand that information from two categories must be used to target the object of the search; or (5) correct selection of the targeted object, but then mislabeling of the attribute in question.

Because almost half of the recursion study errors involved comprehension or production of labels that, at least to humans, sound similar (e.g., "rock" and "block"), I eliminated such confounds in the present task. When a collection contained items with similar labels (e.g., a two-corner yellow rock and a yellow block), I deliberately asked questions about other items, such as the shape of the yellow rawhide. No errors were therefore based on auditory similarity.

As in the recursive task, errors from the second and fifth sources could not be distinguished. In previous studies, however, Alex had labeled object attributes with ~80% accuracy (e.g., Pepperberg 1987c), and his scores on this task were comparable. He thus probably mislabeled the attribute of an object after targeting it correctly (Pepperberg 1992a).

Of Alex's eight errors, five might be attributed to perceptual rather than cognitive abilities.[9] His error on one trial—on "What object . . . ?"—involved confusion between rock (Play-Doh) and hide, which are difficult even for humans to distinguish by sight. Alex does touch the items before a trial, but may have forgotten which object was of which material. Of four errors on "What color . . . ?" one was a red/orange confound and another a red/purple confound. As noted earlier, psittacine and human visual systems differ in color boundaries (Bowmaker 1986; Bennet and Cuthill 1994; Bowmaker et al. 1994, 1996), and Alex's errors on orange are usually "rose" or "yellow," and on purple are usually "blue" or "rose" (Pepperberg 1990c, unpubl. data). Such errors could be related to cognitive processing or chance behavior, but alternative explanations are consistent with psittacine perceptual capacities.

The Implications of Alex's Results

We can again compare Alex's data to data on other species. I focus on abilities specifically evaluated by conjunction (Thomas 1980): flexibility in processing various *types* of information and capacity for processing various *amounts* of information. Parts of this discussion repeat earlier material, but the importance of comprehending conjunction bears review if I am to claim that Alex's aptitude for conjunction was at least as great as that of mammals (Pepperberg 1992a).

A properly designed conjunctive task requires use of the logical operation "and" to relate conceptual categories (Thomas 1980, 1991, 1996). To succeed, a subject must (1) understand the abstract categories that are being represented, (2) recognize that *any* categories can be combined, and (3) not use a conditional discrimination to solve the task. Alex fulfilled these three criteria.

Alex's comprehension of labels and abstract categories was demonstrated earlier (e.g., in Chapters 4, 5, 6); the present task, however, required him to understand categorical concepts at a different level. His task not only involved searching for a single specific collection of attributes, but, unlike some tasks given nonhuman primates, also required simultaneous comprehension of several category and object labels *and* the meaning of entirely different categorical questions (e.g., "How many?" "What's different?") in any session (Pepperberg 1990a, 1992a; note Burdyn and Thomas 1984). Alex thus could not use context to anticipate the topic or type of trial. This study, much like those with marine mammals (Herman 1987; Schusterman and Gisiner 1988), ensured that, to succeed, Alex had to comprehend all conceptual elements in a question and base his responses on his entire repertoire of concrete and abstract labels.

Alex understood the logical operation "and" at a level comparable to that of mammals, although tasks differed somewhat across species (Pepperberg 1992a). Chimpanzees learned, for example, to request multiple objects or actions (banana *and* apple, wash *and* give apple) and were tested on their responses to queries such as "Cookie ? cracker ? pl breadstuff" and "Blue ? green ? pl color" where choices were "and,"

"and," "is," "is" (Premack 1976). Each question type (i.e., category) was, however, tested sequentially and independently; thus flexibility of the "and" operation was not necessarily demonstrated. Also, after learning "red" and "dish," chimpanzees understood "Insert apple (in) red dish" (Premack 1976), but their choices were limited to, for example, red and green dishes, and not red and green dishes and boxes. Monkeys can associate triangles with "same" and heptagons with "different"; the data suggest that they comprehend two conjunctive relationships specific to a set of abstract concepts (Burdyn and Thomas 1984), but do not demonstrate the generality necessary for complete understanding of conjunction (Thomas 1980, 1996; Pepperberg 1992a). Tasks given marine mammals, in contrast, do fulfill Thomas's criterion, by showing responses to novel combinations of different *types* of information: Dolphins and sea lions respond to novel combinations not only of attribute and object labels, but also of action and object labels (Herman et al. 1984; Schusterman and Krieger 1984, 1986; Herman 1986, 1987; Schusterman and Gisiner 1988). Alex's trials thus resembled those of marine mammals: His questions combined any two of three categories, and he could not anticipate the combination (Pepperberg 1992a). His three sets of questions (on color and shape, color and matter, shape and matter), moreover, were intermixed not only with each other but also with those from all studies, *including* the nonconjunctive recursion task. Alex could not, for example, infer that a trial involved conjunction simply because a number of items had colors or materials in common; for that collection he could also be asked, "How many green toys?" (Chapter 7). Instead, he had to determine from the form and content *of the queries* whether his response should be based on a conjunction of attributes; he *then* had to show he understood the conjunctive condition. Alex's flexibility in processing novel combinations of different categories of information was therefore at least as great as that demonstrated by mammals.

Moreover, using a recursive task to study conjunction probably lessened the possibility that Alex processed information conditionally (Pepperberg 1992a). When a task combines conjunction and recursion, subjects must choose among a number of items, each of which has at least one of the criterion attributes. Arguably, Alex may have answered a query about the shape of the green key by the process "If I find green things, I'll give the shape of one that's a key." The condition of color would not, however, be *the* specific criterion used to choose the shape of the key, but would merely limit his choices. He could equally well have begun by limiting his choice based on matter. His processing, although sequential, would not be conditional (Thomas 1980).

Finally, we must examine the demands that conjunctive, recursive tasks place on memory, and how success at meeting those demands can be used to infer advanced cognitive abilities in an animal. Many tasks require two types of memory: reference memory and working memory (Roitblat 1987). Reference memory involves the stable rules of the task (e.g., remembering what *kind* of information the type of question requires), whereas working memory involves information that changes depending on the specific trial of a task (e.g., all the bits and pieces of information necessary for answering correctly). Recursive and conjunctive tasks each require integrating these two types of memory, and their combination increases demands on both types of memory. Adding conjunction to a recursive task adds the logical operation "and" to reference memory and additional categorical information to working memory (Pepperberg 1992a). Although the expected effect of adding to memory load would be to increase the number of errors (see Grant and MacDonald 1986), neither Alex nor marine mammals demonstrated any significant decrease in accuracy with addition of conjunction (Schusterman and Gisiner 1988; Wolfe et al. 1989).[10]

The ease with which novel combinations of different types of information are processed, and whether such processing occurs sequentially or in parallel, may also be reflected in the time a subject takes to respond to a query—its reaction time or response latency. For example, a subject is expected to respond more slowly as the number of signals used to define items, and thus the amount of information to be processed, increases (as when a subject must process vocal labels "color" *and* "shape," in contrast to just "color"; Treisman and Sato 1990). As noted previously, however, Alex's readiness to respond to a given set of items is primarily correlated with his level of interest in their acquisition. Because items in the conjunctive, recursive tasks did not appear to engage his interest, response latencies could neither be meaningfully measured nor compared with those of mammals (Pepperberg 1990a, 1992a).

In sum, Alex performs as accurately on a recursive task that includes conjunction as on one that does not; his data thus demonstrate competence with respect to conjunction and, by implication, comprehension. Because the overall cognitive demands of Alex's task are comparable to those of tasks given to marine mammals, Alex's success suggests that various species have similar capacities and flexibility for processing complex information. These data, however, provide but a small piece of an overall picture. In the next chapter I describe yet another study designed to compare avian and mammalian competencies.

9 Can a Parrot

Understand

Relative Concepts?

> Although we are not yet able to enumerate the elements of which human intelligence consists, we have made some progress toward this end. For instance, if in considering intelligence we focus only on language we will find, very likely, that no infrahuman species, chimpanzee or otherwise, can reproduce human syntax. But if we reject this rather narrow approach and instead analyze intelligence into the several capacities that it encompasses, such as semantic memory, representational capacity, causal inference, intentionality, and concept of self, we will find that the ape, although unable to reproduce human syntax, has many of the other capacities. In contrast, lesser species not only cannot reproduce human syntax, but lack other capacities as well.
>
> David Premack, *Cognitive Processes in Animal Behavior* (1978:423–424)

The ability to comprehend relational as well as absolute concepts (e.g., answer "Is A bigger than B?" in addition to "What color is A?") is frequently used to compare intelligence across species (Premack 1978; Thomas 1980, 1996). The rationale is that relative concepts require more information processing and thus are inherently more complex than absolute concepts: To respond with respect to an absolute concept, a subject merely affirms that something is, for example, "red." The subject forms the concept "redness," then decides if this feature is embodied by a given item, that is, forms a single association between item and concept. To respond with respect to relative concepts, in contrast, a subject must compare two (or more) items based on some abstract concept (see, e.g., Thomas and Crosby 1977). Alex succeeded on one type of relational task, same/different (Chapter 5; see Premack 1978, 1983), and other animals solved relational tasks involving conditional situations (e.g., "if A, then B," Thomas 1980), but as of the early 1980s, no conclusive evidence supported a nonprimate's ability to respond to queries about relative *class* concepts, such as "darker than" (Premack 1978; Thomas 1980).

Determining relational ability, however, presents two problems. First, relative concepts vary in complexity. Supposedly, understanding a relative *class* concept is qualitatively different from, and easier than, understanding same/different (Thomas 1980).[1] Nevertheless, relative class concepts are not simple to understand or use. A subject asked, "Is A darker than B?" must base its answer not on common elements in the two objects or only on physical criteria (e.g., one object's being "grey" and the other "black"), but rather on a comparison; furthermore, "darker" in one trial ("grey" versus "white") may be "lighter" in the next ("grey" versus "black") (Pepperberg 1996). Second, part of the problem in showing animal competence in relative concepts probably involves the abilities of the researchers rather than the animals: Relational tasks often allow organisms to learn something about both absolute and relative concepts concurrently (Premack 1978). Thus researchers using such tasks for cross-species evaluations of cognition must determine the *extent* to which relative information is used in problem solving (Pepperberg and Brezinsky 1991). Despite such potential difficulties, I wondered if, given what Alex had already learned, he could acquire a relative class concept.

Background on Relative Concepts from the Field and the Laboratory

Ethological studies, although not designed to contrast relative versus absolute concept learning in animals, nevertheless suggest that some avian species respond to and use natural stimuli on a relative basis. In the field, eastern wood-pewees may use the relative number of repetitions of a song type sung by a conspecific to assess how the singer will probably act (Smith 1988). Eastern kingbirds seem to proclaim impending social behavior through proportional use of two different song forms: Birds singing more of one form responded less actively to intruders than birds singing more of the other (Smith and Smith 1992). Louisiana waterthrushes *(Seiurus motacilla)* use "primary" (relatively short) songs and "extended" (relatively long) songs when they are, respectively, singing alone versus engaging in approach flights (Smith and Smith 1996). Possibly wood warblers (e.g., chestnut-sided warblers) respond to the relative variation of songs sung in a bout (Kroodsma 1990; Byers 1996). Some species may judge relative size and motivational states of competitors by assessing relative frequencies (Hz) of their vocalizations and whether the ending frequency is relatively higher or lower than the starting frequency (Morton 1977, 1982).

In the laboratory, great tits, after being trained to respond differentially to songs with either a large or a small number of notes, respond to songs with intermediate numbers of notes on a relative basis, and these birds also use, although not exclusively, relative frequencies of notes to discriminate songs (Weary 1989, 1991). Studies of optimal foraging also suggest that birds base feeding strategies on the relative amount of food they can obtain per unit time (Kamil and Roitblat 1985; Kamil 1988). Although birds in these studies appear to be responding on a relative basis, the experimental designs do not prevent subjects from also learning and using absolute cues (Pepperberg and Brezinsky 1991). For example, birds that responded to songs having intermediate numbers of notes might, without necessarily being aware of a relative distinction, clump songs into categories that to *humans* represented "more" versus "less"; birds in other studies may have discriminated frequency (Hz) based on some internal threshold. Thus although all these birds may have made relative discriminations, alternative explanations for their actions are possible, and such studies can at best provide only indirect evidence for the ability to respond on a relative basis. Remember, though, that none of these studies was *specifically* designed to demonstrate understanding of relative or relational concepts.

Somewhat more direct evidence for relative learning comes from transposition experiments and number studies. Transposition experiments were among the first specifically to examine the *extent* of relative learning in animals (see Reese 1968; Riley 1968; Sutherland and Mackintosh 1971; Premack 1978), and are still commonly used (e.g., Schusterman and Krieger 1986). In experiments such as one mentioned above, an animal trained to respond to grey in a choice between grey and white is then presented with black versus grey. Choosing grey in the new situation is interpreted as responding to an absolute value: That is, the animal has previously been rewarded for recognizing "greyness" and continues to respond on that absolute basis; the quality of relative darkness does not seem to influence the choice. In contrast, choice of black in that situation is viewed as a response based on relative darkness (Köhler 1929; cf. Spence 1937; Reese 1968; Riley 1968). Data from such studies suggest that pigeons generally respond on an absolute basis.[2] A nontrivial complication, of course, is that the experiment must be designed extremely carefully so that the animal is given some reason to switch from responding with respect to the absolute basis (see the discussion of proactive interference in Chapter 8). One alternative is to ask the animal to transfer to a different form of the same task (e.g., "brighter than" instead of "lighter than"); a more difficult alternative is to design a task that avoids competition between relational and absolute factors. Unfortunately, researchers are rarely able to implement this more difficult alternative (see below, Premack 1978; note MacDougall-Shackleton and Hulse 1995). Separate evidence for relative judgments comes from studies suggesting that pigeons can respond to relative numerosity, for example, "more" versus "less" (Honig and Stevens 1989, 1993; Honig and Matheson 1995; Emmerton et al. 1997; Emmerton 1998).

Some relational learning appears in every species tested, but comparison of animal capacities is impossible because of the lack of consistency among studies (for detailed discussions of various interpretations of relative concept studies, see Reese 1968 and Riley 1968). In addition, as noted above, tasks that involve relative concepts often allow organisms to learn about both absolute and relative concepts concurrently (Premack 1978); thus researchers using such tasks for cross-species evaluations of cognitive capacity must determine the *extent* to which relative information is used in problem solving. McGonigle and Jones (1978), for example, argue that subjects' responses depend upon which judgment, relational or absolute, makes fewer processing demands.

Difficulties in Designing a Relational Task That Compares Knowledge of Different Species

In relational tasks designed for cross-species comparisons, an animal's performance is controlled by at least three factors (Pepperberg 1990b; Pepperberg and Brezinsky 1991). The first and most obvious involves the animal's inherent capabilities—whether it has the cognitive capacity to perform the task. The second factor is the extent to which the chosen task matches a species' perceptual capacities and natural environment—researchers must ensure that an animal that, for example, has excellent olfactory capacities but poor vision is tested on an olfactory rather than a visual task. The third factor is how well the experimental design allows the subject to learn the task and express its knowledge—many comparative studies may not enable the animal to *demonstrate* the extent to which it has acquired or can base its responses on a relative concept.

The combination of the second and third factors often prevents researchers from determining the weight of the first factor in an animal's performance. Specifically, the belief that many animals, unlike humans, respond preferentially on an absolute basis may be a consequence of perceptual (rather than cognitive) capacities and experimental design (McGonigle and Jones 1978). Data on pitch perception, for example, suggest that starlings respond on a relative basis only as a secondary strategy, *after* acquiring information on an absolute basis (Hulse, Cynx, and Humpal 1984; Cynx, Hulse, and Polyzois 1986; Page, Hulse, and Cynx 1989; Hulse, Page, and Braaten 1990; but see MacDougall-Shackleton and Hulse 1995). (Note also the work on black-capped chickadees, *Parus atricapillus,* Weisman and Ratcliffe 1989; Hurly, Ratcliffe, and Weisman 1990; Weisman et al. 1990; Weary and Weisman 1991; Shackleton, Ratcliffe, and Weary 1992; white-throated sparrows, *Zonotrichia albicollis,* Hurly et al. 1991; and veeries, *Catharus fuscescens,* Weary et al. 1991.) But the starlings' preference for responding on an absolute basis, often referred to as a *frequency range constraint* (Cynx et al. 1986), may derive from ethological priorities (Pepperberg and Brezinsky 1991): For some animals, a vocalization's absolute pitch expresses whether another individual is aggressive or fearful/ready-to-appease; relative pitch changes *within* the absolute range express whether these are increasing or decreasing tendencies (Morton 1977, 1982). Thus shifting absolute pitch changes the overall import of the signal to such an extent that responses to the "fine-tuning" may be overwhelmed. Absolute pitch processing may also be involved in species discrimination, whereas relative pitch processing may involve the identity and dialect of a singer (MacDougall-Shackleton and Hulse 1995). Such findings and proposals, combined with data demonstrating that relative pitch processing is possible, suggest that animals may not demonstrate what they have learned about a relative concept unless experimenters limit the extent to which responses can be made on an absolute basis (Premack 1978; Page et al. 1989).

Finally, as noted in Chapter 4, a task used to measure intelligence or to compare intelligence across species must examine if, and how well, a subject can transfer its knowledge across domains (Rozin 1976; Sternberg 1985; Rogoff 1990). Relational tasks used in this manner must thus not only be ethologically appropriate, but also demonstrate transfer. Many studies, however, do not explicitly test whether relative concepts can be applied across modalities or to objects that differ in various ways from training items (Pepperberg and Brezinsky 1991): Subjects are trained and tested on similar sets of items (e.g., tones: Page et al. 1989; conspecific vocalizations: Weisman and Ratcliffe 1989; balls: Schusterman and Krieger 1986; brightness: review in Reese 1968). Thus the generality of the tested ability is called into question (Macphail 1987; Kamil 1988). Nevertheless, an important first step is determining if the capacity to understand relative concepts exists in any domain.

Why Study Relative Size?

My students and I chose to study relative size for two reasons. First, that task involved physical objects—a modality in which Alex had considerable experience (e.g., same/different judgments, Chapter 5). Second, relative size had been studied with a sea lion (Schusterman and Krieger 1986) that, like Alex, had had considerable language-related training. Our data would thus facilitate cross-species comparisons of abilities.

Rationale for the Experimental Design

Our protocol was designed not only to compare Alex's understanding of relative size with that of other animals, but also to test further several existent skills. We could design questions to examine Alex's comprehension and production abilities in more detail. We could also query him about items that differed in many ways from training objects and thus test his ability to transfer information across domains (Pepperberg 1987a). Moreover, because Alex already responded both to the presence *and* absence of similarity and difference (Chapters 5, 6), we could test for a level of understanding not required of animals in many other studies on relative concepts (see Reese 1968): understanding the *absence* of a size difference. To incorporate these additional tests, our procedures differed in three specific ways from those used in other relational studies (Pepperberg and Brezinsky 1991).

First, in most studies animals indicate by a single physical action (e.g., pointing) the bigger or smaller item. Such a protocol primarily tests comprehension, because the subject need not access or encode additional information after identifying the targeted object (see Pepperberg 1990a). In contrast, because Alex responds to and encodes his responses as vocalizations, we could require him not only to chose an object based on the targeted size dimension (larger *or* smaller), but also encode, as a vocal label, information about a different dimension, specific to that item. Thus Alex was to designate the appropriate object not by pointing, but by *labeling* its color or material—an additional step that significantly increases task complexity (Premack 1978).

Second, most studies on relative concepts lack transfer tests across modalities or domains, or even items that vary significantly from training objects (see Pepperberg and Brezinsky 1991). The sea lion in the study on relative size was, for example, queried about items larger or smaller than the training objects, but questions involved only balls (Schusterman and Krieger 1986). Might Alex demonstrate more generality than that? Because he would be responding with the label for an attribute other than size (i.e., "What *color* larger?" not "Go to the *larger ball*"), we could test his ability to transfer to objects entirely different from pairs used in training: objects he had never before seen, some of materials or shapes he could not label. Note that Reese's (1968) literature review shows that the difficulty of a relational task increases with the number of dimensions that are varied.

Third, unlike other subjects, Alex was queried about objects of equal size. Could he extend his knowledge of absence of similarity or difference in color, shape, and material to report absence of *size* difference? Generally, comprehension of "middleness," rather-than size equality, is examined as the alternative to "not bigger and not littler" in studies of animal intelligence (Reese 1968:156), but we thought absence would allow us to examine further Alex's ability to transfer information across domains. Whether comprehension of the concept of equality rather than middleness might reveal more about intelligence was not the issue; the point was to learn whether Alex could comprehend a relative concept. Evidence that he could respond not only to positive instances of bigger and smaller but also to the absence of such information would strengthen such a claim (note Pepperberg 1988b).

Specifics of the Experimental Design

Alex received M/R training. During sessions, he observed a trainer pick, from a small group of items, two objects of the same shape and material but of different size and color (see below); the trainer showed these objects to the model and asked, "What color bigger?" or "What color smaller?" The model responded appropriately and received the items or the right to request an alternative reward, or erred and was scolded ("No!" "Pay attention!" etc.). As in previous studies, the roles of trainer and model were frequently reversed, and Alex was also given an opportunity to respond. As soon as he would respond reliably with any color label (on five of six consecutive trials), whether or not his response was correct, we began counting trials to criterion (see below); this procedure ensured that he was attending to our queries and, to some extent, the task. If correct, he received the objects; if he erred, trainers modeled the correct response (Pepperberg and Brezinsky 1991).

As ever, we ensured that training would not be associated with specific physical conditions or topics other than relative size. We varied the positions of bigger and smaller objects, asked both questions, and used several different object pairs in a single session to ensure that

Alex learned about the relationships between objects, rather than, for example, to choose "more" or a specific item. To ensure that a large number of objects were available for transfer tests, we limited training to a subset of 18 familiar items: felt circles (diameter of 2.5 and 5 cm), Play-Doh rods (~1.0 cm × 3 and 5 cm), and plastic boxes (3 and 5 cm³) that were green, blue, and yellow. To ensure that the presence of two items did not cue a response associated with relative size, we concurrently reviewed "What's same/different?" for object pairs and "How many?" for various collections (Pepperberg 1987a,b, 1988b). Finally, to ensure that Alex would not associate the task with particular humans, different individuals provided training; these individuals included me and students who were testing other concepts (e.g., number comprehension). Training began in November 1988 and continued until March 1989, excluding breaks for student vacations, finals, and intersession. Sessions occurred two to four times a week and lasted between 5 minutes and 1 hour depending upon Alex's attention span.

Criterion Prior to Formal Testing and Behavior during Training

We began counting trials to criterion on 1 December 1988. Alex was queried about relative size no more than twice in any session. As in earlier studies, we did not want him to expect queries in a given session to involve one topic, and thus to cue into a particular subset of answers. Training ended when Alex reached the designated criterion, after 42 trials, of 24 initially correct responses in 30 consecutive questions (80%) on pairs of training items (Pepperberg and Brezinsky 1991).

During training we tracked changes in Alex's accuracy and type of error. He succeeded on 6 of the first 10 trials, 7 of the second 10 trials, but 8 of the third 10. Of five errors in the first 10 trials (he erred twice on one trial), two were unclear vocalizations ("rue" [a rose/blue mixture], "gree"), one was an identification of object material ("rock"), and two were colors of the incorrect choice. In the remaining training trials, he erred once by labeling object material; all other errors were either the label for the incorrect choice or that for the absent color (one of the three expected colors).

Additional Training That Differed from That for Other Projects

Unlike Alex's previous tasks, this one required him to label one of two simultaneously presented objects (note Pepperberg and Kozak 1986, Chapter 10); it thus required additional precautions against inadvertent cuing. In other tasks that require a vocal response, position cues are irrelevant, as when Alex is queried about the quantity of a random array of objects ("How many key?") or about an attribute two objects have in common ("What's same?"). In contrast, in the two-choice design of the present protocol, an experimenter's direction of gaze or manner of presenting objects could provide inadvertent cues as to which object should be labeled.

To ensure that Alex would not be cued in this manner, before beginning testing we trained him to respond to queries about objects hidden in a cigar box with a reconstructed lid such that the examiner could not see the objects but could still interact with Alex (e.g., meet his gaze; Figure 9.1). I began a trial by leaving Alex's room; I then placed one or more objects in side A of the box, rotated the lid to hide the items, returned to the room and handed the box to the examiner. I told the examiner what type of question to ask (e.g., "How many?" "What color bigger?"); I then sat with my back to both Alex and the examiner. Alex was asked what he wanted as a reward. He watched as that object (e.g., a cork, a grape) was put into side B. The examiner turned the box so that side A faced Alex, rotated the cover to expose the objects(s) to Alex, then asked the appropriate question. The examiner repeated what she heard Alex say. I stated whether Alex was correct; this repetition prevented me from accepting an indistinct, incorrect vocalization that was similar to the expected, correct response (e.g., "gree" for "three"). If correct, Alex was given the items that were the subject of the question and then whatever was in side B; if he erred, he was scolded, experienced a momentary time-out, and the question was repeated. We used this procedure during formal testing.

Only after several months of M/R training did Alex reliably attend to the box and its contents (Pepperberg and Brezinsky 1991). In many sessions he would cease to work, begin to preen, or interrupt with many successive requests for other items (which we put in side B of the box) or changes of location. Because such behavior had previously indicated lack of interest in the training items (his primary rewards; e.g., Pepperberg 1983a), we added objects to the training set; these pairs were then not used in formal testing. And, as always, we found that working simultaneously on several topics improved his attentiveness and the likelihood of his responding appropriately to any question.

Testing

To determine whether Alex had, during training, learned to respond to relative size, we designed four different types of test trials (Pepperberg and Brezinsky 1991; see Table 9.1 for test objects). These different

Figure 9.1. Cigar box with a reconstructed lid used to test relative size. We placed the experimental objects into side A; we put objects that Alex requested, and that would be his rewards, in side B.

trials were presented at random, and, for each type of trial, the particular objects and the colors or materials were also randomized.[3] Each type not only tested Alex's ability to respond on the basis of the concept of relative size, but also often tested other cognitive capacities.

Our main goal was to see if Alex had learned to respond to relative rather than absolute size. Thus one set of test objects included items that were larger and smaller than those used in training. For these sets, the object that was the bigger one during training could then be the smaller of the pair being tested; likewise, the smaller training item could be the bigger of the two test items. Given this possibility for transposing absolute values of objects tested, such trials are called, in psychological parlance, *transposition* trials. We were careful to ensure that the novel-sized item was not always the correct choice so that Alex could not simply learn to respond to the unfamiliar object. Correct responses on these trials would suggest he had indeed learned to answer on the basis of relative size.

To test whether Alex could generalize to objects other than training items, we used two sets of objects that differed in their novelty (Pepperberg and Brezinsky 1991). One involved objects that were familiar from other studies (e.g., toy cars, tin boxes) but that now appeared in additional, unfamiliar sizes. The other included objects, in at least three

Table 9.1. The various objects used in relative size transfer trials with Alex. The sizes refer to the diameter for circular objects or the bases of containers; for such objects as hearts and trucks, the size refers to the longest dimension. (After Pepperberg and Brezinsky 1991.)

Objects	Size (to nearest 0.5 cm)
Woolen felt disks	1.5, 2.5, 5.0, 8.0
Plastic boxes	2.0, 3.0, 5.0, 7.0
Cups (plastic, paper)	1.5, 2.0, 2.5, 3.0, 4.0, 5.0, 6.0, 7.0, 8.0
Wooden plant stakes	3.0, 7.0, 10.0 (0.5 wide); 3.0, 7.0, 10.0 (2.0 wide)
Balls (wool pompons, wood beads, Styrofoam spheres)	0.5, 1.0, 2.0, 3.0, 4.0, 6.0
Blocks (wooden, plastic, sponge)	2.0, 3.0, 5.0, 7.0
Toy trucks	1.5, 3.0, 8.0
Styrofoam stars	6.0, 8.0, 12.0
Erasers	3.0, 4.0
Pasta	5.0, 9.0 (0.5 diameter)
Hearts (wood, paper, wool)	1.0, 3.0, 5.0
Leggo	2.0, 3.0, 5.0 (1.0 wide)
Wooden spools	1.0, 1.5, 2.0
Wooden rings	3.0, 4.0, 5.0
Play-Doh rods ("rock")	3.0, 5.0, 8.0, 11.0 (1.0 diameter)
Polygons (paper, wood, wool, Play-Doh)	2.0, 3.0, 4.0, 6.0
Keys (metal and toy plastic)	2.0, 5.0
Sugar cane (vs. pasta tube)	3.0 × 15.0 (vs. 0.5 × 3.0)

sizes, that Alex had never before seen and often were of shapes and materials he could not label (e.g., wooden hearts, Styrofoam stars). We painted these items various colors, including colors not used in training. In some sense, these trials often also involved transposition, because the sizes of the items varied considerably from those of the training materials. The smallest wooden heart, for example, was 1.2 cm across at its widest point, whereas the smallest Styrofoam star was about 6 cm at its narrowest. Such trials thus involved generalization to entirely novel instances of relative size.

To see how Alex would react to the *absence* of a size difference, some object pairs were of different colors but identical in size and material (Pepperberg and Brezinsky 1991). A response of "none" to a pair of equal size would provide additional evidence both for comprehension of absence of information and for Alex's understanding of a relative concept. We ran a single initial probe trial, followed by a series of such questions.

We also wanted to learn if Alex could, without explicit training, respond with respect to an attribute other than color. We thus designed trials in which item pairs were identical in color but of different sizes and materials (e.g., a blue wooden bead and a blue woolen pompon), and asked, "What matter bigger?" or "What matter smaller?" To ensure that Alex did not learn to provide material labels if objects were the same color, two trials involved items for which colors also differed. Correct responses on these two trials would thus provide additional, albeit limited, data that Alex was attending to, comprehending, and responding to our queries as well as to relative size. We considered using such pairs in additional trials, but decided that the requirements of these trials might not allow us to separate Alex's abilities with respect to question comprehension from those connected to relative size. We thus tested comprehension in the study described in the previous chapter.

Scoring Procedures

As usual, we calculated test scores for first trial and all trial responses and used a conservative value for chance. For all queries except those about equally-sized objects—for which "none" was a correct response—we considered chance 1/2 (to represent the number of colors present), although Alex could have produced any of seven color labels (and often did during initial training) or any other utterance in his repertoire (Pepperberg and Brezinsky 1991). We considered questions for which "none" was a correct response as probes because they had never been part of training; to be conservative, we calculated p using chance values of both 1/2 and 1/3 for these probes.

Alex's Results on the Various Sets of Trials

Alex's overall score was 59/75 (78.7%) on first trials (binomial test, $p < .0001$) and 75/91 (82.4%) for all trials. He erred 8 times each for queries of "What's bigger?" and "What's smaller?" Because our procedure for randomizing questions happened to provide a somewhat greater number of "What's bigger?" queries, his percentage correct is slightly greater for those questions, but both scores are statistically different from chance ($p \leq 0.0038$). Table 9.2 shows the data separately for the several different types of transfer trials.

Transposition did not appreciably effect Alex's accuracy. For queries involving larger or smaller versions of the training items, he scored 8/10 (80%, $p = 0.0439$) for first trials, 10/12 (83.3%) overall. Alex chose

Table 9.2. Alex's responses to questions on relative size. (After Pepperberg and Brezinsky 1991.)

Type of question	First trial			Errors	
	Correct	Total	p	Type	No.
Transposition	8	10	.0439	Color label of wrong object	1
				Random color label	1
No size difference	9		.0269[a]	"What's same?"	1
			.0011	Color label of object	1
What matter bigger/smaller?	8		.0439	Color label of both objects	1
				"None"	1
All other novel objects	34	44	.0001	Color label of wrong object	2
				Random color label	5[b]
				"None"	2
				Matter label of both objects	1

a. Binomial test: The first value is calculated for chance at 1/2, the second is for chance calculated at 1/3.

b. Of these 5 errors, 2 involved a yellow-orange-red confound, a color separation with which Alex has considerable difficulty (see Pepperberg 1990c).

the training item only once in error, but also labeled the training item on two randomly distributed trials for which it was the correct choice. Such data suggest that he did not respond on an absolute basis, and, furthermore, that his responses involved neither avoidance of, nor a predisposition for, novel size; rather, he responded with respect to relative size. Of course, any predisposition Alex might have had to respond on an absolute basis might have been weakened by our protocol: Because we intermingled trials involving size transposition of training items among many trials involving unfamiliar objects of varying size, familiarity effects could have been swamped. Our purpose, however, was to learn whether Alex *could* respond on a relative basis, and our data demonstrate that he did.

Alex's responses on trials involving equally-sized objects was, in one sense, particularly striking. The first time we asked, "What color smaller?" for a pair that did not differ in size, he responded, "What's same?" (Pepperberg and Brezinsky 1991). To our reply, "First you tell us what color smaller," he did indeed say "none." We intentionally did not answer his initial query; a response of a category label might, by association with previous experiments, have cued him to respond "none" (Chapter 6). Alex had not been trained to respond to same-size pairs (i.e., to an absence of size difference), and overemphasizing the importance of his initial, spontaneous responses is tempting. Evidence in favor of giving Alex intellectual "credit" for his behavior is that his first response was not a formal error. He did not, for example, label the color of one of the two equally-sized objects. Moreover, when unable to find a larger or smaller item, he did not ignore our query and respond to one of the other possible questions for which we presented objects pairs ("How many?" "What's different?"); that is, he did not reply "two" or "color," respectively. Conceivably Alex, by asking "What's same?" (a query that at the time he used on average less than once a day), indicated awareness of the similarity in size. And because the number of his possible vocal responses (~100) suggests that his second answer of "none" was not a lucky guess, it is also tempting to claim that his responses demonstrated understanding both of relative size and the appropriate response to absence of information (Pepperberg and Brezinsky 1991).

But behavior consistent with one hypothesis does not exclude others (for examples on avian vocal behavior, see Kroodsma 1990). Because Alex does occasionally respond to queries by requesting other information (Pepperberg 1990c), we believed that it was not Alex's initial responses but rather his overall behavior that would suggest he understood a lack of size difference. To explore the extent of his understanding, we thus needed to see if he could replicate this behavior.

We therefore gave Alex several subsequent trials (interspersed with other types of relative size trials and tests on other topics) to learn whether his initial responses had been a fortuitous accident. On these additional trials on equally-sized items, he scored 9/10 (with chance at 1/2, $p = 0.0098$; with chance at 1/3, $p = 0.0004$). If we count his response on the very first trial (i.e., "What's same?") as an error, his score was 9/11 (81.8%), with p for the two values of chance at, respectively, 0.0239 and 0.0011. His overall score (all trials) was 11/13 (84.6%).

Alex's accuracy suggested that he generalized concepts of both relative size *and* absence of information (Pepperberg and Brezinsky 1991). Such an interpretation is likely even if his initial "none" was a lucky guess and his subsequent, mostly correct, responses were the result of one-trial learning, because one-trial learning is likely only when based on, or easily related to, prior knowledge (see, e.g., Piaget 1952; Bruner 1977; Pepperberg 1990c). Specifically, efficient learning generally reflects the ability—or intelligence—to recognize that information *can* be transferred across domains (Rogoff 1984, 1990). Thus whatever Alex's initial strategy, he accurately responded to queries that required understanding both a relative concept and absence of information.

Alex's generalization ability was also evident from his initial responses to "What *matter* bigger?" or "What *matter* smaller?" (Pepperberg and Brezinsky 1991). Although trained to provide the color of the appropriate object, he correctly responded to all but two first trials involving material for a score of 8/10 (80%), $p = .0493$; his overall score was 10/12 (83.3%). He erred once by giving the color of both objects; the other error was a response of "none." He did, however, provide the appropriate material label on the two trials in which the objects' colors and materials both varied, and respond "none" on the one trial in which the two objects were of identical size. Thus he was not limited to responding within a single dimension and did attend to our questions. Remember, too, that he could have responded repeatedly with the color of the two items or (given that there were two items) with respect to, for example, "How many?" ("two") or "What's same?" ("color," "shape").

Alex was also unaffected by our use of unfamiliar objects (Pepperberg and Brezinsky 1991). His score for queries involving novel pairs of items that were not used in other transfer trials was 34/44 (77.3%, $p = 0.0001$) for first trials, 44/54 (81.5%) overall. These objects were often of shapes and materials he had never before seen and could not label, such as Styrofoam stars. Also, the smaller item in these sets of object pairs was sometimes larger than the training items. And although the relative size difference within pairs was always clear to the human trainers (i.e., we were not trying to learn how close in size two objects must be before Alex could not make a discrimination), on occasion the size difference was somewhat less than what Alex had experienced during training. His accuracy suggests that his concept of relative size was not constrained by the size, shape, or material of the training objects, but could be applied to any physical items.

The Implications of Alex's Understanding of Relative Size

In sum, Alex showed some comprehension of a relative concept, that of size. Although we did not test whether he could (or would) transfer this concept to a different modality (e.g., amount of sound) or a different relationship (e.g., relative darkness), or how close in size two objects must be before he was unable to detect a difference, our study shows that Alex's level of understanding at least matched that of other animal subjects (see, e.g., Schusterman and Krieger 1986). He not only transposed size relationships to objects not involved in training and transferred his knowledge to items of novel colors, shapes, and sizes, and responded to untrained questions about object attributes, but also indicated when items did not differ in size.

As ever, the question often raised in such a study is whether Alex's prior training helped him learn this task. His experience both with numerical tasks and in using a human-based code must be considered. The effects of these two types of training, although related, can be discussed somewhat independently (Pepperberg and Brezinsky 1991).

Alex's familiarity with numerical tasks may have helped with the relative size discrimination. Any individual, animal or human, that has been taught to label quantity probably has formed some concept of more versus less, which is necessarily correlated with relative size (Pepperberg 1987a). We thus do not know if the ability to conceptualize quantity is a prerequisite—or at least provides an advantage—for understanding relative size. We would need to replicate the study with a Grey parrot who is experimentally naive on numerical tasks to determine whether Alex's prior training assisted his learning. Our findings nevertheless show that, with appropriate training, a relative discrimination is within the capacity of this parrot, and by extension, other Grey parrots.

Alex's ability to learn a relative size discrimination was probably unaffected by his training in language-related tasks. In earlier chapters, I discussed how language-based training probably affects only how efficiently and directly we can *examine* his capacities. Note that, according to Lawrence and DeRivera (1954), children can respond on a relative basis before they develop language (see Reese 1968), which suggests that language-based training is unnecessary for acquiring a concept of relative size.[4] Nevertheless, language-related training allowed us to use more complex tasks than were possible with non-language-trained animals: We could ask Alex to respond directly to both presence *and* absence of size differences and, because he responded with labels for attributes other than size (color, material), we could remove many absolute stimulus cues by testing on objects entirely different from training items (see, e.g., Pepperberg 1987a, 1990d).

Overall, these results provide additional evidence that a Grey parrot exhibits a level of information processing often presumed to be limited to mammalian subjects (see, e.g., Premack 1978; cf. Emmerton et al. 1997; Emmerton 1998). Alex's capacity to solve a relative discrimination does not, of course, provide sufficient evidence to claim that his overall cognitive abilities equal those of nonhuman primates or marine mammals. Our data, however, demonstrate the capacities of one avian species, and thus should encourage researchers to engage in additional cross-species comparisons.

10 What Is the Extent

of a Parrot's Concept

of Object Permanence?

> It is not too surprising that animals attribute some permanence to objects according to their localization in space. The adaptive value of this ability is rather obvious, especially for predatory species that constantly monitor the position and movement of their prey. Thus, object permanence is not necessarily the result of ontogenetic construction; it may be preprogrammed in the behavioral repertoire of a given species. This does not mean that object permanence cannot change or become more and more complex as a function of specific training (Etienne 1984).
>
> Jacques Vauclair, *Animal Cognition* (1996:38)

An individual, be it animal or human, who understands object permanence knows (1) that objects are separate entities that continue to exist when no longer in view and (2) that these items have physically distinct identities and properties, fixed in time and space, that are unaffected by simple movement (Piaget 1952). Object permanence is consequently a basic component of perception and cognition and thus an appropriate concept for comparative study (Pepperberg and Kozak 1986). Object permanence was nevertheless largely ignored during almost a century of comparative research (Burghardt 1984; Macphail 1987), possibly because it was considered innate and unitary—so basic that researchers assumed no organism could exist without it (Flavell 1985). Object permanence, however, is neither innate nor unitary. Piaget (1952) showed that newborn humans lack object permanence, that the concept develops in 6 major stages, and that each stage represents a different level of understanding.

The Stages of Object Permanence

Researchers may disagree about the rigidity of Piaget's stages of object permanence, but most concur that development follows six major stages. In Stage 1, subjects do not search for an object they see disappear. As subjects mature, however, eventually they pursue such an object: in Stage 2 by tracking its movement; in Stages 3 and 4, respectively, by recovering a partially and a fully hidden item exactly where they see it disappear. In Stage 5, subjects retrieve an item that has been hidden successively in several locations—hidden, exposed, and rehidden several times. In Stage 6, a subject masters invisible displacements: When an object is hidden in a container, the container is moved behind another occluding device, and the object is then placed in or under this second device, the subject can, after being shown that the first container is empty, successfully infer where the object is.

Such development seems to occur over a child's first 2 years (Piaget 1952, 1954b; Harris 1975, 1983; Sophian and Sage 1983); however, the process by which the concept develops is poorly understood, and few researchers fully agree on how to quantify development (Bower 1982; Harris 1983; Flavell 1985; Baillargeon, Spelke, and Wasserman 1985; Baillargeon 1986, 1987; Baillargeon and DeVos 1991). For animals, the extent of development and how development should be tested are even less clear. The controversies are most easily described from a historical perspective.

Early Object Permanence Studies on Nonhumans

Only after Piaget's writings became widely available in English (Piaget 1971, 1978, 1980) were animals routinely tested for object permanence (see, e.g., Chevalier-Skolnikoff 1976, 1981, 1989). Scientists realized that Piaget's methodology (observations and note-taking about a subject's activity in its natural habitat or the laboratory; see Chevalier-Skolnikoff and Poirier 1977), although designed for humans, paralleled that of early ethologists (e.g., Lorenz 1970) and thus enabled cross-species comparisons of observational and experimental data (Pepperberg and Funk 1990). Moreover, experiments based on the Piagetian framework directly measure performance levels on tasks that are often adaptable for many species (Chevalier-Skolnikoff 1989). Early studies on nonhuman primates and a few nonprimate mammals used a series of tasks based on those developed by Uzgiris and Hunt (1975) for human children; researchers found that the capacities of humans and nonhuman mammals for solving some object permanence tasks were comparable. Nonhuman primates tested with varying success included squirrel monkeys (Vaughter, Smotherman, and Ordy 1972), macaques (*Macaca arctoides,* Parker 1977; *Macaca fuscata,* Antinucci et al. 1982; *Macaca mulatta,* Wise, Wise, and Zimmermann 1974), New

World monkeys (*Cebus capucinus* and *Lagothrica flavicauda,* Mathieu et al. 1976), lowland gorillas (*Gorilla gorilla,* Redshaw 1978) and
language- and nonlanguage-trained chimpanzees (Mathieu et al. 1976;
Wood et al. 1980; Mathieu and Bergeron 1981; Hallock and Worobey
1984). A hamster *(Mesocricetus auratus)* failed to achieve Stage 6
(Thinus-Blanc and Scardigli 1981). Cats *(Felis catus)* appeared capable
only of Stage 4 (Gruber, Girgus, and Banuazizi 1971), but Thinus-Blanc,
Poucet, and Chapuis (1982) claimed Stage 5 for cats and Triana and
Pasnak (1981), using objects of greater relevance (food with control
for odor cues), claimed Stage 6 for cats and dogs *(Canus familiaris).*
Few studies investigated avian object permanence even though natural
behaviors such as food caching and cavity-nesting would make it seem
a likely adaptive trait (see, e.g., Kamil and Balda 1985). Etienne
(1973a,b) studied object permanence in young domestic chickens in
relation to imprinting, but concluded that a chick's capacity did not
go beyond a stimulus-response association, and was unrelated to any
general stage of cognitive development.

These studies were challenged for five major reasons (review in
Pepperberg, Willner, and Gravitz 1997). Researchers questioned
(1) whether standardized tests that ignore environmental enrichment—"enculturation" (Bower 1982; Harris 1983; Flavell 1985)—can
accurately pinpoint the age at which abilities emerge; (2) whether
tasks quantify a subject's "native" cognitive level or, instead, measure
proficiency reached through training, learning, memory (Bjork and
Cummings 1984; Doré and Dumas 1987), or physical development
(Baillargeon 1987); (3) whether success shows that subjects *represent*
the hidden objects (Pepperberg and Funk 1990); (4) whether additional, more complex tasks are required to prove the existence of full
object permanence (e.g., Sophian 1985; Doré et al. 1996); and
(5) whether tasks control for experimental cuing of various types (e.g.,
Pepperberg and Kozak 1986; Pepperberg and Funk 1990; Pepperberg
et al. 1997). All these points merit scrutiny.

Enculturation may indeed affect the rate and extent of development.
Children given enriched environments develop more rapidly than
those without such input (Bruner 1964; Wishart and Bower 1985);
interestingly, chimpanzees given human-like experiences outperform
conspecific-reared controls on imitation tasks similar to those given
children (Tomasello, Savage-Rumbaugh, and Kruger 1993). Object permanence might be similarly affected. The standardized use of Uzgiris
and Hunt's (1975) procedures in different laboratories ensures that any
enculturation implicit in a given task equally affects all species being
tested, but researchers generally do not examine effects of encultura-

tion *prior* to their subjects' exposure to Uzgiris and Hunt's tasks. A subject's prior history, whether enriched or deprived, must be taken into account because competence on different object permanence tasks is affected differently by different histories. Although enculturation probably cannot facilitate a subject's performance beyond its species' inherent limits, enrichment can help a subject reach its maximum potential. Furthermore, even if enculturation does not enable a subject of a given species to perform optimally at a specific age (see Macphail 1987), enculturation that enables *eventual* Stage 6 competence would seem a reasonable level under which to examine relative and absolute rates of development. Thus, if prior histories are taken into account, standardized tests such as Uzgiris and Hunt's tasks, or closely related ones (see, e.g., Dumas and Doré 1989; Gagnon and Doré 1994) that are administered with minimal variation to divergent species, ensure that comparisons are not compromised because of significant variations in experimental design or enculturation across laboratories (see Bateson 1979).

Success at specific Uzgiris and Hunt tasks may indeed reflect training, learning, and memory (Doré and Dumas 1987), whereas failure at a given age may reflect physical rather than cognitive limits (Baillargeon et al. 1985; Baillargeon 1986, 1987). Discussion of the first point is best left until later in this chapter, when results of various studies are evaluated. The second point, however, is straightforward: Very young subjects may fail early tasks because they physically cannot remove a cover (Flavell 1985; Doré and Dumas 1987).[1] For later tasks (or stages), however, success is probably independent of physical competence: Skills needed for such tasks seem related to maturation of specific brain areas involved in cognitive development (Diamond 1990a, 1990b).

Uzgiris and Hunt's tasks indeed do not test representational memory: Subjects need not retain a representation of the hidden item, but may merely search for "something of interest." Researchers thus often add a task based on previous work with nonhuman primates (Tinklepaugh 1929, 1932). The task, which tests whether subjects react with surprise after finding that a researcher has surreptitiously replaced the object initially hidden with a less-favored item, specifically tests whether subjects remember the identity of the hidden item (Funk 1996; Pepperberg et al. 1997).

An area of considerable debate is whether Uzgiris and Hunt's invisible displacements adequately test Stage 6 competence. Doré et al. (1996) propose instead a "shell game" in which an object is visibly hidden under one of three covers and the experimenter then visibly exchanges the position of this cover with one other (Sophian 1985). Whether success at this task, which also tests attention, different types of memory, and spatial cognition (Goulet, Doré, and Lehotkay 1996), should be the criterion for Stage 6 is unclear. Nevertheless, researchers often include this task to provide additional information about cognitive capacities (Pepperberg et al. 1997).

Uzgiris and Hunt's tasks, which involve face-to-face interaction between subject and experimenters, might provide inadvertent cues via an experimenter's direction of gaze or face and body posture (Gardner and Gardner 1975; Wood et al. 1980). Some researchers reject the influence of such cuing for three reasons (e.g., Wood et al. 1980; Triana and Pasnak 1981; Pepperberg and Kozak 1986; Gagnon and Doré 1992, 1994; Funk 1996; Pepperberg et al. 1997). First, more direct cues are inherent in visible displacements. Second, even obvious pointing to the correct cover does not aid subjects who do not understand the task or the indicative nature of pointing, or who simply are unmotivated to perform the task (see Triana and Pasnak 1981; Pepperberg and Kozak 1986). Third, successful completion of Uzgiris and Hunt's Task 15 (a trick; see below for task descriptions) argues against cuing: The task tests whether a subject goes *directly* to where the experimenter *knows* the object exists, even though this site is not the one the subject should logically choose.[2] To control for eye gaze, some experimenters administer trials wearing mirrored sunglasses (Pepperberg et al. 1997).

Odor cues might influence behavior because many tasks hide food to motivate subjects. Odor cues, however, can be controlled, and researchers suggest four reasons why odor cues are unlikely to affect experimental results. First, although olfactory sensitivity varies across species (and particularly across avian species, Cobb 1960; Wenzel 1967; Stettner and Matyniak 1968; Macphail 1982; Harriman and Berger 1986), Gagnon and Doré (1992) argue that odor cues are unimportant even for animals such as dogs that are known for using olfactory information. Second, data on psittacids suggest that odor was probably not important in these birds' searches: Birds made similar numbers of mistakes whether the hidden object was food or nonfood, and often ate little of the food rewards (Pepperberg and Kozak 1986; Pepperberg and Funk 1990). Third, many foods used as rewards are not heavily scented (e.g., seeds), and screens used both to cover the food and as "blanks" almost always previously contained food; also, researchers generally physically handle the screens, which thus pick up human odors. Often crumbs remaining from previous trials or play sessions with different foods act as additional olfactory distractors. Fourth, Uzgiris and Hunt's Task 15 (mentioned above) specifically tests for odor cues: The object is left under one screen, but the investigator gives the appearance of placing the item elsewhere; an animal using odor cues would not be "tricked."

Later Studies on Object Permanence
and Related Behavior in Nonhumans

As a consequence of the above considerations, studies on several species were reinterpreted and sometimes rerun, and at just about this point in the experimental history of object permanence I and my students tested Alex and other psittacids (Pepperberg and Kozak 1986; Pepperberg and Funk 1990). I'll discuss our data later in this chapter; for now, the important point is that I compare our results to those for other animals studied during the same period. Those studies, which also took many if not all of the above considerations into account, involved nonhuman primates (Natale et al. 1986; Natale and Antinucci 1989; Potì 1989; Spinozzi 1989; Spinozzi and Natale 1989; Schino, Spinozzi, and Berlinguer 1990; de Blois and Novak 1994; Dumas and Brunet 1994; de Blois, Novak, and Bond 1998); domestic dogs (Gagnon and Doré 1992, 1993, 1994; Doré et al. 1996); domestic cats (Doré 1986, 1990, 1991; Dumas and Doré 1989, 1991; Dumas 1992; Goulet, Doré, and Rousseau 1994; Doré et al. 1996); and ring doves (*Streptopelia risoria,* Dumas and Wilkie 1995). These studies showed that great apes and possibly dogs reached Stage 6, whereas cats, monkeys, and doves apparently did not.[3]

What intrigued me and my students was that when we began object permanence work, only the previously mentioned studies on chickens had examined, and failed to find, this concept in birds (Etienne 1973a,b). Given that domestic chickens, unlike sturnids, corvids, and psittacids, have little of those neural areas that, in birds, are thought to mediate intelligent behavior (see Chapter 2), a logical conclusion was to examine this concept in other species. Interestingly, observations of several avian species in the wild and in seminatural surroundings detected behavior such as recovery of cached food or cavity-nesting that would require some understanding of object permanence (see studies on nutcrackers, *Nucifraga columbiana,* Vander Wall 1982; Kamil and Balda 1985; European jays, *Garrulus glandarius,* Bossema 1979; jackdaws, *Corvus monedula,* Lorenz 1970; Etienne 1976/1977; marsh tits, *Parus palustris,* Sherry 1982; Shettleworth and Krebs 1982; chickadees, Sherry 1984; Shettleworth and Krebs 1986). Evidence for Stage 6 capacities (representation of images of displacements and absent objects) might even be inferred from certain studies: First, interocular transfer tests show that marsh tits recover caches based on information stored in their brains (Sherry 1982). Second, marsh tits that can engage in multiple recovery sessions avoid caches emptied during their initial recovery session, suggesting that their behavior is based on memory rather than simple rules of movement; this behavior also indicates memory for displacement (Sherry 1982). Third, chickadees that cache two different types of food recover favored foods first, suggesting memory for types of food cached, rather than just for "something of interest" (Sherry 1984). Fourth, nutcrackers continue to search areas where they have cached seeds even when experimenters have removed the seeds to prevent odor cues (a test comparable to Uzgiris and Hunt's Task 15); such recovery is, however, disrupted when local landmarks are changed (Balda 1980; Vander Wall 1982).[4] Parrots, with advanced cognitive capacities, thus seemed good candidates for object permanence studies. My students and I therefore used the following procedures with Alex and other psittacids (Pepperberg and Kozak 1986; Pepperberg and Funk 1990).

General Procedures for the First Object Permanence Studies on Parrots

To hide items, we had to use covers that parrots were capable of removing: small plastic boxes, cups, toy barrels, and pieces of paper towel or construction paper. Paper pieces ($\sim$6.5 cm^2) were crumpled so that birds could not determine the presence of a hidden object by observation. Except for Task 5 (see below), we used identical types of screens in any given trial that required multiple covers; we varied screens between trials to deter habituation and cuing—we didn't want the birds to become bored with the task or learn that the sight of a particular cover indicated a hidden reward. Note that all items used as covers could contain food at various times during the day and were sometimes used to present food in other experiments occurring simultaneously (see, e.g., Chapter 7). We found that a parrot's method of removing a nonrigid screen (extending its head as far as its neck could stretch, grasping a portion of the screen with its beak, and pulling) was unsuited to the cloth covers used in other studies (see, e.g., Pepperberg and Kozak 1986). Birds could, however, grasp a box edge or piece of paper and then remove such covers in a single motion.

We used several food and nonfood items as targets of the search. We chose items based on the preferences of each bird being studied, because we realized that motivation could be a critical factor. We varied these items frequently to avoid "boredom"; as noted previously, Alex quickly stopped responding if we repeated similar tasks with similar rewards, and we suspected that other psittacids would respond similarly. Food items included seeds, grapes, crackers, peanuts, popcorn, walnuts, cashews, ice cream, water, rolled oats, cream cheese, small pieces of banana, bits of egg yolks, jelly beans, cookie pieces, matzoh, and cherries. Nonfood items included mirrors, yarn, corks, a key chain, wooden beads, paper and hair clips, jacks, and rings. Correct responses enabled birds to eat the food, interact with the toy, or pick up (or in Alex's case, request) a different item. Many tasks were repeated at least once with nonfood items; birds' performances never differed between food and nonfood trials. Birds often refused foods they uncovered and interacted with a nonfood item instead. We therefore believed that odor cues were not particularly important (Pepperberg and Kozak 1986; Pepperberg and Funk 1990).

Alex was ~8½ years old when we began his object permanence studies; we thus could not compare his age-related development to that of children or nonhumans. We could, however, see what stage of object permanence an adult psittacid might achieve. Informal evidence suggested that Alex did possess some degree of object-concept (e.g., he visually tracked objects and appeared to remember, over time, favored food items hidden in human fists, in various metal tins, and in overturned cups), but we knew of no formal assessment of object permanence in psittacids (note Lea 1984). According to Uzgiris and Hunt (1975) and Triana and Pasnak (1981), Alex's success at a given level would represent the eventual developmental capacity of the species; in that sense, age would not be a factor.

Our subsequent study also used single individuals, one each of four species (Pepperberg and Funk 1990). Such single subject designs are called "power studies" (Triana and Pasnak 1981). Success by a subject of a power study implies that object permanence is within the capacity of that species. Negative results from one subject, however, need not imply lack of aptitude for the species as a whole (Triana and Pasnak 1981). The four subjects in this second study were all pets. We studied Wok, a 13-year-old Grey parrot, to see if any difference existed between Alex and a bird with significantly less enculturation. To see if differences existed among various psittacids, we tested Fred, an adult Illiger mini macaw *(Ara maracana);* Bruce, an 18-month-old budgerigar; and Yellow Bird, an 8-month-old cockatiel *(Nymphicus hollandicus).*

In all studies, at least two persons administered trials. One person sat behind the apparatus and manipulated the objects. The second person sat behind the parrot, occasionally restraining a bird by placing his or her hands around the bird's body. When we used pets, one examiner was often the bird's owner; pet birds are often ill at ease with strangers, and the owner's presence might be needed to reassure the bird that the experimenters were nonthreatening (Pepperberg and Funk 1990). After the second person released a bird (if necessary), he or she, the first person, and occasionally a third person observed the bird's behavior; one observer recorded each session's events in a notebook. Several different humans were involved, and all humans took all roles (hiding objects, observing actions, and scoring results). Such role transfers helped ensure that birds did not become adept at reading possible cues presented by a single individual.

Neither Alex nor any pet worked for extensive periods. At best, sessions lasted an hour and involved three or four tasks, beginning with a repetition of the last task done in the previous session. This repetition tested a subject's willingness to work: Failure to repeat a previous task usually indicated motivational problems and thus separated such problems from a subject's inability to perform a task (Pepperberg and Kozak 1986; Pepperberg and Funk 1990). Even when we varied objects or foods between trials and allowed birds to eat the food or play with toys they had found (see Doré and Dumas 1987), they generally became uncooperative after three or four tasks. Alex engaged in his standard "boredom" behavior (see previous chapters); all birds often began to ignore us and preen, and flighted birds occasionally took off. When a subject failed to attend, we ended the session and resumed after a period of free-choice requests or a lunch break, or on a succeeding day.

We also found that several pet birds reacted to testing as aversive (Pepperberg and Funk 1990). None of these birds had ever had an item placed within reach that was then hidden before it could be obtained. Two birds (Wok, Yellow Bird) engaged in displacement behavior the first time we completely hid an object (Task 4): They bobbed their heads, plucked at their feathers, or emitted distress calls that they used in other unfamiliar situations. Wok also reacted fearfully when novel items were used as covers (note Zentall and Hogan 1974; Zentall et al. 1981).

Intervals between sessions and the number of sessions needed to complete the 15 tasks varied among subjects. Alex completed all sessions in 5 weeks (Pepperberg and Kozak 1986). Sessions with pet birds were arranged at their owners' convenience, and thus did not necessarily occur at regular intervals (Pepperberg and Funk 1990). Bruce, the budgerigar, was seen 7 times in 6 weeks. Seven visits to Fred, the macaw, occurred over 7 months. Ten sessions with Wok, the Grey parrot, took 5 months. Yellow Bird, the cockatiel, was tested in 10 days, with a brief visit 3 months later to retest 1 task. We could not always impose upon the owners to repeat the 15 tasks 3 times, as suggested by Uzgiris and Hunt (1975); thus only Yellow Bird completed 3 series. Most tasks were, however, repeated at least once because of the task review at the outset of sessions and the retesting on food or nonfood items.

Tasks Used in Our Studies

Despite the criticisms (discussed above) often leveled against Uzgiris and Hunt's procedures, we primarily used their tasks. Our decision was based on three reasons. First, the scale directly measures competence levels via performance on 15 tasks of increasing difficulty; success on specific subsets of tasks directly correlates with Piaget's six stages (Wadsworth 1984). Second, these subsets provide fine divisions of Piagetian stages, and thus yield within-stage comparative data across species. Third, the tasks can be readily adapted for various species; because most studies use either these tasks or closely related ones, we could perform cross-species comparisons that were not confounded by significant differences in experimental design.

Although Uzgiris and Hunt's (1975) tasks are described in almost every publication on object permanence, I repeat these descriptions for readers who are unfamiliar with the tasks and concepts. Tasks are grouped according to the type of skill tested. I also provide a descriptive label for each task for cross reference with the Results section.

Tasks 1 and 2 are generally lumped under the heading "visual pursuit" and respectively test a subject's visual capacity to follow a slowly moving and then a disappearing object (Stage 2). A subject succeeds on Task 1 (arc track) if it tracks an object (determined by eye gaze, head position) smoothly through a complete arc. To succeed on Task 2 (object disappearance), a subject must remain gazing at the point of disappearance or return its glance to the starting point.

The next set of tasks involve simple visual displacement. Tasks 3 (partial hiding) and 4 (complete hiding) test Stages 3 and 4, respectively: the ability to retrieve an object that is partly and then completely hidden under a single cover. In Task 3, a subject can succeed by pulling either the item or the cover. In Task 4, a subject must remove the cover; pulling an item from underneath the cover does not count. Our Task 4, based on Uzgiris and Hunt, corresponds to Gagnon and Doré's (1994) Stage 4b because we never allowed a search for a targeted item until it was completely covered. We did not test Gagnon and Doré's Stage 4a (allowing searching to begin while the item is being covered) to avoid overexposure and possible training on the task (see below).

Stage 5 requires success on many different tasks. Only the beginning of Stage 5 is tested with Task 5 (two sites, visible). In Task 5, two hiding sites are available; an object is visibly hidden in the same place for two or more trials before being hidden in the second site; the task tests whether a subject recognizes when the site changes. Young children commonly fail; this failure is called the "A-not-B error" (see Diamond 1990b). As is often done for children, my students and I used differently marked covers, placed the covers in front of our birds, and then hid the object (see Bremmer 1978a,b; Goldfield and Dickerson 1981; Butterworth, Jarrett, and Hicks 1982; Diamond 1990b, although these studies differ somewhat from ours). Marked covers and this hiding procedure can reduce the likelihood of A-not-B errors (Harris 1973). Nevertheless, very young children err even under these circumstances, and we wanted to make direct comparisons with humans. We used the same hidden objects for all trials to avoid possible perceptual complications.[5]

Tasks 6 and 7 are also part of Stage 5. In Tasks 6 (two sites, alternating visible) and 7 (three sites, alternating visible), the hiding place varies respectively between two and among three covers. Covers are not differentiated in any way *within* a trial. These tasks thus test how well subjects track additional slight complexities in hiding. In both tasks, subjects must search only under the correct cover. We varied rewards or covers in successive trials.

Tasks 8 and 9 involve more complex manipulations of hidden objects, but still test Stage 5. In Task 8 (three sites, successive visible), an object, in full view in the investigator's hand, is passed under successive covers until it is hidden in the designated site. The task tests a subject's ability to track object movement and search where the item has most recently been seen. Success requires searching only the final location; searching the other locations, even if the search replicates the order of hiding, is incorrect. The final Stage 5 test, Task 9 (superimposed covers), gauges persistence in obtaining a hidden item; a subject must remove three superimposed covers to obtain the object.

Like Stage 5, Stage 6 includes several tasks, and all must be successfully completed for full competence. Many subjects succeed on only some tasks, and thus do not achieve human levels. As noted above, some disagreement exists as to what tasks appropriately test Stage 6 (see, e.g., Doré et al. 1996; Goulet et al. 1996). Nevertheless, all Stage 6 tasks prevent subjects from directly viewing transfer of the hidden item between covers; subjects must infer the occurrence of such a transfer. Inferential, rather than perceptual, abilities are thus tested (Pepperberg and Funk 1990).

The first Stage 6 tasks, numbers 10–13, are similar to Tasks 4–7, except that the object is not seen during transfer. The item is, instead, hidden in a small container or the experimenter's hand—the "implementing cover"; the subject sees the implementing cover, but not the item, move to one other screen (Task 10, 11) or a choice of two (Task 12) or three (Task 13) other screens, and the experimenter surreptitiously transfers the object. Only then is a subject shown that the implementing cover is empty. To succeed, a subject must search directly under the correct screen.

Task 14 (successive invisible) is similar to Task 8, but the object is not seen after it is placed in the implementing cover, which the investigator then moves *between* screens. To succeed, a subject may search either the last cover in the path or all covers in order of hiding.

Task 15 is a "trick": An item is hidden, as in Task 14, but under an *intermediate* screen as the investigator's hand continues to move under the other screens; the subject is led to believe the object is under the final screen. The task thus tests whether a subject goes to where the object logically should be found, and also controls for several possible inadvertent cues mentioned above.

We sometimes presented more complex tasks before simpler ones (Pepperberg and Kozak 1986). Task order is important only for developmental studies, and shifting to a different, more challenging task sometimes helped focus a bird's attention. A shift also showed that the bird was not being trained to do tasks in a step-by-step manner, because success at more complex tasks was then independent of experience with the simpler tasks.

How We Scored Trials

Criteria for success were similar to those of previous Uzgiris and Hunt–based studies (e.g., Wood et al. 1980): Birds had to uncover the hidden item or perform another acceptable action within 1–2 minutes of the hiding. For each task, Uzgiris and Hunt lists three to five alternative erroneous actions, from removing the wrong cover to losing interest or searching haphazardly; birds were scored accordingly. Several consecutive correct searches were required for each task. If birds were correct on successive trials after erring midway through a session, however, we occasionally proceeded to the next task (Pepperberg and Funk 1990). We recorded mistrials if birds (1) removed and immediately manipulated covers without searching for the hidden item (i.e., ignored the actual task)[6] or (2) interfered with placement of the object.

Alex's Results

Alex succeeded on all 15 tasks. Some researchers (e.g., Gagnon and Doré 1992) argue that success on these tasks may not ensure full Stage 6 object permanence, but our data were the first to show avian competence at any relatively high level. Subsequent work did demonstrate Alex's full competence (see below, Pepperberg et al. 1997). For now, I focus on results of specific trials that illustrate the importance of enculturation for success on Uzgiris and Hunt's tasks.

Alex initially did not succeed on Task 5, which is the first time a subject must choose between two possible hiding places. On the first trials, we placed a red plastic cup and a grey metal cup in front of him and hid nuts under the red one. He chose the metal cup on several successive trials. He had previously manipulated both cups equally, but had at least once prior to formal testing found nuts under the metal cup; he also saw us remove nuts daily from a metal storage tin. We surmised that such prior experience may have affected his choice. We therefore ended the session and decided to repeat trials at a different time with covers unconnected with nuts (Pepperberg and Kozak 1986).

We repeated Task 5 using red and yellow plastic cups. Alex immediately said "two"; we told him that yes, there were two cups, and asked him to get the nuts. In two trials with nuts hidden under the yellow cup, he chose the red; we switched to red, he chose yellow. The most conservative interpretation was that Alex could not move beyond Stage 4, but we suspected that his failures might be due to inexperience in two-choice tests with physical objects. Until then, Alex's studies had *never* involved *physical* choice between objects— this work predated studies on relative size and the complex number work. Alex was commonly shown one object and asked for information about that item; choice was among numerous possible *vocalizations,* not between physical items. Multiple objects were previously present in only two situations: either in a box containing ~100 items that sat near his cage and from which he could choose anything during limited daily "play" periods (such periods helped determine which objects might motivate him to work; Pepperberg 1981), or as a collection of (usually identical) objects for which he was asked to label quantity (Chapter 7). If he correctly labeled the number, his reward was the entire collection.

To see if Alex's Task 5 failure could be due to lack of experience rather than a conceptual deficit, we proceeded to Task 9, which did not involve multiple choice. Success on more advanced tasks implies understanding of previous ones (Uzgiris and Hunt 1975). He immediately removed the three superimposed covers, thus suggesting that his Task 5 failure was not conceptual.

The problem then was how to provide appropriate enculturation without training Task 5. We did not want to replicate the filmed protocol of Uzgiris and Hunt (1966), in which experimenters demonstrate where an item is hidden in the two-choice procedure; such demonstration provides training and would thus invalidate any subsequent test of this ability (Chevalier-Skolnikoff, pers. comm.; Etienne 1984). Instead, we gave Alex several multiple-choice opportunities during the next day. We presented two cups or two boxes, each of which had a different item placed under or on it; occasionally the cup or box was placed on its side, and the object then placed inside. No "correct" choice existed; instead, Alex simply learned that he could obtain a different (although as best we could design equally reinforcing) reward based on choice of cover.

We then repeated Task 5. Alex responded appropriately on all trials, including his first, and then completed all subsequent tasks. His success on Task 5 after minimal enrichment was not as important as the fact that conventional wisdom suggested that we not test him on more advanced tasks after his initial failure. Only our suspicion that his failure was due solely to lack of a particular type of experience (see Bruner 1964; Held 1965; Landers 1968; Etienne 1973a,b) led us to attempt Task 9, which was beyond his apparent level of ability. Our data might be attributed to a real difference in the sequence of task achievement in mammals versus birds, but we believe our results were more likely the consequence of a specific, if unintentional, human-engineered deficit, particularly because of the speed with which it was remedied (Pepperberg and Kozak 1986). Even children who lack specific enrichment usually experience some two-choice situations early in life; total lack of such practice could affect a particular performance adversely, without affecting overall cognitive ability (see, e.g., Held 1965; Etienne 1973, 1984; for a more general case, see Rubinstein 1979).

Alex also responded in interesting ways on some of the final tasks (Pepperberg and Kozak 1986). Occasionally, at the end of a session, rather than choose a screen himself, he requested that we "go pick up cup." On Task 15, he not only immediately overturned the cover where the object *logically* should have been hidden, but also emitted a clear "Yip" (which he occasionally uses when startled) after his obvious failure to obtain the expected reward. On subsequent Task 14 trials, he occasionally knocked over all covers simultaneously with a single motion of his head, often before we completed the displacement. Such occurrences (four during the last two sessions) were listed as mistrials. We could not help thinking that Alex had devised a "Gordian knot" approach to solving this task.

Data from Other Psittacine Birds

The macaw, the pet Grey, and the cockatiel all succeeded on all Uzgiris and Hunt's tasks; the budgerigar, unfortunately, died before being tested on Tasks 14 and 15. Note that New Zealand parakeets (*Cyanoramphus auriceps*) also achieve Stage 6 object permanence according to these tasks (Funk 1996). Each of our subjects occasionally demonstrated individual differences that provide insights into their problem-solving capacities or the effects of enculturation (Pepperberg and Funk 1990).

For Wok, the Grey parrot that had not received Alex's language-based training, lack of enculturation apparently caused aversion to novel covers: When a novel cover (a paper towel) was introduced (Task 4), he did not remove it, but instead moved as far away as possible, began to feather pluck, and emit distress calls. When we used a familiar paper bag, he responded appropriately to all tasks. We thus introduced other novel covers (e.g., plastic measuring cups) in sessions separate from object permanence trials.

Bruce, the budgerigar, in contrast, seemed unperturbed by any aspect of testing. Even initial use of a cover too large for him to remove— a quarter of a cocktail napkin—did not prevent him from obtaining the hidden object; he simply pulled it from under the cover, or shredded the napkin. He removed covers when we used eighths of napkins or propped paper plate pieces between him and the objects. We must work with other budgerigars and administer additional tasks to determine the extent of object permanence in these birds.

Yellow Bird, the cockatiel, worked more rapidly than other birds, but also spent relatively more time playing with covers or in displacement behavior (e.g., feeding) at the onset of each session. On Tasks 13-15, his initial responses, although inappropriate, were particularly interesting: He moved objects and covers to the table edge, dropped them, watched the descent, and followed experimenters' retrievals. Only after losing interest in these actions did he respond appropriately to trials. His pattern of dropping and observing the descent, and his later repeated hiding of an object *after* retrieving it when we redid Tasks 11-13, are called "tertiary circular reactions" in the Piagetian schema; such behavior indicates Stage 5 competence in another Uzgiris and Hunt (1975) series, "The Construction of Object Relations in

Space." Such behavior was thought to be limited to hominids (Parker and Gibson 1977).

Fred, the macaw, was most cooperative; he was fully flighted and flew to the test site when experimenters entered his house. He provided an arresting interruption during Task 13: Rather than repeat the task, he flew from the test site to the floor to search beneath the coffee table, where he had seen his owner place a seed cup before testing began.

Overall, we had shown that various psittacids are as competent as humans and nonhuman primates that had been tested on comparable object permanence tasks, and more competent than some mammals. But our studies had two drawbacks. First, we had not tracked the age at which Piagetian stages occur or the duration of each stage. Few studies compare ontogeny of object permanence in birds and mammals (but see Funk 1996). Given the differences in psittacine and mammalian developmental timelines, we wanted to assess a Grey parrot's development and compare its rate to that of mammals. Second, as noted above, success on several additional tasks may be required to demonstrate full object permanence, and we needed to see how Alex would fare on such tests.

Why Tracking Development Is Special and How We Proceeded

Ontogenetic comparisons of cognitive capacities such as object permanence are important for two reasons (Antinucci 1990; Gagnon and Doré 1994). First, such studies examine how cognitive mechanisms and their underlying structures become integrated or differentiated during development; comparing various species at different developmental stages may help identify these mechanisms and structures. Second, investigating development may help us understand the function and form of the final mechanisms and structures and possibly provide knowledge of their evolution.

In June 1995, acquisition of a 7½-week-old Grey parrot, Griffin, allowed us to perform a developmental study. At the time, he was not yet fledged (had no wing, tail, or body feathers) and was handfed twice each day, although he ate some foods on his own. We would have begun our study even earlier, but removing him from the breeder prior to this age would have jeopardized his survival. Instead, we had breeders perform the earliest Uzgiris and Hunt tasks on younger birds in their care (see below).

Griffin's environment was quite different from that which Alex had experienced (Pepperberg et al. 1997). Alex entered the laboratory when about a year old, and had spent his previous 6 months with juvenile Greys in a pet store. Griffin, when not napping, in sessions, or being carried by humans, was with two adult Greys and several humans in a large laboratory; birds were restricted to specific areas in this room. When humans were absent and during sleeping hours, Griffin was confined to a brooder prior to fledging and a weaning cage (38 × 71 × 56 cm) thereafter; brooder and cage were in a room separate from the main laboratory. Handfeeding occurred about 9:00 A.M. and 9:00 P.M., but weaning foods and water were always available. In scheduled "play sessions" outside of testing, Griffin sat with humans on the floor and received items (small wooden spools, beads, and plant stakes, plastic barrels, cups, pieces of index cards, etc.) with which he could engage in various activities (e.g., chewing, throwing, chasing). He often received treats and nuts. We concurrently studied his response to mirrors (Pepperberg, Willner, Gupta, and Meister, in prep.).

We knew that Griffin's data might not be typical of Greys. Working with Alex as an adult enabled us to determine an ultimate developmental stage and the capacity of the species. In contrast, claiming that specific developmental stages occurred at a specific age in Grey parrots based on a single, possibly idiosyncratic bird was a much riskier proposition. Nevertheless, my students and I chose to proceed. We knew that recruiting multiple Grey parrots of an appropriate age would be extremely difficult for two reasons: First, reputable breeders do not allow outsiders to handle unweaned birds because of the fragility of chicks < 12 weeks old and the possibility of introducing disease into their aviary. Second, Grey parrots may not adapt to multiple successive owners; thus purchasing birds for research with the intention of re-selling them would be unethical. Nevertheless, if one subject reliably performs a given task at a particular age, the tested aptitude is within the species' capacity at that point in development. As with adult subjects, of course, negative results from one subject do not necessarily imply lack of the aptitude at that age for the species (Triana and Pasnak 1981). Thus although individual differences are likely in Grey parrots, we felt that our single subject study would provide a baseline for future research and permit preliminary cross-species comparisons.

Additional Tasks Used in Griffin's Study

Griffin was given more tasks than Alex or the other psittacids (Pepperberg et al. 1997). First, we used additional controls to ensure that he had not simply tracked experimenters' hand motions rather than the object, and that he followed hidings and did not choose merely where he had last seen the object. We also tested whether he searched for the specific hidden object, rather than "something of interest." Finally, we determined how Greys respond to the "shell game" compared to other species.

To test whether a subject formulates rules such as "go to last place the experimenter touched" or "go to last place object was seen," controls not in the original Uzgiris and Hunt (1975) scale were used (Dumas and Brunet 1994; de Blois and Novak 1995; Pepperberg et al. 1997). In these trials (Task 14a, controls) a researcher places an item in the implementing cover, visits one screen, hides the object, shows the empty cover, visits another screen, then leaves the implementing cover in an accessible site that varies for each trial. A third, untouched screen is present to see if the subject examines only screens handled by the experimenter. A subject shows it is not using a "go to last place object was seen" or "go to last place the experimenter touched" rule by ignoring the implementing cover. This trial was not used in our earlier studies.

Our Task 16 (substitution) is also a "trick," but of a different type: The experimenter repeats Task 14 but surreptitiously hides an item less favored than the one that the subject initially observes (LeCompte and Gratch 1972; Pepperberg et al. 1997). This task tests whether the subject has a representation of the specific hidden object and does not simply search for "something of interest."

We also administered Task S, comparable to that of Sophian (1985) and Doré et al. (1996). S is a form of "shell game": An item is visibly hidden under one of three covers and the experimenter then visibly exchanges the position of this cover with one of two others. Success at Task S may or may not be an acceptable criterion for Stage 6 because it also tests attention, different types of memory, and spatial cognition (Goulet et al. 1996), but the task provides data for cross-species comparisons.

We also gave these three tasks to Alex, who was then 19½ years old. We could thus compare data from a bird with considerable laboratory experience to data from one that was relatively naive.

Differences between Alex's and Griffin's Study

To see if Griffin could succeed on a complex invisible displacement without experiencing simpler ones, we omitted Tasks 10–12. Although mammalian object permanence apparently develops in a fixed sequence (Doré and Dumas 1987; Dumas and Doré 1989), we did not know if the same held true for birds. We also believed that shifting to a more challenging task helped focus Griffin's attention and removed the possibility of stepwise training, because success at the more complex task was then independent of experience with simpler tasks (see below).

Griffin would work only for short periods (Pepperberg et al. 1997). Sessions involved two or three tasks, beginning with repetition of the last task from the previous session. As with other birds, this repetition tested Griffin's willingness to work and separated motivation from an inability to perform a task. Even though we varied items or foods between trials (except for Task 5) and allowed him to eat the food or play with the toy he had found, he usually became uncooperative after about 20 minutes. Thus although we had planned never to end a session until we had reached a task beyond his current level of competence, he often lost interest before that point.

We tested Griffin weekly until he succeeded on Uzgiris and Hunt's Task 15. We could thus track development in a fine-grained manner without providing too many opportunities for learning specific tasks (de Blois and Novak 1994). We could not test Task 16 (substitution) until Griffin had a stable hierarchy of preferences for objects or foods; we therefore waited until he was 33 weeks old.

We performed Tasks 14a and S when Griffin was 51–52 weeks old and simultaneously tested Alex. These trials thus lacked developmental relevance, but the hiatus lessened the chance of Griffin's habituating to or specifically learning the tasks. Despite the delay, the data could substantiate claims for Stage 6 competence.

Data from Griffin and Alex

Griffin's data are in Table 10.1. Here I describe specific trials of interest along with Alex's results. Both birds succeeded on all tasks, including the shell game and thus exhibited full Stage 6 competence, putting them at the level of humans and great apes, and beyond many mammals and nonhuman primates (Pepperberg et al. 1997). We do not know how Grey parrots develop in the wild and thus cannot correlate Piagetian stages and natural behavior, but Griffin's data are particularly interesting because we can compare his development to that of mammals and relate his data to two milestones in captivity, fledging and weaning.

Age at which stages of object permanence develop and ultimate competence both vary with species (Gagnon and Doré 1994). Variation may reflect different physical or cognitive maturation, which may involve genetic predispositions (evolutionary pressures exerted over time), environmental effects (particularly for laboratory animals), or some combination of factors. Teasing apart such factors is not simple and probably not possible, given our current expertise. Behavioral description, however, provides basic data (Pepperberg et al. 1997). Griffin, for example, developed rapidly compared to primates, and overall more rapidly than dogs or cats. He temporarily plateaued at a different stage and time than nonparrots. Like great apes (see, e.g., Wood et al. 1980) and kakarikis (Funk 1996), but unlike dogs or cats (Gagnon and Doré 1994), he made A-not-B errors. What factors might be involved? Let's look at Griffin's progress and his A-not-B errors in turn.

If Griffin is typical, Grey parrots progress differently from mammals. Great apes generally reach the earliest stages more rapidly than humans (e.g., Stage 3 by ~4 months), plateau temporarily at Stage 5, and reach Stage 6 about the same time as humans (Redshaw 1978; Mathieu and Bergeron 1981; Gagnon and Doré 1994). Monkeys develop faster than apes, but slow down as tasks increase in difficulty and fail Stage 6 tasks (Potì 1989; Spinozzi 1989; de Blois and Novak 1994; Gagnon and Doré 1994; de Blois et al. 1998). Cats and dogs develop more rapidly than monkeys, reaching Stage 5 at 7–8 weeks (Dumas and Doré 1989; Gagnon and Doré 1994). Cats, however, develop no further, but dogs master standard invisible displacements after several additional months (Gagnon and Doré 1994). Dogs appear to fail Task S (Doré et al. 1996). Griffin, in contrast, plateaued between Stages 3 and 4 (weeks 9–14) and completed standard Stage 5 and 6 tasks by 22 weeks (an age comparable to another psittacid, the kakariki; Funk 1996); he succeeded on additional Stage 6 tasks, including Task S, at 51–52 weeks. Interestingly, parent- but not hand-reared kakarikis also plateau, and at Stage 4, but not for as long as Griffin (Funk 1996).

Griffin's inability to reach Stage 4 until week 15 is difficult to explain. By week 9 he reacted to object loss, which suggests Stage 4 competence (Piaget 1952); nevertheless, he could not obtain the object. He might have succeeded had we not required physical removal of the cover (Dumas and Doré 1989). His plateau, interestingly, coincided with flight development: From week 11 onward, he accidentally uncovered objects while "helicoptering" (practice flying) and then interacted with these items; he spent considerable time in sessions trying to fly, and at week 14 appeared deliberately to "helicopter" to remove a paper screen from a pen cap and then a toy cup. Thus his plateau may have involved codevelopment or how he apportioned his attention: He may have been more interested in flight practice than our tasks, or neurological establishment of flight connections may have precluded concurrent development of connections needed for Stage 4

Table 10.1. Results of Griffin's trials and the age at which Griffin succeeded in a given object permanence task. (After Pepperberg et al. 1997.)

Task	Stage	Response and order of actions	Age (weeks)
		Visual Pursuit of Slowly Moving Objects	
1	2	Follows object through arc 3× in succession	≤8
2	2	Follows object to disappearance twice, then fails, follows 2× more	≤8
		Search for Simply Hidden Object	
3	3	Objects partially hidden: plays with cover, succeeds by chance, ignores trial	8
		Obtains partially hidden object 3× in succession	9
4	4	Scratches near fully hidden object, reacts to loss 2×, succeeds by chance once	9
		Reacts to loss 2×, succeeds by chance once	10
		Scratches near object 2×, succeeds by chance once	11
		Reacts to loss, fails in his attempt to recover object	12
		Ignores first two trials, reacts to loss 2×	13
		Succeeds by chance 2×, then succeeds after deliberate action	14
		Removes cover to obtain object 3× in succession	15
		Removes cover to obtain object 3× in succession	16
5	5	Choice of 2 sites: fails 1×, succeeds 1×, fails 2×, succeeds 1×	17
		Succeeds 5× in succession	18
6	5	Choice of 2 alternating sites: succeeds 3× in succession	18
7	5	3 alternating sites: haphazard search, mistrial, succeeds 1×, ignores next trials	19
		Succeeds 1×, fails 1×, succeeds 1×, fails 2×, succeeds 3× in succession	20

Search Following More Complex Hidings

8	5	Successive 3 sites: search wrt hiding order, succeeds 2×, search wrt hiding order, succeeds 1×	20
9	5	Superimposed covers: succeeds 3× in succession	20

Invisible Displacements

13	6	3 sites: searches implementing cover 1×, succeeds 3× in succession	21
14	6	Successive 3 sites: succeeds 2×, fails 1×, mistrial, succeeds 2×	21
		Succeeds 5× in succession, final search is wrt hiding order	22
14a	6	Control trials: succeeds 2× with 2 sites, succeeds 3× with 3 sites	51
15	6	"Trick": succeeds but skips a box	21
		Stops looking after last screen, succeeds 2×	22
15a		Repetition with blindfolded experimenters: succeeds 2×	51
16	6	Substitution trial: succeeds 2×	33
S	6	"Shell game": succeeds 2×, errs, succeeds 3×, including sham trial	52

success (cf. Rakic et al. 1986). Whether flight and Stage 4 develop simultaneously in other Greys, in nature or in the laboratory, as well as the possible importance of such codevelopment (such as memory for a nest hole no longer in view or predators that "disappear" into trees), is currently unknown.

Interestingly, other species' temporary plateaus also co-occur with developmental milestones. Dogs' Stage 5 plateaus occur at weaning (Gagnon and Doré 1994); children's final stages of object permanence emerge only as words involving movement and disappearance are acquired (Tomasello and Farrar 1984, 1986; Meltzoff 1988). Kakarikis develop flight between Stages 2 and 3, but weaning occurs just before Stage 4; maybe social behavior that simultaneously emerges plays the same role as flight for Griffin (Funk, pers. comm.). Developmental correlations between object permanence and natural behavior across species might prove important. What happens to wild Greys at 22 weeks, when Griffin succeeded at standard invisible displacements? In the laboratory, he was not entirely weaned and flight developed several weeks earlier (Pepperberg et al. 1997).

Behavioral milestones may be the impetus for, rather than a block to, achieving stages of object permanence. Grey parrot breeders observe Stage 2 at the earliest opportunity for sighted behavior (Carleson, Clyne, Lawson, Parks, pers. comm.). Nestlings follow feeding syringes at 2–3 weeks, essentially as soon as their eyes open. One nestling appeared to track the syringe a day before its eyes opened; whether it tracked by shadow or scent was unclear (Clyne, pers. comm.). By 18 days, nestlings follow syringes through a 180° arc; by 20 days, they track until a syringe is out of sight, and cheep as soon as it disappears (Clyne, pers. comm.). Unlike mammals, who search for relatively stable teats (possibly by scent) and who need not necessarily vie with littermates for nourishment, parrots compete with nest–mates for access to their single food source by tracking a parent's beak. The earlier tracking develops, the more likely the nestling is to survive. Interestingly, such tracking develops later in kakarikis than in Greys (Funk 1996).

Griffin's classic A-not-B Task 5 errors were intriguing because we differentially marked box tops to assist discrimination. Maybe he did not initially notice the colored Xs. The real question, however, is why primates and parrots but not dogs or cats make A-not-B errors. The answer probably depends upon determining the error's underlying cause. Researchers propose at least four hypotheses for children; these possibilities may be relevant for some animals (Pepperberg et al. 1997). First, A-not-B errors may involve memory (Bjork and Cummings 1984;

Diamond 1990b). Possibly, subjects cannot store information from the second hiding (Flavell 1985) or inhibit responding to the previous association of location and reward (Diamond 1990a). Second, the error may involve coordinating object position with respect to the subject's own body, rather than processing spatial cues for the overall surroundings (egocentricity; Bremmer 1978a,b); in other words, the subject uses prior information—"choose to my right"—not current input about the entire situation. This hypothesis may explain why subjects that rely on external reference from a very early age might not exhibit the error (Gagnon and Doré 1994). Third, the error may simply indicate incomplete comprehension of object permanence (Piaget 1952, 1954): The subject sees the object as an integral part of the hiding place, rather than as having independent existence. Finally, the error may reflect an incompletely developed ability to understand object movement in time (spatiotemporal data; Wellman, Cross, and Bartsch 1986).

These hypotheses do not explain why dogs and cats, but not parrots and primates, avoid A-not-B errors (Pepperberg et al. 1997). Memory would seem relevant for all species: Children, for example, err less when reliance on memory is unnecessary—if they can search immediately after the item is hidden in the second location (Diamond 1990b), if cover discriminability is increased, or if cover removal is simplified. Moreover, if given more than two choices of cover, children's errors are not all choices of the previous location, but may cluster in locations near the correct choice (Bjork and Cummings 1984; cf. Diamond 1990b). Nothing in dog and cat ecology requires better memory than other species for alternating specific locations during infancy. Nor are there apparent reasons why infant dogs and cats are less likely to form egocentric associations than infants of other species, or are better at understanding the separation between object and cover or processing spatiotemporal information to the extent that this error would not exist. Interestingly, in children and monkeys, A-not-B errors appear to be associated with incomplete maturation of brain areas related to memory and inhibition of neonatal reflexes (respectively, the dorsolateral prefrontal cortex and the supplementary motor area of the frontal cortex; Diamond 1990a,b). Of course, brain structures in dogs and cats, and even more so in birds, are probably not identical to those in primates that require maturation to overcome A-not-B errors; however, given our knowledge of brain evolution, such analogous (or homologous) areas are unlikely to exist in birds and not in dogs and cats. Conceivably, dogs and cats make the error, but for only a brief period that reflects their overall compression of stages: Griffin, for example, erred in only one session (Table 10.1); possibly, because of individual developmental differences, the error might be missed if the same dogs and cats are not tracked over time.

Task 16 tested whether Griffin and Alex remember (or form a representation of) the specific item that is hidden (Pepperberg et al. 1997). When Griffin was 33 weeks old, his food preferences stabilized, so we could test this possibility. On his first trial, we presented a cashew (a favored item) but hid a less desired Bird Diet nugget during a successive invisible displacement. Griffin upended the final box and stared at the pellet. He immediately turned over the other boxes, then ran to the experimenters. He repeated this behavior on a trial with a different box as the final hiding place. We then replicated the procedure without substitution; Griffin uncovered and ate the cashew without continuing his search. Before testing Alex, we reacquainted him with the procedures because he had not had such tasks in several years. On a standard Task 14 trial, he promptly chose the last screen and obtained a cashew. We then administered Task 16: After upending the box and finding a pellet, Alex turned from the apparatus to the experimenters, narrowing his eyes to slits, a behavior we have come to interpret as "anger."[7] To ensure cooperation on the next trial, we gave him the expected nut. His reaction to finding a pellet on the final trial was similar to his reaction on the first, except that he banged his beak on the table—another sign of frustration or displeasure.

On other tasks not in the Uzgiris and Hunt protocol—Task 14 controls (labeled here 14a) and Task 15a trials (during which experimenters wore sunglasses)—Griffin and Alex succeeded immediately. Alex, however, reprised his "Gordian knot" approach when given Task 15a for the second time; we therefore could not repeatedly test this task (Pepperberg et al. 1997).

Of particular interest were our birds' responses to Task S, the shell game (Sophian 1985), which tests attention, working memory, and spatiotemporal cognition. Both birds passed, which suggests that Grey parrots, unlike dogs and cats, and like humans and great apes, develop a robust sense of object permanence. We did not administer all tests described in Doré et al. (1996) because our parrots succeeded on the most difficult ones.

Possible Environmental Influences on Griffin's Behavior

Two different effects must be examined: those related to training and learning, and those related to general enrichment (Pepperberg et al. 1997). First, because evaluation tasks in a developmental study can conceivably *teach* as well as *test* a targeted concept (Cornell 1978; Hediger 1981; Thomas and Walden 1985; Chapman 1987; Pepperberg and Funk 1990), Griffin's results must be examined for three confounds: Could he (1) learn a task from observing the experimenter's behavior, (2) use repetitions and even mistrials for trial-and-error learning (Doré and Dumas 1987), or (3) use tasks as a stepwise means to learn a concept? Second, given the different ages at which Griffin and Alex entered the laboratory and their initial experiences, their enrichment histories might provide insight into their different Task 5 responses. My students and I believe that general experience, not training and learning, affected our results. For the following five reasons, we believe we did not teach concepts, that Griffin did not learn from the procedures, and that he did not acquire simple rules to facilitate Stage 6 success without object permanence (Gagnon and Doré 1992)—but that general enrichment did affect his success.

First, Griffin could not have learned from the experimenters' behavior because humans never demonstrated correct responses (see, e.g., Pepperberg and Kozak 1986). If Griffin did not react within the allotted time, a cover was not—as in studies by Uzgiris and Hunt (1966) or Sophian (1985)—lifted in his presence; instead, the cover and the hidden object were removed together. Moreover, different items or covers were generally used in successive trials to prevent associations between objects and covers.

Second, although Griffin may have managed one-trial learning of individual tasks, his data were generally inconsistent with trial-and-error acquisition. Had learning occurred, he would have consistently improved *during* a task; instead, he sometimes erred after correct trials (see Table 10.1 for week 20 on Tasks 7 and 8). Also, he might fail on successive attempts one week, but succeed immediately the next without intervening practice (see Table 10.1 for weeks 17 and 18 on Task 5). Conceivably, Griffin's behavior might reflect "insight" (see, e.g., Köhler 1927/1976), but insight is, by definition, different from trial-and-error learning. In an ontogenetic study, moreover, behavior that improves without overt practice is probably a consequence of maturation.[8] And, because we omitted Tasks 10–12, Task 13 was Griffin's first invisible displacement; initial success thus could not have been a consequence of learning about such displacements. Learning may, however, have affected Task 4 if codevelopment of flight was not a factor in his plateau.

Third, if tasks indeed expedited stepwise learning, success on one task should facilitate success on subsequent related tasks. Such was not the case for Griffin (Pepperberg et al. 1997). After finishing Task

4, for example, he immediately made the Task 5 A-not-B error, but succeeded the first time on all trials for the unique Task 9 (uncovering an object hidden in nested covers). Most important, however, is that presentation of and success on visible displacements need not facilitate success on invisible ones (see the rhesus monkey study by de Blois and Novak 1994). Specifically, if visible displacements enable subjects to devise rules such as "go to last place you saw container" or "go to last place the experimenter touched" (Gagnon and Doré 1992), monkeys (and presumably cats; Dumas and Doré 1989) either cannot devise or cannot transfer such rules to new situations; they fail tasks such as 14a, even after performing Stage 5 visible displacements. Learning and training apparently enable a subject to proceed only as far as its species' capacities allow. But even if purported success on invisible displacements in some laboratories was a consequence not of comprehending object permanence but of learning these rules, such success demonstrates an advanced level of cognitive processing unavailable to animals that fail such tasks.

Fourth, Griffin's results were inconsistent with simple rule-governed behavior. Had he used the rule "search last place you saw object," he would, on trial 1 of Task 13, have quit after seeing that the implementing container was empty. Had he used "search screen in most recent contact with container," he would have made no intermediate or sequential searches in Task 14 after successfully choosing the final screen in the first two trials. Moreover, during control trials held 29 weeks after formal testing ended, Griffin and Alex searched neither the last place the object was observed (implementing container) nor the screen last touched by the experimenter or container; both removed covers actually hiding the object. Finally, neither "followed eye gaze" (de Blois and Nowak 1994): Task 15 results did not differ when researchers wore mirrored sunglasses.

Fifth, Griffin's play sessions probably provided enculturation and affected results on some tasks. Play sessions did not provide practice at uncovering items, but did enable him to chew, toss, or throw items subsequently hidden or used as covers, and to interact with multiple items simultaneously. These opportunities may have advanced his dexterity beyond that of birds lacking such experience. Clearly, enculturation helped Griffin in Task 5 compared to Alex: Alex, who had not interacted with multiple separate items prior to Task 5, succeeded only after receiving such exposure. Enculturation also probably prevented tests from being aversive. How Griffin's concurrent exposure to mirrors may have improved his performance is, however, unclear.[9]

The Implications of Success on Object Permanence Tasks

Object permanence tasks examine a constellation of abilities: motor skills, attention, memory, motivation, socialization and enculturation, and the capacity to process spatiotemporal data. Motor skills and motivation are somewhat independent of cognitive capacities that are of primary interest in such a study; attention and memory, in the sense of acquiring and remembering environmental information, are more closely correlated with object permanence. Socialization and enculturation may allow subjects to demonstrate characteristics that have been acquired through interactions within their *specific* social and physical environment—a milieu that may or may not be richer than that in nature (Gollin 1985; Pepperberg and Kozak 1986; Doré and Dumas 1987; Pepperberg and Funk 1990; Funk 1996). The question is: Does success on object permanence tasks indicate general cognitive capacities? Not all animals reach Stage 6, and each stage may be reached by different processes, at different times, or with more or less difficulty (Chevalier-Skolnikoff 1989). What then are the implications of demonstrating a particular level of object permanence?

On the basis of work by researchers such as Rozin (1976), at least one connection arises between object permanence and general abilities. Remember, these scientists expand on Piaget's study of information assimilation and use to suggest that general intelligence *is* the ability to use information acquired in one (environmental) domain to solve problems in another. Thus psittacids' Stage 6 object permanence not only reflects the evolutionary adaptive value of such a trait but also indicates individual ability both to *assimilate* and to *use* available environmental information (Piaget 1954).

Correlations of object permanence and both brain maturation and language development in children are also of interest. Autistic children err more on Piagetian object permanence tasks than mentally retarded children whom they match in global, verbal, and nonverbal developmental age. These data suggest that object permanence tasks require particular types of cognitive ability such as handling abstraction and self-regulating behavior (Adrien et al. 1995); the latter ability may be related to neurological development (Diamond 1990a,b). For normal children, acquisition of words representing the concept of disappearance appears to be correlated with object permanence development, a correlation that also suggests specific cognitive links (Tomasello and Farrar 1984; Gopnik and Meltzoff 1986; see Gopnik, Choi, and Baumberger 1996 for studies on non–English-speaking children). How these findings relate to animals is unclear, but they provide intriguing possibilities for future study.

In sum, ontogenetic comparisons based on Piaget's framework may have limitations (see, e.g., Bower 1982), but this framework offers pertinent comparative guidelines for examining strategies and underlying mechanisms that various species use to solve common problems. Few other systems support valid comparisons between primates, nonprimate mammals, and birds. As many researchers propose, object permanence studies not only demonstrate intriguing parallel development in several evolutionary lines, but may also, if extended to comparisons between closely related species, reveal important information on cognitive, neurobiological, and evolutionary traits (Parker 1990; Pepperberg and Funk 1990; Gagnon and Doré 1994; Funk 1996; Pepperberg et al. 1997).

11 Can Any Part of a Parrot's Vocal Behavior Be Classified as "Intentional"?

Intentional is a philosophical term meaning "aboutness" and reference. It does not mean "on purpose," although "wants it to be the case that" is an intentional phrase.

Carolyn Ristau, *Cognitive Ethology* (1991:92–93)

Intentionality is reserved for special forms of social exchanges for which a number of criteria must be met, such as the sender of a message controlling the content of the message and of its consequence on the receiver.

Jacques Vauclair, *Animal Cognition* (1996:157)

In the operant paradigm, the reinforcer follows a response. The reinforcer was not considered to be a goal which "causes" behavior; rather behavior produces the reinforcer (Greenfield, 1971). How could something that occurred later (the goal) cause something prior (the means)? The answer is that the goal can have mental or cognitive existence not only before its attainment in the outside world, but before the response itself. This psychological existence is called an intention.

Patricia Greenfield, *The Social Foundations of Language and Thought* (1980:255)

The quotations above provide a brief introduction to the many different definitions of intentionality. Researchers debate whether intentionality requires conscious awareness (Vauclair 1996), the extent to which intentional behavior can be reduced to steps in a computer program (ostensibly removing the need to invoke consciousness; see Miller, Galanter, and Pribram 1960; Greenfield 1980), and whether inferences about intentionality of behavior can be drawn from goal-directed actions (Greenfield 1980; Call and Tomasello 1998). Dennett (1971, 1978a,b, 1983, 1987, 1988) tries to resolve the problem by analyzing behavior in terms of *levels* of intentionality, from none at all to the most complex. My goal, however, is not to review various levels or discussions of intentionality, or to claim that all Alex's behavior is intentional. I instead examine a subset of behavior—his acquisition and use of "want"—because data with respect to requesting provides some evidence, at least in terms of certain researchers' definitions, of intentionality (see, e.g., Vauclair 1996).

I had two reasons for teaching "want" and for studying how Alex acquired and used the term. First, I needed to determine whether, when he incorrectly identified objects with labels for more favored items, he was attempting to obtain treats rather than making errors (see Chapter 3). Specifically, if I could separate requests from errors, I would have a better indication of Alex's labeling capacity (Premack 1976). Second, I wanted to determine the extent of his communicative competence—a term generally defined as the ability to convey intent and to respond to the intent of others (Fay and Schuler 1980; see Smith 1991 for a discussion in terms of information processing). Could Alex convey his wants and needs by means of what to him was an artificial communication system? Given that his most frequent identification errors involved labels for treats (e.g., foods not freely available, such as nuts) or items with which he generally interacted for extended periods of time (e.g., corks he chewed to shreds), his behavior suggested some level of intent and thus communicative competence. Also, he would often toss an object he had identified and received from a trainer and immediately produce the label for a more favored object or food. Alex thus seemed a good subject for studying whether a nonhuman might use "want" in a referential, intentional manner.

Some Details of Alex's Previous Behavior

Formally separating Alex's requesting from labeling would be a challenge. His training always used referential rewards—he always received the object he had identified (see Chapter 2). He had never been *shown* that human utterances could be used *either* as labels *or* as requests. Thus my students and I could not determine if his use of labels involved comments on, identifications of, or requests for particular items. Likewise, we were unable, except through contextual inference, to decide if his utterances were comments upon an object not in view or requests for it in its absence. In that sense, his communicative competence was not unlike that of children in the one-word stage (Brown 1973; Bruner 1983; see Griffiths 1985) or animals in other interspecies communication studies (e.g., Savage-Rumbaugh et al. 1980b). Unlike researchers investigating children (Bruner 1983) or even chimpanzees (e.g., Plooij 1978), however, we could not correlate Alex's vocalizations with pointing: Grey parrots do not routinely point to specific objects or places.[1] Also, when he uttered a label we always responded by producing the object to which it referred and then asking him to identify it so he could receive it as a reward, further confounding the problem. Despite all these difficulties, determining whether Alex could indeed signal communicative intentions—separate requesting and labeling—would be extremely important for ascertaining his true representational abilities (see, e.g., Lock 1980). Let's look at why such separation is important.

The Significance of Requesting versus Labeling Behavior, and How Each Develops

Requesting seems to be a basic skill. Parents, for example, interpret childrens' earliest cries and whines as requests; the problem, as any new parent will admit, is often determining exactly *what* the child desires. Early projects to train chimpanzees to communicate with humans encountered few problems in getting the ape to produce a particular symbol to request, for example, M&Ms (Rumbaugh et al. 1973; Rumbaugh and Gill 1977); a chimpanzee so trained, however, could not use the label to *identify* these items (Savage-Rumbaugh 1984a,b). Chimpanzee requesting thus seemed to precede identification, but the training protocol was such that requests were rarely denied, and the *intention* of the ape to obtain M&Ms versus, for example, soda pop was not carefully examined. What the ape actually had learned was unclear. Interestingly, dysfunctional children in intervention programs that use comparable training protocols also fail to develop separate labeling and requesting behavior (see Fay and Schuler 1980; Twyman 1995). A child in an average household, however, sometime in its second year, seems to develop spontaneously the ability to use "denotive symbols," that is, to use utterances not only for demands but also to comment upon its environment (Dore 1985:24) and thus label items. Do some apes and dysfunctional children fail because of inherently limited abilities, or does another reason exist for their failure to move beyond what Dore (1985:24) defines as "indexical signs"—intentional communication of something, but something that lacks lexical meaning?

What indeed have the chimpanzee and dysfunctional children learned, and is the crying child really making a request? What actually is involved in "requesting"? Whether "requesting" is considered a sophisticated form of conditioned behavior (e.g., Skinner 1957) or representative of knowledge about how to use a communicative code to "get things done" (Bruner 1983), the fact remains that a subject that is *truly* capable of requesting not only must verbally discriminate between many different items (must associate the symbol that represents the object with the object itself), but must also spontaneously communicate which object is desired *and* that some other object is not acceptable (von Glasersfeld 1977); that is, some form of labeling is a *prerequisite* for requesting. In that sense, neither a crying child nor the chimpanzee described above was "requesting."

As noted above, however, normal human development includes acquisition of verbal skills that enable intentions to be conveyed (Brown 1973; de Villiers and de Villiers 1978). Dore (1985) argues that development of "denotive symbols" (the ability to use utterances for both requesting and labeling) from "indexical signs" (intentional communication without such separation) is not a strict stage because elements of the former appear gradually in the latter behavior. His most intriguing point (often ignored by some researchers) is that evolution of indexical signs into denotive symbols occurs through social and emotional interaction with caretakers, specifically through dialogues in which both reference and function are explicitly demonstrated. Dore's argument reflects the ideas of others, such as Bruner (1977, 1978) with respect to "scaffolding," Piaget (1954a), Scollon (1976), and Vygotsky (1962), but particularly the social modeling work of Bandura (1971, 1977; see discussions in Chapter 2). The elements that Dore describes are specifically the ones missing from input given the dysfunctional children and chimpanzee described above (for details, see Pepperberg 1997).[2] So, given that training true requesting behavior in a nonhuman would be a significant step in establishing its communicative competence (Savage-Rumbaugh, Rumbaugh, and Boysen 1980a), and that our M/R technique seemed to fit Dore's (1985) criteria for the type of experience needed to engender such behavior, my students and I initiated training on "want" (Pepperberg 1988a).

Training Procedures

One might argue that natural and appropriate input for engendering requesting behavior might not be the same for parrots and children. Little information exists, however, on the natural development of anything resembling requesting in parrots. The only documented "requesting" in wild parrots is food begging, either by the young from parents or between members of a mated pair (Nottebohm and Nottebohm 1969); such behavior more closely parallels that of the crying child, because the bird does not specify a particular object out of a large group. Whether true requests even occur in natural parrot vocalizations is still unknown. My interest, as in all the previous studies with Alex, was simply to learn if such behavior could be trained in a parrot.

My students and I used a variant of the standard M/R technique (Pepperberg 1988a). During sessions on "want," Alex observed the following interactions: The person acting as trainer held two of his favored toys (e.g., paper and cork; we used at least two objects to prevent him from associating "want" with a particular item). The human acting as the model said, "I want cork!" "Want cork!" or, occasionally, "Wanna cork," emphasizing "want" (or "wanna"). The trainer then gave the cork to the model. These actions were repeated with the paper; then the model and the trainer reversed roles. If the model produced only the label, the trainer replied, "Yes, this is a cork . . . Do you *want* it?" Alex, who was accustomed to using object labels as requests, often called "paper" or "cork" during these exchanges. We initially rewarded such utterances to avoid extinguishing the association between label and object. We slowly phased out our positive responses to the one-word vocalizations, and treated Alex just as the model who used one-word responses. Table 11.1 provides an excerpt from a training session. During initial training, we also adapted our protocol to other sessions involving concepts of color, shape, and quantity: After every correct identification of a particular X by the model, the trainer asked, "Do you want X?" For affirmative responses, the object was transferred as usual; negative replies were countered with the query "OK, what *do* you want?" The model responded "Want Y" to receive the more favored Y.

Alex's Acquisition of "Want"

We analyzed Alex's use of "want" two ways. First, we examined the emergence of his use of "want" from transcripts of audiotaping and videotaping sessions held during free periods outside of formal testing or training. Second, we tested his level of intentionality during use of this vocalization over several weeks; that is, we examined whether he indeed "wanted" the object he had ostensibly requested.

Analysis of the transcripts (Table 11.2)[3] shows that Alex's use of "want," "wanna," or "I want" to preface possible requests increased from 0% before training, to 38% after 3 months of instruction, to 70–75% after 9 months. At no time did his accuracy in object labeling decrease (see, e.g., Pepperberg 1983a, 1987b,c), although certain behavior patterns (such as holding on to objects he did obtain for signifi-

Table 11.1. Excerpt of an M/R training session on "want." A refers to Alex, M to Mary Sandhage (a secondary trainer), and I to the principal trainer (me). Note that Alex often uses corks and paper to clean his beak after eating. (After Pepperberg 1988a.)

A: (Has just eaten some banana): Cork.

M: (Looks at Alex): Well, I have a cork. Do you *want* cork?

A: Cork.

I: I *want* cork.

M: (Turns to I, presents a cork): Here's the cork.

I: (Manipulates cork for a few seconds, then starts to manipulate a piece of paper.)

M: *Want* paper.

I: (Presents the paper to M): OK, you can have the paper.

A: Cork.

M: Cork.

I: (Picks up cork, alternates gaze between M and A): I have a cork . . . do you *want* cork?

M: *Want* cork.

I: (Gives cork to M): Yes, you *want* cork.

---

cantly longer periods) suggested that the procedure engendered some minor frustration.

Interestingly, Alex often spontaneously generalized use of "want" to new lexical items without additional training. After learning labels for novel items in which he had some interest, he routinely uttered "Want X," "I want X," or "Wanna X." Objects in which he was not interested, but which we required him to label, were not requested. He rarely took such objects from us after they were identified; note the videotape transcript in the final section of Table 11.2, where he drops the walnut, requests another item, and proceeds to interact with what he requested. A transcript of the first part of a test, which demonstrates some additional differences between Alex's labeling and requesting behaviors, is presented in Table 11.3.

We also tested whether Alex's use of "want" could be intentional. Initially, only anecdotal evidence supported this claim (Pepperberg 1988a). We noticed that Alex appeared to request items even in their absence. At the time, nuts and other staples were in tins stacked in random arrays and thus often hidden by one another; fruits and vegetables were refrigerated and thus absent from view unless a trainer specifically brought them out; toys were jumbled together in boxes placed on a cart, so that the items were generally not in view. Thus unless Alex was responding to a trainer's query about a specific object or was requesting a particular item for a second time (e.g., another slice of banana), he was asking for something not physically present. We found that he generally refused substitute items offered by trainers, often repeated his initial requests, and then ate or manipulated the requested objects. We formalized these interactions, as suggested by Donald Griffin (pers. comm.), into a test: On various occasions, over several months, trainers substituted items for those Alex requested. Alex responded ~86% of the time to presentation of these items by saying "no," and such refusal was coupled (~68% of the time) with a repetition of the initial request (Pepperberg 1987c). These data suggest that Alex had successfully acquired appropriate (functional) use of "want" (see von Glasersfeld 1977).

Some Implications of Alex's Results

Three aspects of our study are noteworthy. First was the specific protocol we used to test intentionality. Second was Alex's flexible use of various phrases involving the term "want" and his acquisition of "wanna go." Third was the relative ease with which Alex learned to request items.

Our protocol to test the intentionality of Alex's use of "want" was designed so that other factors would not affect our results. We were very careful not to present Alex with a simultaneous choice of requested and nonrequested items. In a simultaneous presentation, the quality of individual items might influence the subject's choice. If, for example, Alex had requested banana, and we gave him a choice between a perfect grape and a piece of banana that might not be at the exact stage of preferred ripeness, he might chose the grape simply because it was more appealing. In contrast, by presenting only the nonrequested item, we were testing his actual desire for the object he had requested and, to some extent, his representation of the label for that object.

Alex appeared to use all three forms of request ("I want," "wanna," and "want") interchangeably after being exposed to our modeling procedure. In contrast, other training methods may produce considerably more rigid results (see, e.g., Fay and Schuler 1980). We note, too, that Alex apparently acquired and intentionally used the phrase "Wanna go Y" from his trainers' queries about where he wanted to be situated. If, after he began to use the phrase, a trainer attempted to place him somewhere other than the location requested, he refused to leave the trainer's hand and repeated his initial request. Given that his movement in the laboratory is not often left up to him (because, e.g., he will not

Table 11.2. Examples of Alex's emerging separation of "labeling" from "requesting." The following are representative samples transcribed from taping sessions; we used these transcriptions to monitor Alex's use of "want." A refers to Alex, B to Bruce Rosen, D to Denise Dickson, and L to Leah Schimmel, who were all secondary trainers. Peg wood is a clothes pin. (After Pepperberg 1988a.)

5/4/81: Prior to introduction of "want."

B: Alex, what is this?
A: Ro . . . peg wood.
B: Good boy, that's rose peg wood.
A: That's ro?
B: Yes, that's rose peg wood. Good boy.
A: Banow.
B: What?
A: Banana.
B: You want a banana?
A: Ro.
B: That's a rose peg wood. That's what you have.
A: Banow.

9/10/81: After initial introduction of "want." Note that "shou-er" might have referred to "shoulder" rather than "shower"; Alex was learning to differentiate those two labels at the time.

A: Want shou-er.
D: Wanna shower? OK, you had one already, but we'll give you another one if you want one.
A: No.
D: What is . . .
A: Want corn.
D: Well, which do you want? D'ya want a shower?
A: No.
D: No, you don't want a shower? Let's see . . .
A: Whaddya want?
D: What do . . . well, I want some corn. See . . . mmm . . . corn . . . that's good.
A: Corn.
D: Yeah, did you want some? Hmm . . . Alex? Did you want some corn?
A: Corn.
D: Yeah, it's corn. Good boy . . . that's *corn.* Yeah, yeah.

2/4/82: Approximately 9 months after introduction of "want." We did not know to what Alex was referring when he used the word "kah"; later we learned that a student had used that label for a corn "cob" during the previous weekend.

A: Want kah.
D: Well, this is a *carrot* . . . carrot . . . is that what you want?
A: No.
D: Oh, can you tell me what it is, even if you don't want it? You can tell me.
A: Want kah.

Table 11.2. (continued)

D: Alex, I don't have any "kah" . . . I don't know what "kah" is. But . . . What's this? What is it?

A: Want kah.

D: Alex, I'm sorry . . . I . . .

A: Want kah.

D: Well, I'd get you some kah . . . but . . .

A: I *want* kah!

D: Oh, Alex, I don't know what it is!

7/1/82: "Want" has been under study for a year.

D: Well, what do you want, Alex?

A: Tea.

D: Well, we don't have any tea. How about some pop? Would you like pop? Look, is this what you really want?

A: Want cracker.

D: Oh, you can have some cracker . . . Good boy. [Pause] Now look here . . . What's this? Oooh . . . look.

A: Want banana.

D: Oh, good boy . . . have some . . . Oh, good boy . . . You can have some of that banana. You didn't eat much . . . does it not taste very good? Do you want (it) . . . ?

A: Want some tea.

D: Well, we don't have any.

2/9/83: Note that in our laboratory, "nut" is the label for cashews. Other nuts have their own labels ("walnut," "cork nut" [almond]). This transcript is from a videotape.

L: You can have something if you want it. (Shows Alex tin of nuts.) Do you want some of these? Hmmm? Do you want some of these? This what you want?

A: Wanna nut.

L: Yeah, you're a good boy. You can have a nut. Yeah, you're a good parrot. (While Alex is eating nut): You're a good boy. Do you want anything else, Alex? Hmmm?

A: Wanna nut.

L: Want another one? How about a different kind? What kind of nut is this? (She holds out a walnut.)

A: Walnut.

L: That's right. It's a walnut. Good boy. (She hands over walnut.)

A: (Takes walnut, drops it immediately.)

L: You don't like walnuts? Do you want anything else? Hmmm? D'ya want something else?

A: Want cork.

L: OK, you can have a cork. Here you go. (She tosses cork onto top of cage.) Go pick up cork.

A: (Walks over to cork, picks it up and starts chewing.)

L: Good boy. Yeah.

Table 11.3. Transcription of Alex's responses to the first three questions of Test 344, given on 1/12/82. K refers to Katherine Davidson, the examiner, I to the principal trainer (me), and A to Alex. Peg wood is a clothes pin. Details of test procedures are presented in Chapter 2. (After Pepperberg 1988a.)

K: OK, Alex, let's start. What's this? (Holds up clothes pin.)

A: Peg wood.

I: (Sits facing wall, observing neither A nor K): That's "peg wood."

K: That's right—here's the peg wood, Alex. You're a good boy. (Hands over the clothes pin.)

A: (Takes clothes pin, but drops it immediately): Want cork.

K: OK, here's cork.
   (A plays with the cork for about one minute.)

K: Enough, Alex. Gimme cork (holds out her hand; Alex relinquishes cork). What's this, what shape? (Holds up a red, triangular piece of wood.)

A: Cor-er wood.

I: I think he said "corner wood."

K: (Briefly turns away, then reestablishes eye contact with parrot): Alex, what shape? Talk clearly!!

A: *Three*-corner wood.

I: "Three-corner wood."

K: You're right—the *shape* is three-corner wood. (Hands wood to parrot.) (A takes wood, chews off one of the corners, drops it.)

A: Wanna nut. (K gives him a nut, which he eats.)

K: Look, what's this? (Holds up a piece of grey rawhide.)

A: Wanna nut.

K: *No* nuts . . . first tell me, what's this?

A: Grey hi'.

I: "Grey hide."

K: Good parrot . . . here's the grey hide.

A: (Refuses to take the hide): Wanna nut!

attend to testing or training if he is on a student's shoulder), we frequently overrode such requests, in contrast to our usual compliance with requests for objects and foods. Thus his perseverance in these locative requests waned relatively quickly. Note, however, that he never substituted object labels for location terms or vice versa in "Want X" and "Wanna go Y."

Although Alex was trained for approximately 6 months before he uttered "want" reliably in situations correlated with requesting, the ease with which he acquired apparent intentional use of the term contrasted strongly with the difficulty experienced by some chimpanzees.

Like Alex, these chimpanzees had also always previously received the object that had been "named." The procedure used to teach them to separate requesting from labeling, which did not involve modeling, actually led to a temporary breakdown in food-symbol correlations (i.e., short-term loss of labeling ability; Savage-Rumbaugh 1984b, 1986). Researchers tried to show the chimpanzees that a label was not necessarily a request by substituting either a different food for the one the chimpanzee had "named" (Savage-Rumbaugh 1986) or a single food for the animals' responses to all objects (note Greenfield 1978). Reinstating previous levels of labeling accuracy required a series of steps. Simply switching between providing the food that had been labeled and the common food was ineffective, as was providing plastic inedible "foods" followed by the common reward (Savage-Rumbaugh 1986). Finally, after the researchers presented both labeled and common foods, faded out the labeled food by giving smaller and smaller portions, used hugs, pats, and so forth to indicate approval, and added a "showing" versus an "offering" gesture ("What's this?" versus "Want this?")[4] to denote whether the situation required labeling or requesting, the chimpanzees learned to separate the two functions of their labels (Savage-Rumbaugh 1984b).

My students and I encountered no such problems with Alex. Our relative ease in training "requesting" versus "labeling" by using the specific term "want" and demonstrating its use via interactive modeling suggested that the technique might be particularly potent, and possibly useful even for humans with communicative deficits. I am claiming not that mechanisms of learning in an intact animal are homologous to those of humans, either with or without communication problems, but rather that a method that alters the behavior in one species might alter behavior in another (Pepperberg 1988a). In particular, many intervention programs in use when I performed this study on Alex heavily relied on behavior modification and operant conditioning methods that were initially derived from nonverbal studies on learning in pigeons and rats (Tinbergen and Tinbergen 1972; Lovaas 1977; Moerk 1977; review in Fay and Schuler 1980). At the time, I suggested that a successful instruction program for teaching a vocal, interactive, intentional and referential form of human communication to a species that does not normally acquire such behavior might provide insights into critical variables for human learning. If Alex is any indication, the characteristics of optimal instruction identified by researchers such as Bandura (1971, 1977) are also similar to some proposed by Krashen (1982) for acquisition of a second language by humans: that input must be comprehensible and relevant, and that not only must the correlation between a vocalization and its referent be clearly demonstrated, but teachers must also *show* how a vocalization is to be used and why it is to be used. Interestingly, a colleague has recently had considerable success in using our modeling technique to assist developmentally delayed children (Pepperberg and Sherman, in prep.).

In sum, an organism that has learned to communicate with other organisms learns (1) not only individual signs but usually also sign combinations, whether vocal or gestural or combinations of both (see, e.g, Smith 1988); (2) appropriate contextual use of signs so that the signs acquire some level of reference and become more symbolic than iconic; and (3) rules of conversational interaction including some degree of intentionality. Clearly, the extent to which natural animal communication codes adhere to these criteria may vary among species (see, e.g., Smith 1977, 1991), and even the existence of such behavior in animals can be hotly contested (see, e.g., Bickerton 1990), but these criteria are meant for comparative purposes. For animals that are being taught to communicate with humans, these criteria have led to arguments about the extent to which the codes that have been learned correspond to human language (see Chapter 3). My point is not to rehash these arguments, or to claim that Alex's use of "want" is isomorphic with that of humans. Rather I wish to emphasize that modeling protocols have aided overall communicative competence by providing the means (1) to add a vocabulary item such as "want," which is often otherwise difficult to train, to a repertoire; (2) to expand production from labeling to predication, which implies an extremely simple form of syntax but a form nonetheless; and (3) to introduce a shared (and effective) way to signal what kind of speech act ("labeling" versus "requesting") a speaker is making. This system can thus be used to train at least rudimentary forms of the three aspects of a communication code—semantics, syntax, and pragmatics—that are critical for acquisition of functional competence (Bruner 1983).

12 Can a Parrot's Sound Play Assist Its Learning?

> Why does a child engage in monologue? What is its function for her? What role does it play, if any, in the development of her language or thought? What role does it play in the construction of her view of self in relation to others? How does it differ in function and form from dialogic speech?
>
> Katherine Nelson, *Narratives from the Crib* (1989:1)

Until this point in my studies with Alex, I had consistently reported data from experiments designed to train and test for specific types of learning. But my students and I had also noted two forms of vocal learning we had not formally presented to the scientific community. First, although Alex's new labels usually appeared in sessions initially in a modified, rudimentary pattern—first as a vocal contour, then with vowels, and finally with consonants (see Chapter 15)—he occasionally uttered completely formed new labels after minimal training and without any overt preliminary "practice." Second, he often engaged in a form of sound play outside of sessions in which he recombined labels or label parts; these innovations quickly became part of his repertoire if we provided a corresponding object. Some of this behavior was probably based on natural avian capacities. Compare, for example, how an immature Bewick's wren or white-crowned sparrow, still in subsong, may produce full crystallized song during interactions with an adult and occasionally maintain that form in subsequent interactions (Kroodsma 1974; Baptista 1983). Alex's behavior might also parallel processes involved in human language acquisition, particularly those in which childrens' babbling occasionally results in adult-like utterances (Dore 1985). Alex's data also parallel data on a child's responses when adults react as though the child's initial labeling attempts are meaningful (e.g., "fast mapping," Cary 1978; see Locke and Snow 1997). Thus, for two reasons, my students and I began examining these aspects of Alex's behavior. First, his actions seemed relevant to his developing communicative competence; second, we felt an analysis might determine how much these patterns function in animals. In this chapter I will describe our study of Alex's monologue speech; in the next chapter I will examine how our "mapping" of speech affected his development.

Monologue Speech in Children and Possible Parallels with Birds

Many researchers, myself included, are fascinated by and have concentrated on studying how social interaction affects vocal learning (e.g., Snowdon and Hausberger 1997), but other influences clearly are relevant. One effect is how much an individual, human or nonhuman, actively practices its communication code in the absence of overt social stimulation. Why do babies babble and manipulate sounds in monologues in their cribs (see, e.g., Weir 1962)? Why do immature birds seem to warble to themselves not only in isolation boxes in laboratories (Marler 1970; Marler and Peters 1982a,b) but also in the wild (Baptista 1983)? Baldwin (1914) even reported that a Grey parrot occasionally engaged in private practice before emitting an utterance in perfect form for its trainers. Some researchers argue that these practice periods are an integral part of development (Piaget 1966; Vygotsky 1978), and that because such practice occurs in diverse species, it supports an evolutionary theory of language play (Kuczaj 1998). Did Alex also exploit some form of private practice? My students and I decided to determine whether he did engage in such behavior, and if so what effect it might have on his development. A brief review of the phenomena in children will help put our study in perspective.

What researchers call "monologue speech," to distinguish it from interactive vocalizations, does not seem essential for human language acquisition, but has been observed for most children (Weir 1962; Kuczaj 1977, 1983; Nelson 1989). Monologue speech actually has two components: *private speech,* which is produced in the absence of any possible receivers, and *social-context speech,* which is produced in the presence of potential receivers but which does not appear to serve any obvious communicative purpose (Fuson 1979; Kuczaj and Bean 1982). Monologue speech thereby differs from *social speech,* which is directed toward other individuals. Because the purpose of a communication code is generally to transfer information among individuals (Searle 1969), the function of monologue speech, which has no communicative intent, clearly needed investigation.

Some researchers suggest that monologue speech permits language play: practice that, like most play, facilitates learning by allowing immature individuals to experiment with the adult system without experiencing the consequences of failure (see, e.g., Piaget 1966; Vygotsky 1978). Monologues, unlike other forms of vocal practice, do allow a speaker complete freedom to choose topics and contexts, attempt novel forms, and examine similarities and differences between familiar and novel forms (Piaget 1966; Britton 1970; Ryan 1973; Reynolds 1976; Scollon 1976; Black 1979; Kuczaj 1983; Nelson 1989). Several examples exist of speakers being trained in adult usage through formal, programmed routines whose development suffers because their practice is public.[1] Such speakers face the consequences of incorrectly communicating something; they may be interrupted and corrected for inappropriate usage (see Rice 1991), or be forced to engage in possibly boring drills (Kuczaj 1983). Conceivably, these negative consequences inhibit further practice and hence delay development (see, e.g., Krashen 1976; Kogel, Dyer and Bell 1987; Yoder and Kaiser 1987; Rice 1991; note Salmon et al. 1998). Debate exists, for example, as to whether caretakers who recast a child's error or expand elementary attempts at communication help or hinder development (e.g., Morgan, Bonamo, and Travis 1995; Nelson et al. 1995; Bohannon et al. 1996; Morgan 1996; Bloom et al. 1998). Even in ostensibly unstructured dialogues with caretakers, the constraints of constructing replies with appropriate word meaning (semantics), grammar (syntax), and functional use (pragmatics) could inhibit experimentation by speakers (Feldman 1989). In contrast, absence of negative feedback or the need to use correct semantic, syntactic, and pragmatic forms during monologues might encourage practice and accelerate learning.

Testing this hypothesis about the function of monologue speech is difficult for at least three reasons. First, data collected in the absence of communicative partners are difficult to interpret: Word or phrase meaning cannot be determined by, for example, environmental cues (Nelson 1989). Second, not all children engage in overt monologue speech, and those who do vary considerably in the overall amount used (Kuczaj 1983) and usually stop by about the age of 2½ years, while language is still developing (Nelson 1989). Third, some children engage instead in covert forms such as subvocal or nonvocal (mental) practice that cannot be quantified or analyzed (Vygotsky 1962; Kuczaj and Bean 1982). Consequently, analyses of the styles and learning strategies of various children often yield conflicting conclusions about the value of monologue play/practice (Bowerman 1982; Nelson 1989). Researchers have nevertheless performed such analyses; the apparent consensus is that monologues enable children to integrate what they have recently heard, practice these new forms, and thus progress toward adult communication. But how might such practice work for a parrot, who is not necessarily learning human language, but *is* acquiring functional use of human speech?

As noted earlier, such behavior is probably part of avian development: Laboratory-raised, isolate oscine birds that learn songs from audiotapes engage in solitary vocal practice (e.g., emit monologue-like routines) prior to developing adult song forms (Marler 1970; Marler and Peters 1982a; Hultsch 1990). These birds, as well as those in the wild, may practice the order of notes in their songs and the order of songs if they have a repertoire, recombine elements from different tutors if given multiple input sources, and practice different songs in different contexts (e.g., under different lighting conditions); if they are not reared in social isolation, they may also change the song types they practice after noting how different songs affect other individuals (Lemon 1975; Kroodsma 1981, 1988; Hultsch 1990; Nelson 1992; Margoliash, Staicer, and Inoue 1994; King, Freeberg, and West 1996). Such behavior roughly parallels human practice of syntax, semantics, and pragmatics (West and King 1985). Similar behavior has been informally observed for parrots that were learning either their natural vocalizations (Nottebohm and Nottebohm 1969; Nottebohm 1970) or to mimic human speech (Baldwin 1914; Todt 1975b). The function of this avian behavior and the *extent* to which it occurs as monologue in the wild are not yet determined, but the existence of such practice suggests that a bird could provide data comparable to that of children who engage in monologues.

Alex's behavior suggested he would be a good candidate for study. Preliminary data demonstrated certain parallels with children with respect to level of practice and type of material practiced; additional data might illuminate the role of monologues in his learning processes.

What Aspects of Monologue Speech Should Be Studied in Alex's Behavior?

The earliest forms of language play/practice involve sound play: imitation, modification, variation, and recombination of parts of words that are being acquired, often with little apparent regard for meaning (Stern and Stern 1928; Hurlock 1934; Piaget 1962; Garvey 1977a,b; Weeks 1979; Kuczaj 1983). Such play differs from noise play (e.g., phonemic babbling) because sound play occurs only after a child begins to acquire adult use of words (i.e., limited reference as well as association; Lock 1980). Sound play, however, occurs well before anything resembling mature language exists and decreases as language development proceeds (see, e.g, Weir 1962; Scollon 1976; Weeks 1979). Sound play is most likely in private or social-context speech (Garvey 1977a,b).

Prior to this study my students and I had found that Alex spontaneously produced English monologues under the same conditions as children in the early stages of learning: occasionally in his trainers' presence, but mostly when alone (Pepperberg and Matias, unpubl. data obtained from tapes). He also used English labels and phrases in a referential, contextually applicable manner but had not acquired (nor was he being trained to acquire) a communication code isomorphic with human language (see, e.g., Pepperberg 1981, 1987c). Consequently, we concentrated on studying Alex's sound play while he learned to produce and comprehend various labels, even though portions of his monologues, like those of children, appeared to contain more complex material, such as lexical substitutions (Pepperberg, Matias, Brese, and Conn, unpubl. data) recombinations of word patterns in phrases and sentences (Weir 1962; see Gallagher and Craig 1978; Craig and Gallagher 1979; Pickert 1981; Furrow 1984).

What Labels Should Be Studied in Alex's Practice?

The question then arose as to which labels would provide interesting information about Alex's use of monologues. Because much of his sound play seemed, at least from preliminary data, to involve recombinations of items already in his repertoire, we decided to track his progress while training a label that could be formed from parts of existing labels. The choice of label was influenced by Alex's existing conceptual knowledge (Pepperberg, Brese, and Harris 1991), because some researchers believe that children learn most if they can relate new information to previously acquired knowledge (Piaget 1962; Rummelhart and Norman 1978; see also Vygotsky 1962, 1966). Thus to benefit from novel input, children need both to assimilate and to accommodate this input in terms of existing data, and learning is optimal when new material is only slightly more difficult than that already learned (Nelson 1978; Krashen 1982; Kuczaj 1982a). Several studies have shown that children are more likely to imitate behavior patterns which are only moderately different from existing forms and of which they have partial understanding, than entirely novel behavior patterns (Piaget 1954a; Krashen 1976, 1980). With respect to language, new labels that are modifications of existing labels and that encode only slightly novel information are thus likely candidates for practice (Weir 1962; Ryan 1973; Scollon 1976; Kuczaj 1982b, 1983).[2] Would a parrot respond like children?

My students and I thus began part one (Experiment I) of our two-part study by training Alex on "none." "None" contained phonemes already in his repertoire (e.g., in "nut," "no," "one") and, at the time, acquisition of "none" would allow us to extend the recently completed same/different study (Chapter 5) to include nonexistence (Chapter 6).

We also wanted to see if the extent and timing of Alex's practice might be affected by the degree of phonetic similarity between new and existent labels (Pepperberg et al. 1991). We thus monitored Alex's monologues while training two sets of new labels. One set ("nail," "bread") had little phonetic similarity to existent labels and only the "n" of "none"; the second set ("sack," "green bean") had more similarity than the first set to existent labels and equal similarity to "none" (Experiment II, 1988). "Sack" shared phonemes with existent labels "back" and "cracker"; "green bean" combined the existent "green" and "be." We included food labels to examine possible motivational effects.

Our study would also specifically test whether a subject trained by programmed, sequenced routines with correction procedures was likely to use monologues for "no fault" practice of novel utterances prior to their social use. Remember, Alex was being *trained* to communicate—to use labels referentially—rather than being exposed to an environment that *allowed* consequence-free "acquisition" without necessarily teaching meaning (Krashen 1976; Lambert 1981). Even though training incorporated few formal drills, and attempted to recreate a relatively natural learning environment (e.g., observation of and interaction with individuals who were themselves engaging in referential communication; Pepperberg 1981), Alex always experienced negative consequences for erring in his trainers' presence: Trainers scolded him and removed training objects from sight. Such negative events did not occur during monologues. We might therefore discover if Alex was, like some children, "practicing" in a solitary, consequence-free context the communication code he was being taught in a formal, interactive procedure.

How We Studied Acquisition of "None"

We used standard M/R training (Chapter 2) to demonstrate the use of "none." We produced "none" in response to "What's different?" or "What's same?" if nothing was different or same between two objects (Chapter 6). We also demonstrated connections between "none" and "one," "no," and "nut," which were already in Alex's repertoire. Models initially produced "none" as "nnn-won," followed by "none" (Pepperberg et al. 1991). Such a procedure is similar to Bruner's (1983) scaffolding for young children: demonstrating old forms upon which new learning can be built. During training on "none," we continued to test other concepts and train Alex on tasks that used existing labels, but did not introduce other novel vocalizations. During training sessions, we recorded observations on Alex's progress in a notebook.

To obtain monologue samples, we audiotaped Alex when he was alone but active and vocal (Pepperberg et al. 1991). Unlike children, he did not vocalize if isolated in a dark room (Pepperberg and Matias, unpubl. data). We thus taped in the early evening after trainers left but while ambient light still existed, and in the morning before trainers arrived. When dusk preceded trainers' evening departures, we provided 1–2 hours of artificial light with a time-activated switch. Because we taped without visual observation (that particular laboratory layout did not allow a one-way mirror), we did not know if Alex's utterances were contextual, that is, if he talked about items with which he played, or whether, like some children, he "free associated" to produce utterances based on structural, phonological, or semantic similarities (Piaget 1962; Bloom 1970; Britton 1970). Taping in this manner, however, was comparable to the procedure used in most child studies (e.g., Weir 1962; Nelson 1989), and meant that our data would not be affected by an observer (Fuson 1979). Inconsistencies in the behavior of a subject who is drifting in and out of sleep, a matter of concern for children's studies (e.g., Weir 1962; see Gallagher and Craig 1978), were not a problem: We could hear Alex moving about, which meant that his monologues occurred when he was fully awake.

We documented Alex's utterances with a Marantz PMD 221 tape recorder controlled by a Lafayette Instruments Voice-Activated Relay (VAR) set to begin after an unadjustable delay of 0.008 s from the onset of sound levels comparable to those of human conversation (Pepperberg et al. 1991). Recording stopped after 1.5 s of continuous silence. The microphone for the switch was at cage level about 4 feet from Alex. We used the recorder's internal microphone, and set the recorder on a chair also about 4 feet from the cage (Figure 12.1).

Taping sessions and transcriptions covered three time periods. During the first period, 7/15/86 to 8/27/86, prior to any training, we taped Alex to collect baseline data and make final adjustments to the equipment. The second taping period, 9/5/86 to 10/21/86, started with training on "none" and ended with Alex's first production of "none" in our presence. The third period, 10/22/86 to 10/24/86, ended when Alex used the label reliably. We taped every other working night/morning on a Monday-Wednesday-Friday-Tuesday-Thursday 2-week schedule,

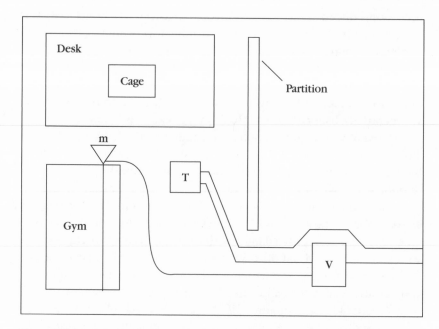

Figure 12.1. Schematic diagram of the laboratory for recording Alex's solitary sound play (other furniture, such as chairs and benches, was present, but is not diagrammed). The parrot is always on either the desk or the cage. T is the tape recorder with its internal microphone; it is placed slightly below the level of the cage. V is the voice-activated relay switch, and m is the microphone for the switch. The gym is the stand used by the parrot when trainers are present. The vertical line represents the perch (a wooden dowel rod) to which the microphone is attached. The perch is about 4 feet above the level of the stand so that the microphone is at the level of the cage. The thin lines connecting the pieces of equipment are electrical wires. (After Pepperberg et al. 1991.)

and altered taping dates only when absolutely necessary in an attempt to provide a random sample.

We obtained considerable data (Pepperberg et al. 1991). In 35 sessions, we recorded 10,246 sounds, 51.5% of which were English labels or related vocalizations. Other sounds included parrot calls and whistles, cage noises, ringing phones, and so forth; a complete listing is in Pepperberg et al. (1991). The 10 initial tapes (9 sessions) provided 4 hours (about 1,800 sounds) of baseline data. The 26 tapes (26 sessions) during training provided another 9 hours of data (about 8,500 sounds).

To check transcription accuracy, a student and I initially worked independently. Katherine Brese transcribed data into symbols and English words or sounds. I retranscribed two tape samples, one taken before and one after training on "none" had begun, without knowing the order of taping. These retranscriptions checked for interobserver reliability and "expectation cuing."[3] Our transcripts agreed to within 88.9% for the early sample of English and English-related vocalizations and 92% for the later sample; disagreements did not involve targeted vocalizations (Pepperberg et al. 1991).

We analyzed the data using two BASIC programs on an IBM PC. One program searched transcripts for a specific label or phrase targeted by the user and gave counts and percentages for each night and for the total number of times this particular vocalization occurred in the data. The second program listed occurrences of all words per night, broke down the data into non-English and English sounds, and calculated nightly and total percentages of English labels or sounds.

The Development of Alex's Production of "None"

Alex produced "none" in his trainers' presence after 7 weeks of modeling. Four more weeks of training were needed, however, before "none" could be recognized aurally by all trainers. During these periods, he practiced in private. Analysis of his monologues showed the following (Pepperberg et al. 1991):

1. Alex's overall percentages of "none"-related utterances ("one," "un"), although ≤0.6%, differed markedly before and during training. We counted both "one" and "un" as attempts at "none" because we actively modeled "none" from "one" and the VAR may have cut off initial bits of utterances. Table 12.1 (part A) shows percentages of "one" and "un" separately and together in baseline and training data compared with the total number of English vocalizations. If any English utterance was equally likely, we would have expected 5 "one/un" during baselines, but recorded 0; similarly, we would have expected 18 such utterances after training began, and recorded 20 overall. Thus Alex's use of "one/un" after training began became at least as likely as any other vocalization in monologues, and clearly contrasted with the absence of such vocalizations prior to training.[4]

2. "One/un" occurred most (12 times) in the first session after training began (9/5/86).

3. Alex used "one/un" in monologues (beginning 9/5/86) several weeks before attempting "none" in his trainers' presence (10/21/86).

4. Alex's average overall number of English vocalizations per session increased after training began. Moreover, the highest total utterances per session (1,021) occurred immediately after we initiated training on

"none." Several possible explanations (e.g., novel input, additional attention from trainers, etc.) could account for this increase.

5. During training, Alex's use of "um" in monologues also changed, which was intriguing because "um" is not as closely related to "none" as "one" or "un." Before training, "um," if present, was used ~66% of the time in monologues as part of a phrase (e.g., "go on the gym um geepers"), whereas after training it was used only by itself. At the time, "um" had never been recorded in any context in social speech.

These data, although intriguing, were clearly limited to "none," which had been chosen because it was likely to engender monologue speech. We thus proceeded with Experiment II (1988). That experiment would, we hoped, show how labels with different degrees of phonetic similarity and different levels of connection to current cognitive knowledge might be practiced in monologues.

Alex's Acquisition of "Nail," "Sack," "Bread," and "Green Bean"

We again used M/R techniques (Pepperberg et al. 1991). We began by training "nail" and "sack" simultaneously, and introduced "bread" and "green bean" at subsequent 1-week intervals after Alex first uttered "nail." We used screws of different sizes as the referent for "nail" because Alex manipulated ribbed screws more than smooth nails. For "sack" we used variously colored paper bags (2″ × 4″). To interest him in sacks and introduce the similarity between "sack" and existing labels (to provide scaffolding), we talked about "cracker" that was in the "sack," or suggested that Alex could go "back" to his cage (something he occasionally requested) if he identified a "sack." "Bread" referred to pieces of whole wheat or rye, and "green bean" to the vegetable. Training occurred two to four times per week and lasted 5 to 60 minutes, and we simultaneously tested other concepts. We noted when each label was first produced in a trainer's presence and when all trainers could distinguish a label reliably.

We used different recording equipment in Experiment II. To avoid possible data loss from the inherent delay of the VAR switch, we obtained a Sony TCM-5000EV cassette recorder with an automatic voice-operated recording (VOR) system (Pepperberg et al. 1991). This "bucket brigade device" VOR did not cut off the starting point of Alex's utterances. Sensitivity was set so that recording began at sound levels comparable to those of human conversation. By connecting a unidirectional AKG CK8 microphone to the recorder, we focused on Alex and cut out some extraneous room sounds. We placed the microphone where the VAR microphone had been and put the recorder on the gym (Figure 12.1).

Our taping schedule and environment also differed slightly from those in Experiment I (Pepperberg et al. 1991). We now also recorded while trainers were at lunch. Because the laboratory had been relocated in autumn 1987 to a windowless room, all lighting (VitaLite fluorescent lamps) was timer-activated. Lights were on for 1-2 hours after trainers departed, and for about an hour before trainers arrived in the morning. Lights in an exterior room provided additional low illumination in the evening and morning ("dusk" and "dawn") to avoid abrupt onset/offset of darkness. Taping again occurred without visual observation by human trainers.

Experiment II generated considerable data. In 53 sessions, we recorded 24,157 sounds, 71.0% of which were English labels or related vocalizations. The 7 initial tapes, made prior to training (8 sessions), provided 4⅔ hours (about 1,900 sounds) of baseline data. The 35 tapes (45 sessions) made during and after the training provided another 27 hours of data (about 22,000 sounds).

As before, two of us transcribed independently and compared our results. Barbara Harris transcribed all data and I retranscribed two samples (one taken before and one after training on "nail" began) without knowing the taping dates. Transcripts agreed to within ~94%. Disagreements did not involve targeted vocalizations. We analyzed data with the same programs used in Experiment I.

The Development of Alex's Production of the Second Set of Labels

The amount of training required before Alex produced each new label varied considerably (Pepperberg et al. 1991). He uttered "green bean" for the first time in our presence less than 1 week after its introduction and "nail" in less than 2 weeks, but "bread" took considerably longer and he never produced "sack" reliably. In all cases, Alex required additional training after initially uttering a new label in our presence before all trainers could identify the vocalization.

Data analyses showed that, as with "none," overall percentages of monologue vocalizations related to novel labels were low ($\leq 6\%$), and marked differences occurred in monologues for periods before, during, and after training (see Table 12.1, part B). We noted the following (Pepperberg et al. 1991):

1. Before training (during baseline), few of Alex's vocalizations resembled targeted labels. No utterances resembled "nail" or "bread," which (except for the /n/) had no closely related precursors in his

Table 12.1. Average occurrence of labels and attempts at labels before and after training as a percentage of English utterances and related vocalizations. (After Pepperberg et al. 1991.)

	Percentage of English vocalizations		
A. Experiment I	one	un	one + un
Baseline data (before training)	0.0	0.0	0.0
During training			
9/05/86–10/21/89[a]	0.29	0.16	0.45
After initial training			
10/22/86–11/24/86[b]	0.60	0.0	0.60

	x + nail[c]	nail
B. Experiment II		
Baseline data (before training)	0.0	0.0
During training		
7/04/88–7/14/88	0.09	0.0
After initial training		
7/18/88[a]–7/26/88[d]	3.67	0.30
7/27/88–8/01/88[d]	0.26	0.0
8/03/88–8/16/88[b]	0.31	0.0
8/18/88–8/24/88	0.05	0.0
8/26/88–9/28/88	0.0	0.0
9/30/88[e]–10/13/88	0.07	1.21

	back[f]	bach	cracker	"ss-words"	sack
Baseline data (before training)	1.27	0.49	0.0	0.07	0.0
During training					
7/06/88–7/22/88[g]	1.37	0.56	0.0	0.23	0.0
7/26/88–8/01/88[d]	2.19	0.73	0.04	0.29	0.0
8/03/88–8/16/88	2.14	1.41	0.0	0.05	0.0
8/18/88–8/24/88	0.52	0.19	0.05	0.05	0.0
8/26/88–9/28/88	3.15	0.13	0.0	0.0	0.0
9/30/88[e]–10/13/88	1.71	0.37	0.0	0.03	0.0

repertoire. Labels in his repertoire, for example, "green" and "bok" (Alex's frequent mispronunciation of "box"), that were similar to "green bean" and "sack" appeared somewhat less often or as often, respectively, as would be expected by chance. If any single utterance was as likely as any other for the period before training, we would have

Table 12.1.   (continued)

			Percentage of English vocalizations	
			braa, graed	bread
Baseline data (before training)			0.0	0.0
During training				
7/20/88–7/27/88			5.74	0.0
After initial training				
7/28/88[a]–8/16/88[d,h]			1.12	0.0
8/18/88[b]–8/24/88			0.66	0.0
8/26/88–9/28/88			0.0	0.0
9/30/88[e]–10/13/88			1.41	0.0
	green	be[i]	related	gb[j]
Baseline data (before training)	0.03	2.58	0.04	0.0
During training				
7/28/88–8/03/88[a]	0.12	0.49	0.04	0.0
After initial training				
8/05/88–8/16/88[a]	0.38	1.15	0.04	0.0
8/18/88–8/24/88	0.14	3.27	0.71	0.66
8/26/88–9/28/88	0.20	4.53	0.07	0.20
9/30/88[d]–10/13/88	2.0	3.39	1.04	2.05

a. First appearance of the targeted vocalization—or one very close to the target—in the presence of a trainer.

b. Date at which novel vocalization became reliable.

c. x + nail represents utterances related to "nail"—e.g., "awail," "benail," "banail," "blail," "lenail," "chail," "mail," etc.

d. Period during which another novel vocalization first appeared in the presence of a trainer.

e. Date of return of regular students.

f. During taping, "back" always occurred as "Wanna go back," "I'll be back in 2 hours," etc.

g. Date at which trainers heard something like "sock."

h. Alex started to eat less bread on 8/16/88. By 8/26, he ignored it, but began to eat a small amount again after students returned.

i. "Be" occurred often as part of "You be good!" "I'll be in tomorrow," etc.

j. Refers to any "bean"/"green bean" combination.

counted 11 vocalizations such as "green" and "bok," but found only 2 and 7, respectively.[5] "Be" and "back" appeared more frequently than expected, but may have been special cases: Just before leaving for lunch or dinner (just before taping began), trainers always told Alex, "You be good!" and "I'll be back in 2 hours" or "I'll be back tomorrow."

2. In all cases, vocalizations related to targeted labels increased in monologues during and just after training. We also found utterances

we had never heard in either private or social speech or during training, such as "awale," "sock," "graed," and "ch-bean." Certain labels already in Alex's repertoire, such as "green" and "cracker," also increased in private speech at this time, such that "green" (but not "cracker") was as frequent as expected by chance (Pepperberg et al. 1991).[6]

3. Alex uttered labels related to targeted vocalizations in monologue periods that differed from those of Experiment I. He produced most "none"-related utterances in the first session immediately after training began; in contrast, he produced the greatest number of "nail"-related utterances ("banail," "benail," "blail," "chail," "jemail," "lemail," "lobanail," "loobanail," "mail," "shail," "wgail") for a single session (55 observed; 23 expected by chance) on 7/18/88, *after* the first production of "nail" in our presence. No single session had a noticeably larger number of "sack"-related utterances, although the highest percentages were in sessions just after these utterances were first heard by trainers, on 7/22/88. Only for "graed" did Alex behave as in Experiment I: The most "graed" in a single session (76 observed; 3 expected by chance) occurred 6 days after "bread" training began and 2 days before it was heard by a trainer, on 7/26/88. Only three taping sessions occurred between introduction (7/28/88) and production (8/03/88) of "green bean," and the single practice vocalization ("ee/ea") during that period could have been a corruption of "green" as well as an approximation of "bean."

4. In all cases, Alex practiced new labels in monologues before producing something *reliably* for all trainers, but practiced only some novel vocalizations in monologues before his *first attempt* in a trainer's presence (Pepperberg et al. 1991). Both "awale" and "graed" appeared on tapes 2 days before we heard anything related, respectively, to "nail" or "bread" in training, and "ss" sounds were added to several existent vocalizations (e.g., "ssreally") in monologues 6 days before we heard "sock" (his first real attempt at "sack") in social speech.

5. Monologue sessions with the most vocalizations related to targeted labels also included the longest uninterrupted strings of these sounds (Pepperberg et al. 1991). Thus for 7/18/88, we found "banail chail mail," "mail chail banail," and "shail banail," as well as "jamail good boy banail" and "banail all right blail." Strings were present in other sessions, but not to the same extent. Similarly, on 7/26/88, Alex produced strings of up to 11 repetitions of "braa" mixed with "graed." Moreover, in the monologue after we deemed "green bean" reliable (8/16/88), he produced uninterrupted strings such as "keen green graeen," "bean green graeen bean," and "keen green bbbbb."

6. As in Experiment I, Alex's overall use of English utterances per session increased after training began on the first targeted label. This increase was maintained during initial training for each label, then decreased until students returned for autumn classes (see below). The most English utterances in a single session (940) occurred on 7/14/88, just before Alex produced the first targeted vocalization, "nail," in a trainer's presence (Pepperberg et al. 1991).

Clearly, Alex's training affected his nightly monologue patterns. He often used training labels or related vocalizations in monologues before producing them in his trainers' presence, and always practiced in private before using the labels *reliably* in social interactions. Monologue speech also appeared to increase overall soon after training began on a new concept or label, although this increase could have been caused simply by additional caretaker attention. Several explanations are possible for Alex's behavior; by comparing the amount and type of play/practice in Alex's monologues with that of children, I could determine whether hypotheses for children's behavior might be relevant for Alex.

Amount of Practice in Alex's and Children's Monologues

The average percentages of targeted utterances were low (0.0–5.74%, Table 12.1, parts A and B) compared to percentages of other vocalizations in Alex's monologues. Five factors may be responsible for the relative representation of these utterances. Each factor may have affected Alex and children in similar or different ways (Pepperberg et al. 1991).

First, Alex may not have needed extensive monologue practice because, like children, he was already facile at varying established patterns. Even though he often delayed practice until he incorporated some initial form of a targeted label into his repertoire, he then experimented, like children, with sounds that differed only slightly from those he already produced. First approximations to "nail," for example ("awale" in a monologue and "banail" in our presence), were immediately followed by numerous variations. Similarly, "graed" and "braa" were common soon after "graed" appeared in monologues. Experimentation led to the targeted form fairly rapidly and thus may have removed the need for extensive monologue practice.

A second possibility is that Alex did not practice extensively until he, like some children, "assimilated" and "integrated" new material into his repertoire (Piaget 1954). This possibility might explain differential practice of "nail" and "sack": The former was distinct from most labels in his repertoire, whereas the latter resembled "back," "cracker," and, to some extent, "bok" (how he pronounced "box"). Before practicing "nail," Alex appeared to devise his own "scaffolding" (Bruner 1983; Pepperberg et al. 1991): Early in training, he occasionally responded to a nail with "I want banerry" ("banerry" is his label for apple; see Pepperberg 1990c), but refused the fruit. Although "banerry" and "nail" are related neither physically nor closely phonologically, Alex, in social speech, progressed from "I want banerry" through "I want ban-ail" to "I want nail." He began practicing after producing "banail." Lack of practice on "sack," in contrast, is difficult to explain. Despite the phonetic similarity of "sack" to labels in his repertoire, and our em-phasis of these similarities, he seemed unable to use these labels as a basis for either practice or an initial approximation. Instead, he added "ss-" to utterances such as "really" and "truck," possibly working from the /k/ in the latter case; even then "sstru" and "sstruck" occurred only three times. Conceivably, Alex did not consider "back" as a separable unit in "Wanna go back" or "I'll be back in 2 hours." This inability to produce a "sack" precursor, for whatever reason, might have been responsible for his lack of sound play/practice on that label—and, pos-sibly, his ultimate failure to produce the label reliably.

A third similarity between Alex and children that might account for the small percentage of sound play in his monologues is related to his repertoire size when this study began. Once children have acquired a large number of labels, sound play in their monologues decreases com-pared to other forms of verbal play (Craig and Gallagher 1979; Nelson 1989). By now, Alex had a moderate repertoire (Pepperberg 1987c),[7] and his monologues contained other forms of practice comparable to those of children with repertoires of similar size (see below).

A fourth reason for the small amount of practice is that Alex may have needed less practice to produce labels containing familiar sounds (e.g., "none," "green bean") than labels with entirely new sounds or sound combinations. This interpretation suggests that Alex, unlike chil-dren (see, e.g., Piaget 1954), experiments least with familiar sounds. Interestingly, Alex's percentage of "bread" attempts ("braa," "graed") was among the highest for any training period. To support such an interpretation, however, we would have to discount production and variation of labels that existed prior to training, and consider practice of intermediate forms (such as "banail") as practice of unfamiliar sound patterns. This interpretation thus seems unlikely, but cannot be dis-counted.

Fifth, conceivably, we recorded only a portion of Alex's actual prac-tice. The first piece of a short utterance or phrase such as "un" might have been cut off by the VAR in Experiment I, but not by the VOR in Experiment II. The higher overall practice in the second experiment thus might have been a consequence of the improved system, but the continued small percentages of targeted labels suggest that technology was not the major factor responsible for the small amount of recorded practice.

Type of Practice in Alex's and Children's Monologues

Although the amount of Alex's practice was important, of greater interest are analyses of how periods and types of practice affected label acquisition. Specifically, Alex practiced *during* training, and in ways that suggest intriguing similarities to and differences from the behavior of children. These comparisons are clearest for labels that he practiced most: "none," "nail," and "bread" (Pepperberg et al. 1991).

Alex, much like children, practiced "none" in consequence-free monologues almost 7 weeks before actively using the label in social speech. His manner of practice, however, seemed to differ from that of children. During monologue play, children notice differences *and* similarities between adult utterances and their own imitative attempts and actively modify their vocalizations toward the adult standard (Kuczaj 1983 and refs. therein). Unfortunately, we can make no strong claims for modifications in Alex's monologues: In all but one instance, "one/un" activated the VAR and, because of the onset delay (sound level and the 0.008 s setting), attempts at "none" might have been recorded as "one/un." In social speech, however, Alex did not modify "one/un": His first social attempt at "none" (in response to his trainer's "What's same?") was "one." He used "one" rather than "none" in training for another week, then continued using "one" interchangeably with "none" for 4 more weeks before using "none" reliably with all trainers. Although Alex first used variants of "none" in a consequence-free context, he might have been reviewing or resurrecting an existent related form rather than, like children, actively modifying this form or practicing the targeted vocalization. Had he been actively upgrading his utterances, we would have observed productions like "nnnn-won" or "nnnnnn" in social speech.

Alex's behavior was noticeably different for "nail." Little practice of "nail" existed in monologues (0.09%) prior to an attempt in our presence; however, immediately after uttering "banail" in social speech, he began both to practice in private and to modify this utterance toward the standard form. His vocalization strings (e.g., "banail chail mail") were comparable to those of Weir's (1962) son in monologues. Alex practiced for 10 days before producing "nail" reliably for all trainers. After 6 more weeks he used the label itself, rather than variants, in monologues; he produced "nail" with greater frequency (12 expected, 36 observed) than expected for a random label and with a percentage frequency (1.21%) comparable to that of other novel vocalizations.

Alex's practice of "bread" at times resembled that for "none" and at others that for "nail." As with "none," Alex practiced, in monologues, intermediate forms ("braa," "graed") before using them in social speech, but he practiced for only 2 days. As with "nail," practice sessions were relatively long, but for "bread" Alex modified intonation more than phonology. Unlike his production of "nail," which eventually matched standard English, "bread" stabilized for a time halfway between "braa" and "graed," with "graed" used most often. Decreased interest in eating bread may, however, have affected his motivation for acquiring its label. After we substituted muffins as the referent, his production matched (at least aurally) that of his human trainers (note Chapter 15).

Alex's Forms of Practice Other Than Sound Play

Although our analyses concentrated on sound play, Alex's monologues in both experiments contained advanced behavior similar to patterns observed by Dore (1987) in at least one child's crib speech. The child, taped between 22–36 months of age, engaged in private "dialogues": She did not reproduce caretakers' vocalizations exactly or as isolated units, but instead recreated entire scenarios that had occurred with her caretakers (cf. Kuczaj 1983). Alex's behavior was not identical to that of Dore's subject; Alex did not integrate into his monologues the context or form of training dialogues (Pepperberg et al. 1991). But, both in private and social-context speech during Experiment I, he uttered "What's same?" for which "none," as well as other category labels ("color," "shape," "matter"), was an appropriate response. Unlike the child, however, Alex did not use question and answers in the same utterance set. (An utterance set consists of vocalizations separated by < 1.5 s silence, the VAR offset criterion.) Only once, on 11/3/86, after he had begun using "none" in his trainers' presence, did he even use "What's same?" and "none" (and, later, "matter," "shape," and "none") as a sequential set of utterances (separated by ≥ 1.5 s). Of course, Alex's exposure to adult language was not as extensive as that of the child in Dore's study, and he was not at the linguistic level at which children employ private "dialogue" in monologue speech.

Alex did, however, reconstruct and reinvent scenarios not involved in training the novel labels (Pepperberg et al. 1991). In 1986 and 1988, monologues included recognizable utterances from the daily routine (e.g., "you go gym," "want some water") and strings similar (but not identical) to longer patterns he often heard just prior to being left alone (e.g., "you be good, gonna go eat lunch, I'll be back tomorrow," "okay, you be good, gonna go eat, I'll tomorrow bye"; 9/22/86, 7/12/88). Question and answer dialogues, such as "(snap, snap, snap) How many? Four" (8/05/88), were not present in 1986 monologues, but did exist, if infrequently, in 1988. Although we had not altered our interactions with Alex in those 2 years, his additional laboratory experience and training may have affected his communicative level. Pickert (1981), for example, found that both the percentage of private speech devoted to "dialogue" and the complexity of this "dialogue" increased as her child's language and social competence developed. Interestingly, a juvenile Grey parrot, Kyaaro, subsequently included in our studies (Chapter 14), not only engaged in monologues, but also included entire dialogues in his practice: He reproduced voices of two different trainers, as well as his own and a synthesizer, in exchanges such as "Listen, Kyo (trainer 1)." "Click, click, click, click (synthesizer)." "How many? (trainer 2)." "Four (Kyaaro)." "Good boy!!! (trainer 1 or 2)." At the time, his linguistic level was far less sophisticated than that of Alex.

Even in 1986, Alex experimented with "none" in the context of familiar labels if not familiar situations. Such behavior suggests that he used our scaffolding—modeling "none" on "n" + "one"—as do children whose caretakers introduce new vocalizations within the context of existent forms (Vygotsky 1962; Cross 1977; Kuczaj 1982b). Use of scaffolding was clear in 1988, when Alex seemed to devise connections if we did not provide them. Even during training on "sack" (which he never acquired), he appeared to engage in scaffolding by adding "ss-" prefixes to existent labels.

Overall, Alex's monologues seemed less similar to those of children for whom sound play is most important, and most similar to those of children who are decreasing sound play in favor of private "dialogue" and social-context speech. He often, but not exclusively, practiced vocalizations first introduced in a setting that contained consequences for errors in the consequence-free context of private speech (Pepperberg et al. 1991). He privately practiced "none" well before he produced the label socially. For other labels, sound play/practice occurred just before the label was first used in social speech or was reliably understood by trainers. The percentage of sound play in monologues was small, as it is for children who have similarly acquired a fairly large repertoire but not adult competence. And like children in the early stages of language acquisition, Alex was beginning to reproduce particular training scenarios (e.g., imaginary "dialogues" with trainers) in monologues.

Implications of Our Findings for Work with Children and Other Species

Any correlation between an increase in Alex's vocal play and an increase in task variety and trainer input might be important given ongoing disagreements about the effects of caretaker input on human language development (see, e.g., Snow 1979; Collis 1981; Shatz 1982; Gleitman, Newport and Gleitman 1984; Furrow and Nelson 1986; Harris et al. 1986; Velleman 1987; Morgan et al. 1995; Nelson et al. 1995; Bohannon et al. 1996; Locke and Snow 1997; Goldin-Meadow 1997; Bloom et al. 1998; Huttenlocher 1998). Separating specific training effects from those of general increased interaction is, however, difficult (Pepperberg et al. 1991). Although Alex's monologue speech increased 30.8% overall during training on "none," which necessarily increased caretaker input, his increase in utterances per night also coincided with students' return from summer vacation. Their return meant more people in the lab and a greater variety of tasks, and could have provided stimulation equal to that he received in training. Thus we carefully analyzed the 1988 data, which included material after training on novel labels had been mostly completed, and material from sessions conducted both before and after the return of students.

Data from 1988 show that training and the return of students each doubled the number of English utterances in Alex's monologues (Pepperberg et al. 1991). Increases in monologue speech might thus be related to general vocal environmental enrichment rather than something specific to label training. Given the assumption presented earlier that vocal play/practice facilitates acquisition (e.g., Kuczaj 1983), increased practice could imply that Alex's acquisition and use of labels is related to the overall level of vocal interaction he experiences (see also Dodd 1972).

Further study of Alex's and also our juvenile parrots' monologues might suggest additional research topics. If monologue speech is indeed one of the few opportunities for consequence-free language practice, then utterances experienced without the negative consequences of training (e.g., those heard informally but repeatedly from trainers) are more likely to appear in our birds' repertoire without prior monologue practice than labels we specifically train. Such a result would suggest correlations between a learner's use of monologue and the style of caretaker input (Velleman 1987; Yoder and Kaiser 1987). Reexamination of data from studies on chimpanzees trained in American Sign Language (ASL) might thus prove interesting: Did Nim, who first received repetitive, corrective classroom drills (often with trainers not fluent in sign; Terrace 1979b) but later simply observed interactions between fluid and fluent signers (O'Sullivan and Yeager 1989), engage in monologues on labels taught in the former case but not the latter? Similarly, although dolphins may learn to reproduce human-designed whistles more readily in a situation without specific training (Reiss and McCowan 1993) than in one involving operant methodology (Richards, Woltz, and Herman 1984; Sigurdson 1989),[8] no data have been presented on monologue practice. Similar studies might be useful for children: Dialogues using "scaffolding" might at first seem helpful for learning (e.g., Bruner 1983), but the value of this scaffolding might need reevaluation if input also included negative feedback that engendered an intermediate step of consequence-free monologue practice prior to acquisition (Feldman 1989; Nelson 1989). Might learners skip monologue practice (and progress even faster) if caretakers refrained from rigid instruction and instead provided a consequence-free environment of carefully graded input? Such studies remain to be done, as well as studies involving the effect of introducing new labels that sound similar to existent ones (e.g., do they encourage acquisition, or cause confusion because of their phonological similarity?), and of introducing new labels that involve concepts similar to those already understood (e.g., is it easier to introduce shape terms after establishing the concept "color"?).

Of interest, too, is whether motivation affects practice as well as acquisition. Acquisition may be motivating merely because it allows increased interaction with caretakers, but Huttenlocker (1974) proposed that the presence of specific labels in a child's lexicon may reflect the salience or interest value of their referents. If motivation for label acquisition is reflected in increased practice, a correlation could exist between salience/interest and amount of practice. We tested for such a correlation in Experiment II by training labels for referents that were probably of differential interest (Pepperberg et al. 1991).

Given that Alex's preferences and dislikes were clearest for edibles (see, e.g., Pepperberg 1981) and that his reward for identifying an object is generally that item, his willingness to identify edibles was usually related to his momentary desire for that food and was why we had never previously trained food labels. Such differential interest was, however, necessary in Experiment II, and we thus trained "green bean" and "bread." Our data (Table 12.1, part B) suggest that motivation may indeed affect the frequency with which Alex practiced a label in private speech: He stopped producing "braa" and "graed" about the same time he stopped eating bread, but he continued to eat green beans and to practice sounds related to that label; he also later produced clear renditions of "bread" when we introduced muffins. Reexamination of monologues for emotional content might provide additional insights on use and frequency (note Dore 1989).

Moreover, parallels and differences between human and psittacid monologues may exist for the monologues of other birds, and at levels other than sound play (Pepperberg et al. 1991). Not only Alex (Pepperberg, Matias, Brese, and Conn, unpubl. data), but various songbirds engage in behavior seemingly comparable to social-context monologues, which supposedly represents a more advanced developmental stage than sound play. As noted earlier, juvenile Bewick's wrens and white-crowned sparrows may produce adult versions of song in the presence of adult conspecifics before all adult functions of these songs could reasonably be used by the singer (Kroodsma 1974; Baptista 1983). No one, however, has examined specific *functions* of this behavior. Given that many oscine birds practice and refine the order of notes in their songs as well as the structure of these notes and the order of their songs (see, e.g., Todt and Hultsch 1998), and that for some (but not all) birds note order is important for meaning (see the review in Pepperberg and Schinke-Llano 1991), the purported roles of various forms of monologue speech in human language acquisition (e.g., as a means of analyzing and refining syntax; see Craig and Gallagher 1979; Kuczaj 1983) may provide a framework within which to investigate the role of similar practice in birds (see, e.g., Marler and Pickert 1984).

In sum, Alex clearly engaged in vocal practice that probably has origins in the behavior of wild birds and that also demonstrates some parallels with young children's behavior. Our data thus suggest that vocal play is relatively widespread, and that such behavior may be basic to many forms of communication, possibly even forms that are nonvocal.[9]

13 Can a Parrot's Sound Play

Be Transformed into

Meaningful Vocalizations?

Interpreting infants' word-like vocalizations as intentional and meaningful is a normal, almost unavoidable adult activity. Adults also universally provide infants with simplified forms that typically serve as models for the first conventional words, forms that often serve little communicative function other than allowing children to participate in social interactions. Adults are also very likely to respond to child vocalizations with imitation or with "expansions" into more conventional adult-like forms.

John Locke and Catherine Snow, *Social Influences on Vocal Development* (1997:274)

In the previous chapter, I noted two forms of learning that occurred in the laboratory outside of formal training. The second form involved what happened to Alex's sound play—spontaneous recombinations of existent labels or parts of labels—when he was given objects to correspond to these utterances. Under such circumstances, he quickly incorporated novel vocalizations into his repertoire. He did not appear to use these spontaneous utterances to describe or request novel items or situations, but I could not assess his intentionality. Of interest, however, was the similarity of his behavior to that known in children as "fast mapping" (Cary 1978) or "referential mapping" (Rice 1980): the ability to connect a novel object with a novel utterance (note Clark 1987; Dollaghan 1987; Tomasello, Strosberg, and Akhtar 1996). I thus decided that detailed examination of this behavior, much like that of monologue speech, might provide insights into Alex's learning processes and provide comparisons with children's behavior.[1]

Had Alex restructured elements of his repertoire in spontaneous, innovative (untutored) ways to describe or request entirely novel situations or items (see, e.g., Fay and Schuler 1980), he would have attained the "Holy Grail" of competence in animal-human communication studies (see Savage-Rumbaugh et al. 1993), because intentional creativity is one of the most difficult aspects of communication to train. Clinicians helping dysfunctional children, for example, may successfully train an understanding of language elements and rules for their combination (Parkel and Smith 1979; Leonard et al. 1982; Reichle, Rogers, and Barrett 1984; Schwartz and Leonard 1985; Dollaghan 1987), but fail to teach intentional creativity, and some subjects in intervention programs never learn to use spontaneous, novel communication. Such creativity, however, is common even in the early stages of normal human language development (see, e.g., Marschark 1987; Clark 1998).

Whether Alex or any animal can engage in such behavior is a matter for study and debate; at present, little evidence exists for intentional *creativity* in the wild or in the laboratory (cf. Griffin 1984). Animals do combine signals, apparently intentionally and in rule-bound ways, to express different messages: Eastern kingbirds use one vocalization in connection with different displays or locations to connote different messages (Smith 1986); syntactic combinations in pigeon guillemots *(Cepphus columba)* relate to message content (D. Nelson 1985; see Snowdon 1988 for material on primates). At present, however, little evidence exists that these combinations are *intentionally innovative*—that *novel* combinations are created for *novel* situations. Species have not been tested for their ability to combine existent responses to the appearance of a "novel" happenstance, such as a normally flightless predator seen overhead. Only single instances have been reported of what appear to be intentionally creative uses of material already in a repertoire. A Florida scrub jay *(Aphelocoma coerulescens)* once combined hawk and snake alarm calls in the presence of a perched owl (Hailman and Elowson, pers. comm., December 1984); a vervet monkey emitted an alarm call when losing a territorial confrontation with a neighboring troop, causing participants to scatter (Cheney and Seyfarth 1990). Even under controlled laboratory situations, however, data are not much clearer.

Most reported instances of innovative, productive combinations involve abstract symbols manipulated by laboratory animals taught to use human-based communication codes (Gardner and Gardner 1971; Premack 1976). Bottlenose dolphins that learned to associate individual computer-generated whistles with specific objects and activities (e.g., ring, ball, rub) produced novel "ring-ball" combinations while engaging in novel concurrent play with both objects (Reiss and McCowan 1993). Other examples involve chimpanzees and are few in number, such as the combination of ASL signs to form "water bird" for swan and "cry hurt food" for radish (Fouts 1974), and the combination of computer-generated symbols to form "Coke which-is orange" for

Fanta (Rumbaugh and Gill 1977). Pygmy chimpanzees *(Pan paniscus)* also combine symbols, but most combinations are of abstract symbols with iconic gestures (come, chase, pat), rather than combinations of two or more abstract symbols (Savage-Rumbaugh et al. 1986). In many cases, determining these signals' degree of intentionality, innovation, and referential content (the degree to which signals stand for, refer to, or represent some object or event) is extraordinarily difficult. Interpretations vary: Some researchers dismiss novel (re)combinations of items in an animal's repertoire as "babble-luck" (Thorndike 1943); others suggest that the behavior reflects intentional, innovative productions (see Ristau and Robbins 1982; Dennett 1983; Harré 1984; Harris 1984). Unfortunately, detailed documentation is often impossible in the course of fast-paced training or testing sessions. Even when sessions are well documented, as in the dolphin study, continuous videotaping outside of sessions to ensure that, for example, ring-ball combinations are never produced in the absence of concurrent ring-ball play, is prohibitively expensive in both cost and transcription time, and sampling cannot, by definition, provide complete documentation (Ristau and Robbins 1982).

On first glance, the above material implies that the combinatorial creativity observed in all children learning language (e.g., the two-word stage, Brown 1973), which seemingly reflects or coincides with certain levels of cognitive development (Bates et al. 1991; Nelson 1996), is not widespread in nonhumans. Nevertheless, the limited examples given above suggest some capacity for intentional creativity. Interestingly, the frequency of laboratory animals' spontaneous behavior appears to be correlated with positive responses from their trainers (Rumbaugh and Gill 1977; Pepperberg 1990c); language-deficient humans similarly attempt more intentional communication if their initial spontaneous actions, whether or not intentional, are referentially rewarded (Calculator 1985; see also Carr and Kologinsky 1983). Menyuk and colleagues (1995) review comparable data for both premature (at-risk) and full-term (presumably normal) babies. Even more interesting may be data that suggest that the *type* of response given to spontaneous combinations might critically affect development (see Chapter 12): Caretakers who actively train communication skills (whether in children or nonhumans) may unwittingly extinguish a nascent capacity for spontaneity by focusing too closely on production of targeted words or actions and by actively discouraging untargeted responses. Such a focus overlooks the possibility that spontaneous, untargeted actions might help expand the range of (or even establish) functional communication: A novel, untargeted utterance might not have any clear immediate communicative intent, but might be a human word whose function could be demonstrated. Caretakers who appropriately respond to subject-initiated actions, that is, help "map" a novel label onto a specific object or action, might thus teach the referentiality of such material, increase the frequency of its production, and eventually encourage intentional interactions in individuals who are in the process of learning to communicate (see Fey 1986).

Types of "Mapping"

To determine how best to respond to Alex's novel combinations, I studied child language acquisition data (Pepperberg 1990c). I was intrigued by circumstances described by Carey (1978) as leading to "fast mapping" versus "full mapping" during acquisition of word meaning by normal children. Interactions leading to "fast mapping" provide initial information that the new vocalization is a word and demonstrate something about its semantic and syntactic properties (see, e.g., Snow 1979; Rice 1980; Dollaghan 1987). Such interactions establish "associative speech" (Chapter 3), for which only a correlation exists between an utterance and (usually) the specific object or action to which it refers (Lock 1980). Rice (1991) refers to a similar situation as QUIL (quick incidental learning). In contrast, interactions leading to "full mapping" provide experience in using novel utterances in many contexts and lead to "representational speech" (Carey 1978), in which comprehension and production are comparable to adult usage. Representational speech occurs only after a subject learns to integrate linguistic and nonlinguistic contextual information about novel utterances, and is able then to use these utterances with respect to a large assortment of referents and situations (Chapter 3). The process by which children (and probably nonhumans) acquire representational speech is slow and not preordained because it requires learning outside the context of first use (Nelson 1996).

Procedures I designed for Alex would probably produce a situation somewhere between "fast" and "full mapping," and I use "referential mapping" (Pepperberg 1983b, 1990c; after Rice 1980) to refer to both the technique and the training outcome. Full mapping was unlikely because Alex's laboratory experiences were not adequate to allow him to acquire word meaning comparable to that of adult humans. Extensive interactions with trainers, however, ensured comprehension and production beyond fast mapping (associative speech), and appeared to encourage spontaneous experimentation. Referential mapping thus supplements other techniques for training functional label use and has been the primary means of managing Alex's and now our other birds' spontaneous interactions.

Referential mapping enabled Alex to attach meaning to two forms of novel utterances: (1) spontaneous combinations of two or more labels already in his repertoire, and (2) recombinations of parts of such labels. In humans, the latter vocalizations are called phonemic recombinations. Whether Alex comprehends and actually divides his vocalizations at the level of human phonemes is unclear (Pepperberg 1983b; Patterson and Pepperberg 1994, 1998), but the divisions appear as such to human listeners.[2] The following sections describe the mapping procedure and report on that portion of Alex's repertoire acquired through referential mapping.

The Referential Mapping Procedure

Referential mapping involves three procedures. First, we respond to Alex's utterances as though they were all intentional; second, we use a modification of M/R training to extend the contexts in which he experiences the vocalizations; third, we use "sentence frames" to demonstrate how new utterances might be used. The value of these procedures is best appreciated by examining their use in the laboratory (Pepperberg 1990c).

Data on both humans and birds suggest that experiencing appropriate consequences of an utterance may assist learning (i.e., "contextual applicability" or "functionality"; see Chapter 3); thus all trainers respond to Alex's novel expressions with appropriate objects or actions—as if he indeed understands the significance of his utterance. Whether the utterance *is* intentional is unimportant. Our aim is to demonstrate that these phrases *can* be meaningful and that vocalizations *can* be used to control, or at least influence, Alex's environment and his caretakers' actions.

Our work is based primarily on studies on children, although similar strategies probably hold true for birds (see below). For children, language competence advances when adults interpret and respond to utterances as intentional even before evidence exists to support such intentionality (see, e.g., R. Clark 1978; Lock 1978; Furrow and Nelson 1986; Girolametto 1988; cf. Gleitman et al. 1984). Snow (1979), for example, describes how children's first words develop reference in part through a caretaker's assignment of meaning to spontaneous utterances. Similarly, Veneziano (1988) proposes that a child's utterance that lacks language value at the moment of production can acquire value later if caretakers interpret the utterance as meaningful and the child reacts positively to the interpretation. Repeated interactions then "conventionalize" the sounds (phonemic patterns) and sound-meaning connection (referential aspects) toward standard communication. Blank (1974) even suggests that children use incompletely understood terms to get adult feedback about the exact meanings of these utterances (note Leonard et al. 1983). This strategy is related to Brown's (1958) "original word game," in which a child designates an object and caretakers provide the object label and correct the child's attempts to produce this label. More recently, Rice (1991) and Nelson (1996) describe the extents (and limits) of learning that such interactions engender.[3]

Analogous mechanisms are apt to exist for birds. Young birds, through interactions with adults, probably learn not only what to sing but also how song is to be used: Juvenile Bewick's wrens and white-crowned sparrows may produce adult songs in the presence of adult conspecifics before the songs could reasonably be meaningfully used (e.g., for territorial defense or mate attraction). Also, inexperienced singers apparently alter their songs on the basis of the behavioral input of adults (Kroodsma 1974; Baptista 1983; West and King 1988; Todt and Hultsch 1998; see Pepperberg 1985, 1988c).

Alex's mapping procedure also included a variant of M/R training (Chapter 2), whereby my students and I demonstrate further the possible relevance of his spontaneous combinations (Pepperberg 1990c). We thus model an interaction corresponding to the novel and now-targeted vocalization: One of us produces the novel utterance while the other produces an object or an example of the term (e.g., for colors) or demonstrates the action to which it refers. We then reverse roles, to demonstrate that the utterance is neither specific to nor controls only a particular individual's actions. If Alex emits the vocalization during our demonstration, he is shown (and may receive) the object or action (e.g., a "chain" of paper clips). Trainers model object or action identification (by responding to "What's this?"), and, when possible, also demonstrate varied applications of the utterance (e.g., show Alex that "box" comes in several colors). Using a variety of concrete examples and social modeling positively affects human language learning (Brown 1976; Nelson 1996), and appears to facilitate specific types of avian vocal learning (review in Pepperberg 1985, 1988c). Conversely, programs that do not referentially reward spontaneous imitation or that encourage nonreferential imitation (e.g., "say after me, 'ball'!") do not teach a subject how to use these imitations in subsequent referential speech (Courtright and Courtright 1979; McDade,

Simpson, and Lamb 1982; remember Mowrer 1950; Grosslight and Zaynor 1967).

Finally, human trainers provide additional cues about the appropriate context in which to use novel object or action labels (Pepperberg 1990c). While the targeted item is manipulated or the action demonstrated, either by a trainer or by Alex, humans produce "sentence frames" such as "You're eating a *green nut!*" "I'm holding a *green nut!*" "Want another *green nut?*" in which only the label for the action or object remains fixed and stressed. New vocabulary items are often presented to children in much the same way (Berko-Gleason 1977; de Villiers and de Villiers 1978; K. Nelson 1985; Tomasello et al. 1996).

Testing

Before testing Alex on labels we had mapped, we required that he produce the utterance, whether or not its referent was present, without phonological variation for 3 weeks. We then formally tested him to see if labels for nonfood items had indeed been mapped to the appropriate object, that is, had referential meaning (Pepperberg 1990c). We asked Alex to identify both training objects and somewhat novel items: familiar objects that, for example, featured newly acquired colors. We did not test food labels: Alex's willingness to identify food is related to momentary desire; his food preferences change rapidly, and this changeability precludes testing labels of edibles. Circumstantial evidence (refusal of foods other than those requested and subsequent requests for original items, Pepperberg 1988a) nevertheless suggested that food labels had acquired some contextual meaning. We used test procedures described previously.

The Results of Our Mapping Procedures

We successfully mapped several spontaneous vocalizations onto specific objects or actions. These mapped utterances account for 2 of 7 (28.5%) color labels, 12 of 15 (80%) food labels, and 11 of 25 (44%) nonfood object labels then in his repertoire.[4] He did not, however, acquire any abstract labels (e.g., for categories or numbers) through mapping. Formal tests on nonfood items (Table 13.1) showed that Alex had indeed learned to use their labels accurately. Although we did not test food labels, I list the food labels we mapped in Table 13.2. Test objects could vary from training items; thus we asked Alex to identify as "rock" stones of various shapes and sizes, "chain" constructed out of different numbers, colors, and sizes of both paper clips and shower

Table 13.1. Test results for labels of nonfood items trained with referential mapping. Scores are the number of correct identifications Alex gave divided by the total number of presentations. Chance values varied from 1/30 to 1/50 because Alex could have produced any of the 30–50 labels in his repertoire during any of these tests. He could also have produced any number of combinations of these labels or any number of other, contextually applicable vocalizations (e.g., "I want X"), but these were not included in the calculation of chance. Peg wood is a clothes pin. (After Pepperberg 1990c.)

Object	Score	%	Erroneous i.d. (no. of times made)
Showah (shower)	28/32	87.5	Water (4)
Rock	28/31	90.3	Ring (1), truck (1), chalk (1)
Grate	25/31	80.6	Grey (3), grain (1), truck (1), chalk (1)
Cup	23/29	79.3	Key (2), box (1), grey box (1), key chain (1), kah (1)
Chain	20/23	87.0	Grain (1), key (1), key chain (1)
Key chain	19/22	86.4	Truck (2), chain (1)
Box	16/20	80.0	Cup (2), sock (1), truck (1)
Chalk	12/15	80.0	Bock (1), truck (1), rock (1)
Wool	8/10	80.0	Wood (1), blue (1)[a]
Truck	8/11	72.7	Chalk (2), box (1)
Tray	6/7	83.4	Key (1)
Grey wood	22/28	78.6	Wood (3), green wood (2), U wood (1)[b]
Grey hide	22/28	78.6	Hide (4), U hide (2)[b]
Grey peg wood	22/28	78.6	Grey wood (3), green peg wood (1), peg wood (2)
Orange wood	10/15	66.7 (81.8)[c]	Wood (2), rose wood (1), yellow wood (1), U wood (1)[b]
Orange key	9/15	60.0 (76.9)[c]	Key (4), yellow key (2)
Orange hide	8/11	72.7 (80.0)[c]	Hide (2), yellow hide (1)
Orange peg wood	8/10	80.0	Rose wood (2)

a. "Blue" was at the time pronounced as "boo."
b. U represents an unintelligible label.
c. Numbers in parentheses represent scores if "generic" errors are not considered. Generic responses are correct identifications of the object without the color marker.

curtain rings, and "cup" and "box" of any size or color and made of paper, plastic, or metal. Alex therefore demonstrated some level of referentiality with respect to these labels.

Alex's spontaneous utterances relating to food were often difficult to map, and we could not judge in advance when mapping would be

Table 13.2. Labels for food items and action-related phrases produced by Alex without training. (After Pepperberg 1990c.)

Still in repertoire	No longer in repertoire
Food	
Rock corn (dried corn)	Carrot nut
Wheat	Green nut (pumpkin seeds)
Grape	Banacker/banana cracker
Grain	Corn nut
Carrot[a]	Parrot nut (pecan)
Popcorn	Chirwi (kiwi)
Citrus (orange)[a]	Rock nut (Brazil nut)
Banerry (apple)	
Grey nut (sunflower seed)	
Cracker	
Cherry	
Action phrases	
Wanna go X	
Go away	
Come here	
Go chair	
You tickle me	

a. Label acquired via Alex asking, "What's this?"

successful (Pepperberg 1990c). How frequently he produced *any* spontaneous vocalizations and how quickly these utterances stabilized in his repertoire did, however, seem to be correlated with the rapidity of his trainers' responses and the desirability of the item or action chosen as a referent. We could not quantify these correlations because Alex's preferences, particularly for foods, changed rapidly. What follow, therefore, are subjective descriptions of how some labels were mapped. The descriptions demonstrate the flexibility of mapping and are not intended as interpretations of Alex's behavior.

Alex's Spontaneous Combinations of Object Labels

Although we could not determine whether Alex's spontaneous combinations were intentional (see, e.g., Dennett 1983), referential mapping allowed us to expand his repertoire without engaging in debate on this issue. Some behavior patterns were nevertheless intriguing (Pepperberg 1990c). For example, during winter 1982 we began training Alex on concepts of "hard" and "soft," using, respectively, dried and fresh corn. Alex had recently acquired "rock" for lava-stone beak conditioners (see below). Subsequently, in the absence of any type of corn, he spontaneously uttered "rock corn," and we gave him a dried corn kernel. He ate the corn, has since requested dried corn in that manner, and refuses dried kernels if he requests "corn" (see, e.g., Pepperberg 1988a). Had he abstracted the concept of hardness from the "rock" and transferred that concept to the dried corn, or had he merely uttered a fortuitous combination of sounds, which, when rewarded, became part of his repertoire (e.g., the "babble-luck" concept of Thorndike 1943)? The reason for his behavior was, however, not important; our goal was to map such vocalizations. Other utterances spontaneously entered Alex's repertoire and remained for various periods, apparently depending upon his interest in the item chosen as referent. He spontaneously produced "rock nut" and received a Brazil nut. He was unable to crack it by himself, did not eat the nutmeat when we cracked it for him, and lost interest in the item fairly rapidly. He rarely produced "rock nut" after that incident and, unlike "rock corn," the term is gone from his repertoire. We provided (1) shelled pumpkin seeds in response to "I want green nut," (2) dried banana chips for "I want banacker/banana cracker" ("banacker/banana cracker" were used interchangeably), (3) a candy-coated peanut for "carrot nut," and (4) commercially available "corn nut" snacks when he uttered that combination. He rarely ate any of these foods, and the vocalizations vanished (Pepperberg 1990c). We responded to "pah-corn" (a variation of "pah-uh" [pasta] plus "corn") with puffed corn and a distinctly enunciated "popcorn"; the food was a favorite, and "popcorn" is now in both his diet and his repertoire. Sunflower seeds (a portion of his standard feed at the time) were given in response to "grey nut," and the term remained in his repertoire for as long as we fed seeds.[5]

Interestingly, children, like Alex, often exhibit a correlation between interest in an item and retention of its label. Huttenlocher (1974) proposes that a child may discard words from his or her lexicon because their referents subsequently decrease in salience or interest, not because of inadequate storage ability or lack of mental representation. (Note studies of language loss because of disuse, such as Berman 1979.) Alex may react similarly, but proving that underlying mechanisms in child and parrot are homologous is unlikely.

Alex's Spontaneous Production of Action Phrases

Alex also produced spontaneous combinations involving phrases related to actions, although far fewer than those related to objects (Pepperberg 1990c). Children also exhibit this bias toward object labels in spontaneous combinations, though the overall proportions of object versus action words in children's repertoires probably reflect the extent to which caretakers use object versus action labels (see, e.g., Gentner 1982; Maratsos 1991; Hampson and Nelson 1993; Caselli et al. 1995; Nelson 1995, 1996; note Choi and Gopnik 1995 for studies on Korean, and Tardif 1996 for studies on Chinese). Studies on *unsolicited* lexical imitation, however, showed that both normal and language-impaired English-speaking children spontaneously produce significantly fewer action words than object words (Leonard et al. 1982; Schwartz and Leonard 1985; Conti-Ramsden and Jones 1997). Because of our small sample size and the difficulty in devising and evaluating tests for action-related words (see Premack 1976; Leonard et al. 1982; Nelson 1996), I did not test Alex's action labels. Thus I cannot claim that action-related labels acquired the same referentiality as nonfood labels. His use of action terms nevertheless appeared contextually applicable: He used such labels and phrases in appropriate situations whether or not he necessarily comprehended each of their specific meanings (Pepperberg 1990c).

To form most action-related variants, Alex excerpted part of a standard phrase, which he then used as a request; trainers responded *as though* he had made an intentional demand (Pepperberg 1990c). Thus, during a long training session, Alex uttered "go chair"—a portion of the (by then) commonplace phrase "wanna go chair" (see below). The student sat, temporarily discontinuing the session. Alex subsequently used this utterance to call a halt to whatever was happening. Alex also excerpted "go away," which he uses to break contact with trainers. This phrase is closely related to two situations in which it was first routinely used in his presence. The first situation occurred when he refused to comply with our requests for object labeling: He routinely said "no," turned or walked away, and began to preen. We responded with a brief time-out, accompanied by the phrase "I'm gonna go away!" Now he asks ("go away") that the trainer depart. The second situation involved scheduled departures of trainers (e.g., at lunchtime or the end of the day). At those times, we added "I'm gonna go eat lunch/dinner," and placed Alex inside his cage. Now, during long or difficult sessions, Alex may request students to "go away" or he occasionally states, "I'm gonna go away," climbs off his training chair, and tries to leave. After such an interaction, attempts to continue training are usually fruitless.

Alex's Use of Word Segment Recombinations as Innovative Labels

Alex also produced new vocalizations by combining parts of existing utterances. For humans, such (re)combinations are called phonemic.

Again, some parallels may exist with young children's behavior: Both normal and language-impaired children are more likely to reproduce spontaneously words that are consistent with their existent phonology than words involving new phonemes (Leonard et al. 1982; Schwartz and Leonard 1984; also Menn 1983; Messick 1984; McCune and Vihman 1987; Studdert-Kennedy 1991; Baddeley et al. 1998), and these reproductions often have little correlation with the children's levels of comprehension.

Alex spontaneously recombined phonemes to produce new utterances most often during free periods outside of testing and training (Pepperberg 1990c); our juvenile Greys behave similarly (see, e.g., Neal 1996). Such utterances thus appear in contexts similar to those of children's vocal play (Chapter 12). Alex spontaneously emitted utterances that rarely, if ever, were used by trainers but that closely resembled both existing labels and separate human vocalizations—for example, "grain," which appeared after he acquired "grey." In this instance, trainers gave Alex parakeet seed (an item not normally in his diet), talked about this "grain" and identified it for one another; later we substituted granola and finally sprouted legumes. Similarly, we gave him a ring of interconnected paper clips when he produced "chain," provided the appropriate fruit for "grape," and gave him wire mesh (later a nutmeg grater) that could be used to trim his beak after he uttered "grate." "Cup" (from "up") was mapped to measuring cups and plastic mugs, "copper" (first produced as "cupper") to pennies, and "block" to cubical wooden beads. "Chalk" (apparently from "talk") was mapped to blackboard chalk of various colors and lengths, and "truck" to toy cars and trucks. Alex rapidly integrated these vocalizations into his repertoire and used them routinely to identify or request appropriate objects (Table 13.2). He also devised "cane," "shane," and "cheenut," to which we did not respond; he abandoned these utterances fairly quickly (Pepperberg 1990c).

Alex's Single Instance of Possible Intentional Creativity

As noted earlier, in neither human children nor animals can we be certain that creative combinations are intentional. E. Clark (1982) suggests that children routinely create or intentionally reform existing labels to fill gaps or express meanings for which they have no words, but Blank (1974) argues that children produce novel vocalizations merely for the general attention such behavior attracts from caretakers (note Schuler 1979; Sundberg 1983, 1985). Determining animal intentionality is even more difficult (see Dennett 1988). We thus simply repeated Alex's innovative vocalizations and presented novel, possibly interesting objects to increase his overall arousal level and encourage vocal play (Pepperberg 1990c). Also, Alex's innovative labels were (and still are) generally produced in the absence of any specific object or action to which they could refer. He does not create specific, novel labels for novel objects we present, but may instead use the label of an item that shares some perceptual features, for example, call an unshelled almond a "cork" (note E. Clark 1973; Bowerman 1978; Kamhi 1982; Mervis, Mervis, Johnson, and Bertrand 1992; Mervis and Bertrand 1993).

One incident, involving his response to apples, nevertheless suggests that Alex has some capacity for intentional creativity (Pepperberg 1990c). We were examining the effect of another parrot's presence during training and were limited to using a colleague's pet—one that did not talk and would attend only if we used her favorite food, apples. We thus made an exception to our rule against training food labels, and in fall 1984 began training "apple." At that time Alex already used the labels "banana," "cherry," and "grape." During formal sessions, he began to produce a /p/. At the end of the season for fresh apples, he refused these fruits, and his vocalizations remained at this level. We thus removed apples from training and the laboratory. Apples were reintroduced in the spring, were eaten, and /p/ reappeared in the first training session. During the second week of training, however, Alex looked at the fruit, said, "Banerry . . . I want banerry," and snatched a bite. He not only persistently identified the fruit as "banerry" in subsequent sessions, but also slowed production and sharpened his elocution ("ban-*err*-eeee"), much as trainers do when teaching a new label (Pepperberg 1990c).

For five reasons, I suggest that "banerry" was an instance (albeit the only one ever recorded) of intentional creativity (Pepperberg 1990c): (1) "Banerry" was not a vocalization Alex had heard from any human. (2) All previous elisions in his speech ("banacker" = "banana" + "cracker"), had been produced in the absence of a specific object. (3) No one had ever referred to an apple as anything but "apple." (4) Alex had never emitted the utterance prior to training with apples, at least in the presence of humans or during recorded monologues. (5) The label made some semantic sense: Apples taste somewhat like bananas and look like very large cherries.

Referential Mapping and Alex's Queries for Information

My students and I have also successfully mapped many of Alex's requests for information (Pepperberg 1990c). Thus whether or not his queries "What's that?" "What color?" and so forth, are intentional, we treat them as such. Although Alex repeatedly asked about the shape of wooden plant stakes ("long" wood), round objects ("no-corner"), or the label for the board above his gym ("shelf"), he never acquired the appropriate labels. He did, however, learn "grey" by querying a student about the color of his reflection in a mirror (Pepperberg 1983b), and began uttering "rock" after querying us about a lava-stone beak conditioner he repeatedly tossed from the top of his cage. We answered his query ("What's that?") about covers in the Piagetian object permanence study (Chapter 10) with "box"; he produced "bock" (Pepperberg and Kozak 1986). "Bock" and "box" are now used interchangeably to label square or rectangular containers. After asking about the vegetable we were eating and its color, he began to ask for "carrot" and acquired functional use of that label and of "orange." He identified novel "grey" and "orange" objects, on first trials, without subsequent training (Pepperberg 1990c); few errors were made on later identifications (Table 13.2).[6]

Might Referential Mapping Encourage Additional Creativity?

Some studies with children suggest that referential mapping may encourage experimentation with the communication code (i.e., additional language play) and help develop lexical creativity (e.g., Menyuk et al. 1995; Nelson 1996). Such creativity is also called "segmentation" (Marler 1970) or "combinatorial productivity" (Miller 1967). In its strongest sense, lexical creativity, as discussed earlier, is the spontaneous, intentional, innovative combination of items in a repertoire: It reflects the understanding that individual portions of a repertoire can intentionally be combined in novel, untutored ways, based on explicit rules, to describe or request entirely novel situations or entities (E. Clark 1982). Such behavior is the antithesis of "babble-luck" (Thorndike 1943): an accidental collection of single words in an apparently appropriate context (see Ferrier 1978). Separating lexical creativity from less complex combinatorial behavior is difficult. Numerous interpretations exist about the actual intention of the young child who produced the famous example "Mommy sock" (Bloom 1970), and the argument that the child had simply commented upon the contiguity of mother and sock cannot be excluded. True lexical creativity, in contrast, is the ability of an individual to create "Polly want chocolate cookie!" after learning "I'm Polly," "I want chocolate," and "That's cookie" (not chocolate). Lexical creativity, however, is not a unitary phenomenon, and the term often refers to a number (and range) of different behavior patterns (see Pepperberg 1990c). Projects to encourage combinatorial communication in animals and language-impaired humans thus often study the simplest case, that of "lexical substitution."

Lexical substitution requires much less knowledge of the communication code than does lexical creativity (Premack 1976); the former merely involves substituting one element with a related item in a stock phrase. An example is the behavior of a subject who, having learned "Polly want a cracker," subsequently learns "cookie," and spontaneously produces "Polly want a cookie" (Fromkin and Rodman 1974). The subject has merely segmented the original sentence into "objects wanted" (cookie, etc.) and a stock phrase, "Polly want . . ." Lexical substitution is nontrivial because it requires semantic *and* limited syntactic knowledge (Schuler 1980); that is, a subject must, for example, know to substitute a noun for a noun and to maintain word order. An individual capable of lexical substitution, however, is not necessarily capable of lexical creativity. Lexical substitution is an educable skill in nonhuman primates (Premack 1976; Rumbaugh and Gill 1977), marine mammals (Herman et al. 1984; Schusterman and Krieger 1986), and handicapped humans (e.g., Karlan et al. 1982). With respect to two stock phrases, Alex's lexical substitution may, like those of children, have been encouraged by referential mapping. I refer to his spontaneous placement of new labels in the X and Y slots of the stock phrases "I want (wanna) X" and "Wanna go Y."

Alex constantly heard "Wanna go gym?" (a structure of wooden dowels and ropes), "Wanna go back?" (to his cage), and variations such as "You have to go back!" used in conjunction with being moved about by trainers. "Gym" and "back" were, until summer 1980, the only labels used in the "Wanna go . . ." routine because his cage and his gym were the only sites where he would remain calm for testing and training. Trainers responded appropriately to his first ever utterance of "Wanna go gym" when Alex was on his cage. Later, if he used the phrase when on the gym, they responded with comments like "Well, you're already *on* the gym!" Alex began using the phrase predominantly when he was *not* on the gym; he often uttered the phrase while stretching his body in its direction (Pepperberg 1990c). If he was on the gym and stretching toward another location (e.g., his cage), he would alter the phrase somewhat: to "Wanna go gym—no" or "Wanna go . . . Wanna go . . ." If trainers asked, "Wanna go back?" his reply was often a squeaky "yeah." During this period, Alex also began to allow us to place him at other sites (e.g., on a chair or on a trainer's knee or shoulder). We concomitantly uttered sentence frames such as "I'm putting you on the chair," "Here, sit on my shoulder," and avoided using the phrase "Wanna go Y" each time we moved him to these sites. We wanted to see what he would do. Within 3 months, he began using "Wanna go Y" with various Ys, and was always moved to the requested site. Although we did not at first examine the intentionality of these utterances, a limited study showed that he responded to three out of four subsequent attempts to place him elsewhere with "no" and a refusal to leave our hand (Pepperberg 1988a, 1990c).[7] Admittedly, the correlation between our referential mapping of "Wanna go Y" and Alex's development of lexical substitution is not straightforward. Nevertheless, our techniques probably helped establish both the initial association between the phrase "Wanna go Y" and movement and the initial association between the labels and various locations to which they referred.

We also noted likely instances of lexical substitution in Alex's use of "want." Although he was explicitly trained to use "want" to separate requests from identifications (Chapter 11), training involved very few labels. Alex, however, spontaneously placed additional object labels, some even as they were trained, into the X slot in "I want X" or "Want X." Again, I can merely suggest, but not prove, that the positive experiences Alex received in response to his innovative utterances (e.g., increased attention, novel objects) encouraged novel behavior.

How Useful Is Referential Mapping for Expanding a Repertoire?

Although my studies, unlike, for example, those on marine mammals (e.g., Herman 1987, 1988, 1989; Schusterman and Gisiner 1988, 1989), never emphasized syntax—not even word order—Alex's behavior involving "I want X" and "Wanna go Y," and his consistent use of what appeared to be adjective + noun to combine color and object labels, suggest he had learned some elementary patterns. Could analysis of his label combinations for rulelike constructs lead to other techniques to expand his communication? His label combinations did appear to reflect simplistic rules like Braine's (1976:8) "positional productive patterns" or at least a form of analogy discussed by E. Clark (1982): Many innovations were, in addition to lexical substitution in phrases, formed by combining X + nut. Arguing for or against syntax based on these findings would be both premature and unwise, but the prevalence of certain forms in Alex's spontaneous utterances suggests ways (like those used for dysfunctional humans) to expand his repertoire: by basing labels for novel items on combinations of existent labels (Leonard et al. 1982). Given that other innovations involved phonemic rearrangements (Chapter 12), introducing labels similar to existent labels might be useful.

Analysis of Alex's identification errors may also provide insight into both the mechanisms of label acquisition and the effects of referential mapping (Pepperberg 1990c). Alex could have responded to queries of "What's this?" for the various items we tested with any of 30–50 object labels, or any of 50–80 possible utterances—including labels for colors, shapes, and quantities—but all errors were either other labels acquired through referential mapping or mispronunciations of these labels (e.g., "sock" for "bock" [box]; Table 13.1). Might Alex be classifying labels to some degree according to how they were acquired? A subsequent study on his capacity to label photographs of a select group of objects he could already identify (Stanford 1995; Pepperberg 1996) supports this idea: Alex made a new category, "picture toy," for these labels, and almost all errors were labels in this category. Data from songbirds also may support this hypothesis: The conditions under which blue-winged warblers *(Vermivora pinus)* acquire their songs affect, to some degree, the context in which they later use the songs (Kroodsma 1988). Additional studies of context-dependent learning would be intriguing, as would other experiments on how nonhumans form categorical and phonological classifications.

Interestingly, referential mapping may not encourage production of action and descriptive labels (Pepperberg 1990c). In contrast to the many object labels Alex produced in spontaneous utterances and we mapped, few spontaneous utterances involved human adjectives or action labels. Conceivably, Alex, like children, is less likely to produce spontaneously action and descriptive labels than object labels (see the references above; Tomasello et al. 1997). His dearth of action labels may, however, be a consequence of training and his environment. Alex's input is mostly limited to what I deem necessary for experiments, which primarily require object or category labels. Providing situations for action labels might be helpful, but such situations are difficult to design (Rice 1991): Even for children, emerging verbs may be used to label events (cf. Pinker 1989), and planned activities can be interpreted in ways not expected by a researcher (see, e.g., Gleitman 1990): "Give," for example, may also demonstrate "take." Nevertheless, we might obtain data on nonhuman learning strategies by studying how attempts to map actions or adjectives might affect Alex's spontaneous combinations.

The Implications of Our Success in Using Referential Mapping to Expand Alex's Repertoire

Referential mapping procedures enable me to attach meaning to Alex's spontaneous utterances and thus readily expand the number of items and actions about which we communicate. The extent to which referential mapping facilitates intentional communication is unknown, but the technique may significantly affect overall vocal productivity: Alex is rewarded not only for each new viable vocal pattern he creates, but also for the *act* of creating such patterns. Some supporting evidence comes from a language-based learning paradigm for a pygmy chimpanzee (Savage-Rumbaugh et al. 1985, 1986, 1993): The ape's combinations of signs and symbols appeared to increase when humans responded appropriately.

In sum, acquiring the conventions that guide use of a communication code is an important part of learning any code, be it conspecific or allospecific, for any species. Full functional use of a code thus requires not only associating labels with objects and actions, or learning a small set of rules to guide comprehension, but also acquiring general strategies that enable further learning (Pepperberg 1990c). Devising methods that teach such strategies is an important first step, particularly in helping a subject gain competence in an allospecific code. Mechanisms underlying the development of conspecific communication in species as diverse as parrots, pygmy chimps, and humans are unlikely to be homologous; however, examining apparent analogies may enable researchers to develop procedures specifically aimed at expanding communicative competence in an *allospecific* code.

14 What Input Is Needed to Teach a Parrot a Human-Based Communication Code?

In order to teach language, the practitioner must accomplish three major goals. One is to identify the specific tasks to be taught and the sequence in which to teach them. The second is to design an intervention program. The third is to place the teaching activities in a broad context of social policy. In doing so the practitioner must go far beyond the available basic literature, and take risks usually avoided by theoreticians and empirical scholars. For example, in order to identify the tasks and sequencing (i.e., what to teach and when to teach it), a language teacher must adopt certain general principles of language acquisition, identify specific components of the process, decide which components are likely keys to successful teaching, and assume a particular direction for the cause-effect relations among components.

Mabel Rice, *Biological and Behavioral Determinants of Language Development* (1991:448)

With the addition to the laboratory of two juvenile Grey parrots, Kyaaro and Alo, in spring 1991, I began to answer two long-standing questions: Why had M/R training enabled Alex to learn so many communicative and cognitive skills when other researchers' paradigms had failed? Was Alex particularly adept? That I had succeeded with Alex, and that other projects had failed, was not in dispute. But was it the technique or the bird? I needed to determine if some specific element of the M/R technique was crucial for engendering learning. Given that Alex was chosen at random from among many parrots at a pet store, I doubted that I had had the good fortune to have picked the psittacine Einstein. If Kyaaro and Alo replicated Alex's success with M/R training, but failed with other protocols, I would have some answers.

A Brief Review of the M/R Technique and How Its Effects Could Be Tested

As noted in Chapter 2, reference, functionality, and social interaction are the main elements of M/R training, although other factors (intensity of interaction, a subject's motivation to learn, etc.) clearly play important roles. Reference and functionality refer to real-world use of input; social interaction highlights various components of the input. My goal was to design input that varied with respect to these elements and then evaluate the effects of such variation. Furthermore, I believed that by using different conditions to train Alex and the two juveniles, Kyaaro and Alo, I could determine the *relative* importance of these elements for allospecific learning in a mimetic species (Pepperberg 1994b). To answer my main question I also needed to determine how varying training conditions might affect learning not only to reproduce sound, but also to comprehend and use sounds appropriately: My goal had always been to teach *referential* communication, not merely mimicry of human speech. To provide input that varied with respect to reference, functionality, and social interaction, I contrasted sessions of M/R, videotape, and audiotape tutoring; each of the last two omitted some aspects of input. M/R and videotape sessions, moreover, could each have varying levels of the aspects of input (Table 14.1; Pepperberg 1997, 1998).

Variations in Training

I began work when Kyaaro and Alo were, respectively, 6½ and 10 months old, and also used data from a study conducted when Alex was 12 years old. The juveniles had arrived from their breeder about 3 months earlier. Neither had had any training while with the breeder. Although my students and I talked to them about foods and items in the laboratory, we spent most of those months scaling their relative preferences for objects we planned to use in training (Pepperberg 1994b). They had no formal training during this period. I also used this time to design the input categories, "M/R-variant 1," "M/R-variant 2," "basic video," and "audiotape" sessions, each of which included different amounts of reference, functionality, and social interaction (see, e.g., Pepperberg 1994b). As a control, I contrasted these sessions with standard M/R training. I designed video-variant 1 and video-variant 2 after finding that birds failed to learn during basic video sessions. Briefly, the similarities and differences of the various techniques are as follows.

In M/R-variant 1, Alex's input lacked as much reference and functionality as possible compared to what was available in the usual M/R procedure (Pepperberg 1994b). While Alex watched, one human asked a question and the other uttered a string of foreign number labels without reference either to specific objects in the laboratory or to

Table 14.1. Components and results of various types of tutoring regimes. (After Pepperberg 1997.)

Regime	Reference	Functionality	Social Interaction	Parrot(s) tested	Learning occurred?
M/R	Yes	Yes	Yes	All	Yes
M/R-variant 1	No	No	Yes	Alex	Partial
M/R-variant 2	Yes	Minimal	Minimal	Alo, Kyarro	No
Basic video	Yes	Partial	No	Alo, Kyaaro	No
Video-variant 1	Yes	Partial	Minimal	Alo, Kyaaro	Minimal
Video-variant 2	Yes	Potential	No	Alo, Kyaaro	No
Audiotape	No	No	No	Alo, Kyaaro	No

Alex's previously acquired English number labels (Pepperberg 1994a); Alex thus did not initially observe any modeled connection between labels and their referents. These labels were part of a study on ordinality, counting, and serial learning (Silverstone 1989). The set, *il ee bam ba oo yuk chil gal,* was derived from Korean count labels both to permit comparisons with children (Fuson 1988) and to be maximally different from English numbers Alex already knew. *Bam* (pronounced /baem/) and *ba* were substituted for the Korean *sam* and *sa* because of Alex's occasional difficulty in producing "ss." Correct replies received vocal praise and the opportunity to request anything desired (Pepperberg 1987b); errors in the string (omissions, interchanging the order of elements) elicited scolding and time-outs (Pepperberg 1994b).

In M/R-variant 2, Kyaaro's and Alo's input lacked some functionality and as much social interaction as possible compared with what was available in standard M/R training (Pepperberg and McLaughlin 1996). A single human sat with her back to the bird, who was seated on a perch within reach of an object (e.g., a key) suspended by a pulley system. The trainer consistently repeated various phrases about the item, for example, "Look, a shiny *key!*" "Do you want the *key?*" and so forth, thus replicating what is often heard during language learning in young children (note the use of sentence frames, Pepperberg 1981; Chapter 13; see de Villiers and de Villiers 1978). Conceivably parrots, like humans, most readily remember ends of word strings (Lenneberg

1967; Silverstone 1989). The trainer did not, however, make eye contact with the bird or ever hand over the object. If the bird attempted to produce the targeted label, the trainer provided only vocal praise. This procedure also examined how lack of what child language researchers call "joint attention" (mutual focusing of child and caretaker on an object the caretaker is labeling; see, e.g., Baldwin 1991; Baldwin et al. 1996) affected learning in a parrot.[1]

In basic video sessions, Kyaaro and Alo received input that paralleled what was available in the M/R procedure but that avoided social interaction and minimized functionality: I videotaped Alex's M/R sessions and exposed the juveniles to those tapes (Pepperberg 1994b). Although Alex already knew the targeted labels (Pepperberg 1990a,d), his M/R sessions were not structured as reviews but replicated actual training. Alex occasionally erred or interrupted with requests for other items and changes of location (see, e.g., Pepperberg 1987b), behavior that enabled us to demonstrate "corrective feedback." As in regular live M/R sessions, trainers switched roles and occasionally deliberately erred. A zoom lens enabled us to include life-sized images of Alex and the targeted objects in addition to the somewhat smaller images of the entire training scenario (the object, Alex, and two humans; Pepperberg, Naughton, and Banta 1998). The tapes also retained the normal patterns of breaks for nonvocal exchanges (e.g., periods when trainers preened Alex) and trainers' departures by using, respectively, scenes of such nonvocal interactions or a blank screen. We analyzed the audio portion of the video (Kay 5500 DSP Sona-Graph) to ensure that the sound was not degraded compared to that of Alex's "live" performance (Pepperberg, Naughton, and Banta 1998). During sessions, each juvenile parrot was placed on a perch in a separate room so that the videos were viewed in isolation; thus no direct social interaction with any humans occurred. By watching a tape of a human or Alex produce a particular sound and either receive an object or be scolded, the juveniles saw but did not experience directly the effect of a vocalization. Videos, therefore, demonstrated reference but lacked clear functional meaning.[2]

Interestingly, several studies have examined how well children learn from videos, and provide some parallels with our work (Pepperberg, Naughton, and Banta 1998). Researchers find that young children learn best from video (e.g., *Sesame Street*) when they watch with an *interactive* adult who calls attention to objects ("See the trees?"), responds to the child's questions ("[It's a] man with a ball"), corrects the child ("No, it was a doggie"), requests information ("What's that?" "What *kind* of toy?"), expands the child's utterances (child: "Balloon";

mother: "Balloons. *Three* balloons. One, two, three"), or repeats the child's utterances (Ball and Bogatz 1970; Lemish and Rice 1986). The extent of interaction may be critical, because not all children learned in the presence of adults (St. Peters et al. 1989), and children with language impairments (and who seem to lack relevant social skills) learn significantly less than normal children of the same age (Rice 1991); also, 3-year-olds learn less than 5-year-olds. Interestingly, when the results of training language to mentally retarded students with video-only, therapist-only, or therapist-and-video methods are compared (Watkins, Sprafkin, and Krolikowski 1990), students trained via therapist-only and therapist-and-video methods learned significantly more than students trained solely via videos. Possibly older and normal children already know how to extract information from videotaped input (see below). I tested whether limited social interaction would help Alo and Kyaaro learn from videotaped input in video-variant 1 (Pepperberg 1997; Pepperberg, Naughton, and Banta 1998). Video sessions thus were repeated with "co-viewers" who merely ensured that the birds attended to the input. Trainers provided social approbation for viewing and pointed to the screen, making comments ("Look what Alex has!"), but did not repeat new labels, ask questions, or relate content to that in other sessions. Thus trainers provided only limited assistance with extracting information from videotaped input. A bird that attempted the label received vocal praise but not the object. Thus the amount of social interaction was limited and the amount of functional meaning was the same as in basic video sessions.

Because neither juvenile used labels from video sessions in the vocal "practice" that follows M/R training (see Chapter 12), I was concerned that their attempts to produce the labels had been extinguished by lack of reward during sessions (Pepperberg 1994b, 1997). Thus I designed video-variant 2 (Pepperberg, Naughton, and Banta 1998): To the basic video protocol I added a system that, in the absence of social interaction, rewarded a parrot for attempting the label. The system was controlled by a student in another room who monitored a parrot's utterances through headphones: If the bird attempted the label, the student activated a pulley, lowering the object from a container to a spot within the bird's reach. We audiotaped sessions to test for (inter)observer reliability.

Finally, to test how learning might be affected if input totally lacked reference, context, and social interaction, I trained the juveniles via audiotapes. To ensure that sessions paralleled M/R and video procedures, the audiotapes consisted of the audio portion of basic video presentations. In these audiotape sessions, juveniles sat on a perch and listened to tapes in isolation; no objects were associated with the sounds presented over the speaker (Pepperberg 1994b). Note that this procedure resembled that of early studies of song learning in oscine birds in social isolation (e.g., Marler 1970).

Some Details of the Training Sessions

Two parts of my basic question about the efficacy of M/R training that could be answered by using altered input were (1) whether a bird that had previously succeeded through M/R training would fail if certain aspects of input were removed, and (2) whether birds that had not previously received M/R training would preferentially learn with that procedure or would learn equally well under more impoverished conditions. To answer (1), Alex, who had already learned to produce and comprehend labels and concepts through basic M/R training, received M/R-variant 1 training. To answer (2), Kyaaro and Alo received labels in M/R, M/R-variant 2, basic video, video-variant 1, video-variant 2, and audiotape sessions so that we could compare their behavior; neither received M/R-variant 1 training (Pepperberg 1994b; Pepperberg and McLaughlin 1996; Pepperberg, Naughton, and Banta 1998).

Although my students and I attempted to train Alex in M/R-variant 1 without any reference, he would not attend until we included a minimal point of reference: a sheet of paper with the symbols 1–8 traced along the diagonal. (He did not know that his English number labels corresponded to these symbols; Pepperberg 1994b.) In a typical session, a trainer showed the paper to the model and stated, "Say number!"; all previous queries about quantity had been "How many?" The model uttered the altered Korean labels and was allowed to request a toy or food, or erred and was scolded. As in basic M/R sessions, we routinely reversed the roles of model and trainer; Alex was also given a turn. Although Alex's reward usually is either the object he has labeled *or* the opportunity to request a favored item ("I want X", Chapter 12), here we used only the latter. Training, therefore, lacked functional meaning and all but minimal reference. The procedure did, however, maintain joint attention among bird, humans, and the pictured numbers.

I chose labels for the juveniles' training based on three criteria. First, all labels were ones Alex could clearly produce (see, e.g., Pepperberg 1990d), so that any failure to learn would not be caused by a Grey parrot's inability to produce a particular sound. Second, I chose labels that referred to objects in which birds had demonstrated interest to ensure that differential motivation to obtain an item would not affect whether or how quickly they acquired a label (see Chapters 12, 13). Third, I counterbalanced labels, so that, with a few exceptions, labels used for one bird with one technique were used for the other bird with another technique (e.g., Pepperberg 1994b). Thus Alo was trained on "cork" in M/R sessions, whereas Kyaaro initially had "cork" in basic video. Both birds, however, were trained on "paper" via live tutors, on "rock" via audiotape, and on "key" and "block" in M/R-variant 2 to compare their speeds of learning for a given condition. For each juvenile, I repeated one of the two labels from basic video in the video-variant 1 condition to test the effects of co-viewers, and repeated some labels (e.g., "cork") that had not been learned in nonbasic M/R procedures in subsequent M/R sessions (Pepperberg, Naughton, and Banta 1998).

The Results of Training with Various Types of Input

Although Alex eventually acquired the modeled string of number labels,[3] his results differed in two striking ways from those for utterances learned in M/R sessions (Pepperberg 1981, 1994b, 1997, 1998; Silverstone 1989). First, he needed 9 months of training, an unusually long period of instruction, to learn the string. Second, and most interesting, was that he could not immediately use, or subsequently learn to use, these labels referentially either for serial labeling or for quantity. He could merely repeat the labels by rote; that is, he did not see any 1:1 correspondence between a set of eight objects and the string of labels. Even after we modeled this correspondence with various sets of items, he could not use elements in the string to refer to smaller quantities, for example, to say "il ee bam" when presented with three items and asked to "say number." Alex had thus learned to reproduce—that is, mimic—but not comprehend these human vocalizations. I doubted that the task was too difficult, given his previous success on both production and comprehension of human labels after M/R training (see, e.g., Pepperberg 1990a, 1992a). I believed that his failure was a consequence of the training and not caused by a lack of general cognitive capacity.

Kyaaro never learned to produce, either for trainers or in private practice (Pepperberg 1994b, 1997; Pepperberg and McLaughlin 1996; Pepperberg et al. 1998a), labels trained via audio ("wood," "rock") or any form of video ("cork" [initial training], "truck" [initial training], "key," "block," "chalk," "bear"), except for video-variant 1 ("truck"). Even his attempts at "truck" during video-variant 1 sessions were not clearly recognizable. His early results for labels taught in M/R sessions ("nail," "paper") were also difficult to understand: When 11 months old, he ran them together ("ail-er") in a manner that trainers could not distinguish for testing (Pepperberg 1994b).[4] In contrast, he produced clearly differentiated "nail" and "paper" during monologues. After additional training, his labels were at the criterion for testing (see Chapter 2). On identification tests for items trained under nonbasic M/R procedures and video-variant 1, Kyaaro scored 0 on every trial, including those on "truck." On tests for basic M/R-trained labels, however, his first trial scores were 34/40 for paper and 35/40 for nail, wood, wool, and cork (Table 14.2, see Pepperberg 1994b; Pepperberg and McLaughlin 1996; Pepperberg, Naughton, and Banta 1998 for details).

Similarly, Alo never clearly produced, in the presence of trainers, any labels taught via M/R-variant 2 ("key," "block"), basic video or video-variant 2 ("wood," "nail" [initial training], "chalk," "chain") or audio ("key," "rock") (Table 14.2). Tapes of monologues also revealed a total lack of practice of these labels, but frequent practice of labels trained in M/R sessions. Like Kyaaro, Alo attempted to produce a label trained in video-variant 1 ("nail"), and also failed to identify the object or even approximate the utterance on tests for labels trained in this and all other nonbasic M/R procedures. In contrast, on tests on basic M/R-trained labels, her first trial scores were 34/40 for cork and paper, 35/40 for truck, and 38/40 for wool (Table 14.2).

The Implications of the Results for Learning in Parrots

These results show that social interaction, reference, and full functionality are all important training elements if Grey parrots are to learn both to produce and to comprehend an allospecific code. Absence of even some of these elements affects whether and how acquisition occurs. Detailed examination of each condition will clarify what may have contributed to the birds' failures to learn.

For audiotaped input, all three factors are missing, and birds failed to learn anything, even simple mimicry of labels they had heard (Pepperberg 1994b, 1997). Intermittent monitoring of sessions via an intercom showed that Kyaaro and Alo whistled and squawked while listening to the audiotapes, but did not attempt to reproduce any human utterances they heard; they also failed to practice the utterances in private or for trainers. Possibly audiotape input was processed as background noise, corresponding to environmental sounds that, in the wild, would be inappropriate to reproduce and would thus be "tuned out" (Pepperberg 1998). Even environmental sounds in the wild, how-

Table 14.2. Results from various training regimes with different amounts of reference, functionality, and social interaction. Labels not learned in non-M/R sessions were subsequently learned if switched into M/R sessions. (After Pepperberg 1994b; Pepperberg and McLaughlin 1996; Pepperberg, Naughton, and Banta 1998.)

Object	Training	Test score	Erroneous identifications (no. errors)
			Alo[a]
Paper	M/R	34/40	Cork (4), unintelligible (2)
Cork	M/R	34/40	Paper (1), unintelligible (5)
Truck	M/R	35/40	Cork (2), paper (1), other (2)
Wool	M/R	38/40	Cork (1), paper (1)
Wood	Video	0/20	Paper (3), cork (8), four (4), unintelligible (5)
Nail	Video	0/20	Paper (2), cork (4), four (8), unintelligible (6)
Nail	V-v 1[b]	0/30	Truck (9), wool (8), paper (8), cork (4), other (1)
Chain	V-v 2[b]	0/30	Truck (7), wool (7), paper (2), cork (2), other (12)
Key	Audio	0/20	Paper (3), cork (4), four (5), unintelligible (8)
Rock	Audio	0/20	Paper (1), cork (10), four (2), unintelligible (7)
			Kyaaro
Paper	M/R	34/40	Nail (5), unintelligible (1)
Nail	M/R	35/40	Paper (1), unintelligible (4)
Cork	M/R	35/40	Nail (1), paper (3), other (1)
Wood	M/R	35/40	Nail (3), paper (2)
Wool	M/R	35/40	Cork (3), wood (1), paper (1)
Key	M/R 2	0/30	Wood (5), nail (15), paper (4), other (6)
Block	M/R 2	0/30	Wood (4), cork (13), wool (10), other (3)
Cork	Video	0/20	Paper (8), nail (6), unintelligible (6)
Key	Video	0/20	Paper (4), nail (13), unintelligible (3)
Truck	V-v 1[b]	0/30	Nail (13), paper (2), cork (1), other (14)
Chalk	V-v 2[b]	0/30	Cork (18), paper (6), wood (2), nail (1), other (3)
Bear	V-v 2[b]	0/30	"Chain"[c] (10), paper (3), other (17)
Wood	Audio	0/20	Paper (6), nail (7), unintelligible (7)
Rock	Audio	0/20	Paper (6), nail (9), unintelligible (5)

a. By the end of these experiments, Alo had developed behavioral problems that prevented my students from testing her on labels trained in the M/R-variant 2 condition. Alo had worked well in our laboratory (see Pepperberg 1994b; Pepperberg, Naughton, and Banta 1998) until she had a series of operations and until two of the students, to whom she had closely bonded, graduated. She became progressively more difficult to handle (e.g., the student performing M/R-variant 2 sessions on "block" had to have other students place Alo in position), and she has now been retired from our laboratory. See Pepperberg and McLaughlin 1996 for details.

b. V-v 1 is video-variant 1, V-v 2 is video-variant 2.

c. The label "chain" was in the process of being trained; Kyaaro used his approximation to it.

ever, might have relevant referents (e.g., the sound of wind in the trees signaling an approaching storm); in audiotape sessions, however, birds were given no opportunity to deduce explicit meanings for the sounds they heard, nor were they shown any purpose for which the sounds could be used. Their vocal or physical responses to the sounds did not affect what they subsequently heard or received. Thus they were given no obvious reason to acquire the sounds. Had they acted like some songbirds (e.g., marsh wrens, *Cistothorus palustris*, Kroodsma and Pickert 1984a; blue-winged warblers, Kroodsma 1988), they might have learned the allospecific sounds from the tapes and then produced them either at random or in connection with some relevant or irrelevant cue (Pepperberg 1997). Had the birds engaged in some type of information processing, they might have transferred skills from M/R sessions to audio sessions: Given that other human utterances were presented along with objects in M/R sessions, birds might have attended to the tapes without overt learning but then made some association between the novel sound and the novel object subsequently presented to them in tests; they might then have attempted the targeted label (see Pepperberg 1997). Also, had they not completely ignored the tapes, they might initially have attempted some utterances, possibly because the sounds reminded them of other interactions with humans (Mowrer 1954); lack of impact of their vocalizations on their situation could have subsequently extinguished their behavior. Overall, no net learning would have occurred, but my students and I would have found some practice. The birds did not, however, respond in any of these ways (see, e.g., Pepperberg 1998). Whether such responses to allospecific taped input are similar for all parrot species is, of course, unknown, but researchers have found that some oscine birds can learn allospecific vocalizations from tapes in social isolation (e.g., chestnut-sided warblers, Kroodsma, Meservey, and Pickert 1983; see below).

The juveniles also failed to learn from basic video training, which included reference and limited meaning in the absence of interaction (Pepperberg 1994b); they attended to the videos but did not acquire the modeled sounds. They could have failed for at least four reasons: First, they did not realize that the interactions they observed could be transferred to their own situation. Second, they could not determine exactly what aspect of Alex's behavior caused transfer of the desired objects. Third, they, like children, may have found visual aspects of the video more salient and memorable than audio aspects (Pezdek and Stevens 1984; Watkins et al. 1980; cf. Anderson and Collins 1988), or they may have been confused by the physical separation of visual and audio signals on video (see Radeau 1994) and thus failed to recognize the label-referent connection.[5] Fourth, they stopped responding to the tape because they received no reward for what could have been their first approximations to the targeted vocalization. The first three reasons suggest that they could not or did not process the presented information, or that the input was somehow impoverished and thus incapable of being processed; the third and fourth reasons suggest that a potential association between object and label either did not occur or was extinguished (Pepperberg 1998). The fourth reason was unlikely to be valid, given that they failed to learn in video-variant 2, where attempts would be rewarded. In fact, no attempts occurred that could have been rewarded (Pepperberg, Naughton, and Banta 1998). Clearly, merely watching Alex and human trainers receive objects for producing particular sounds was insufficient to enable the juvenile birds to acquire the targeted labels. Even having a trainer to maintain the birds' interest and direct their attention to actions on the screen did not help: Although Kyaaro made a few attempts ("tru," "uh") at the label he heard during co-viewer training ("truck"), and Alo produced some "nail"-like vocalizations, neither bird ever produced a clear utterance during sessions, nor did they apparently understand that these utterances represented an object label, or how these labels could be used outside of training (Pepperberg, Naughton, and Banta 1998). Remember that the birds received no reward other than social approbation for their attempts in the co-viewer condition. Thus even though they vocalized, they received no direct feedback on how an utterance could be used nor were they given any reason to work toward comprehension, and both birds scored 0 on tests of labels trained in any video session (Table 14.2).

Results of video-variant 2 (and, to some extent, M/R-variant 2; see below) show that physical rewards will not engender referential learning of an allospecific code in the absence of social interaction (Pepperberg 1997). Even in basic M/R training, reward cannot be the only factor important for learning because a bird is rewarded only *after* it attempts the targeted label (Pepperberg 1981; see Chapter 2). The reward demonstrates and reinforces the *referentiality* of the utterance. Some parallels probably exist for humans. Bandura (1977), for example, argues that reward cannot create novel behavior but can effectively shape behavior acquired in an approximate form or regulate existent behavior, and that under natural conditions, people rarely learn exceptional (i.e., inhibited) behavior patterns they have never seen performed by interactive models (e.g., videotapes can train a noninhibited behavior but not one that is inhibited; Bandura 1976).[6] I suggest that even if our birds had acquired the targeted utterances from videotapes and obtained some reward, they would have been unlikely to use those utterances referentially. An early study on nonvocal allospecific learning in a chimpanzee, although quite different in specific training from our video-variant 2, also exemplifies the importance of interaction if *referential* learning is to occur: The ape, taught in the absence of trainers to use a computer keyboard to produce symbols based on human language to answer specific questions or make requests, did not comprehend the symbols: She reliably signaled "M&M" to obtain observed candy, but could not choose M&M in the presence of the symbol (see Savage-Rumbaugh et al. 1980a,b). Although she, unlike our birds, did acquire some associative use of symbols through trial-and-error learning, she had no real comprehension of what she had learned.

Clearly, the contrast between Alex's data from the M/R-variant 1 experiment and data from previous studies in which he received basic M/R training shows that social interaction conjoined with severely limited function and meaning engender, at best, production but not comprehension of allospecific utterances (Pepperberg 1994b). In M/R-variant 1, Alex received positive feedback merely for making particular sounds in response to a specific cue. He was given no reason to learn either the meaning or the appropriateness of these sounds. Subsequently, he could not transfer his acquired behavior to related situations. Thus only the simplest form of processing was probably involved in his learning—associating a specific situation with reward, and possibly recognizing that no additional processing was needed to achieve the desired end. Such training represents the input for most pet mimetic birds, and explains why parrots were once thought incapable of doing more than mindlessly reproducing human speech (see, e.g., Lenneberg 1973): Pets did what was necessary for reward (be it the attention of their owners or a food reinforcer), but no more.[7] Such learning may, in fact, be an artifact of captivity (Pepperberg 1997, 1998). That a parrot might be rewarded for such *specific* allospecific vocal associations in the wild is unlikely: After learning, for example, to reproduce a particular predator call in one situation, the parrot would have to be able to *transfer* such learning to different situations in which the predator was present for the vocalization to be useful.

The results of M/R-variant 2 training suggest that Grey parrots are also unable to acquire or comprehend elements of an allospecific code if input is referential, fully functionally meaningful, but not interactive (Pepperberg and McLaughlin 1996; Pepperberg 1997). Thus shared visual attention between trainee and trainer (i.e., joint gazing and directional pointing) probably is as important for animals that are learning allospecific codes as for humans acquiring language (note Baron-Cohen 1995 for general learning). Arguably, the need for shared visual attention for learning in animals might be an artifact of captivity: All four great ape species that used referential pointing to direct the attention of human companions and thus create joint attention were raised in human environments (gorilla, Gomez 1990; chimpanzee, Boysen and Berntson 1989; bonobo, Greenfield and Savage-Rumbaugh 1990; orangutan, Miles 1990; Call and Tomasello 1994); little evidence exists for such behavior among nonhuman primates in the wild (Menzel 1973; Goodall 1986; cf. Boesch 1991). Interestingly, capuchin monkeys *(Cebus apella)*, who apparently do not obtain information from human eye gaze, use information from joint attention in the form of a human point to solve human-designed problems (Anderson, Sallaberry, and Barbier 1995).[8] Clearly, shared visual attention plays a significant role in language acquisition in humans (Murphey and Messer 1977; Schaffer and Crook 1979; Bruner 1983; Tomasello and Todd 1983; Rocissano and Yatchmink 1984; Baldwin 1991, 1993, 1995; Dunham, Dunham, and Curwin 1993; Pine 1994), and in most cases in which apes were taught elements of a human-based language, pointing and joint gaze with humans were an integral part of the training (e.g., ASL: Gardner and Gardner 1978; Miles 1990; Yerkish: Greenfield and Savage-Rumbaugh 1990; review in Call and Tomasello 1994). Our results clearly demonstrate comparable requirements in a nonprimate. I suggest that similar mechanisms involving shared attention may indeed occur in the wild (Pepperberg 1998), although no data yet exists about vocal learning in Grey parrots in their natural environment.

My studies provide some information about conditions that enable Grey parrots to learn *referential,* allospecific communication (Pepperberg 1997). Although pet parrots acquire extensive repertoires of allospecific sounds (see, e.g., Amsler 1947; Boosey 1947), these utterances are not usually used referentially (Chapter 3). Indeed, the term "to parrot" has come to *mean* mindless mimicry. My research suggests that pet birds' mimicry results from lack of reference and functionality (and often interaction) in the input they receive. Grey parrots (at least) seem to learn human utterances most readily under certain conditions and to acquire fully functional, referential use of such a code under even more limited conditions (Pepperberg 1994b, 1997; Pepperberg and McLaughlin 1996; Pepperberg, Naughton, and Banta 1998). Even the use of live video rather than taped input, to avoid possible habituation effects, failed to engender referential learning (Pepperberg, Gardiner, and Luttrell 1999).[9] Various training conditions remain to be tested (e.g., those involving reference and full functionality in the absence of social interaction; those involving reference and limited functionality with full interaction; and those testing the effectiveness of two- versus three-dimensional referents), but input that is fully referential, functional, and socially interactive ensures that these parrots not only can produce but also can fully comprehend allospecific vocalizations. Lack of some or all of these aspects affects the course and extent of allospecific acquisition. Also, comparisons of my data with that from other species lead to interesting insights into how communication, both allospecific and conspecific, is learned.

Additional Parallels with Other Species pertaining to Taped Input

Earlier, I noted intriguing parallels among birds, apes, and humans with respect to the need for reference, functionality, and social interaction for learning communication skills. Additional parallels may provide insights not only into how Grey parrots learn to communicate with humans but also into how other species learn both allospecific and conspecific communication. Given that few studies, other than those that train animals to use human codes, have examined how absence of reference and functionality affect what is learned and whether learning occurs (see Pepperberg 1997), I'll concentrate on the effects of social interaction after a brief review of other research using audiotapes, videotapes, and comparably impoverished input.

Researchers have shown that most songbirds can learn *conspecific* song via audiotapes in isolation (i.e., from input lacking all three factors; see, e.g., Marler 1970), but experiments have discovered little about the effects of such training: Specifically, do birds trained in that manner understand the function and meaning of their songs—know how to use their songs in the contexts of mate attraction, mate guarding, or territorial defense? Some studies have shown that birds must learn not only what to sing but also when to sing and the appropriate context for song, and that such learning appears to occur through social interaction, that is, that socially learned contextual aspects of song are critical for communication (see King and West 1983, 1989; Kroodsma 1988; King, Freeberg, and West 1996; review in West, King, and Freeberg 1996).[10] Wild birds respond less vigorously to laboratory-reared conspecific isolates' songs (which are generally aberrant) than to wild conspecifics' songs (see, e.g., Thielcke 1973; Shiovitz 1975; Searcy, Marler, and Peters 1985), but comparisons have not been made with songs of laboratory-raised birds trained with conspecific audiotapes. Moreover, little information exists on the use of *allospecific* song learned from tape, because fewer oscines engage in such learning. White-crowned sparrows, for example, efficiently learn allospecific song only from a live tutor (see, e.g., Baptista and Petrinovich 1984,

1986). Note that wild chestnut-sided warblers' repertoires include both territorial and courtship songs (Byers 1992, 1996), but tape-tutored hand-raised birds learn only allospecific "territorial" songs (Kroodsma et al. 1983): Is this result simply a consequence of the bird's viewing its isolation box as a territory worth defending, or of social isolation? In contrast, a wild chestnut-sided warbler that apparently learned song from an indigo bunting *(Passerina cyanea)* seemed to use this vocalization in courtship (Payne, Payne, and Doehlert 1984). A song sparrow that learned allospecific songs from live tutors in the wild actively used those songs in territorial defense (Baptista 1988). We simply do not know if an allospecific vocalization acquired by an oscine without contextual, referential training can be used contextually and referentially (Pepperberg 1997). No isolate learning experiments exist with respect to conspecific communication in apes or, for obvious reasons, in children, but work described earlier on socially isolated apes trained via a computer interface to use an allospecific code clearly reinforces the importance of social interaction for referential learning.

Another interesting case is that of the zebra finch, for whom even conspecific learning from tapes is difficult (Price 1979, but see ten Cate and Houx 1998). Under some conditions (birds were caged in auditory contact with conspecifics and could peck response keys to hear taped songs), some zebra finches may learn some conspecific vocalizations from audiotapes (Adret 1993). Many birds, however, learned songs that were only partial imitations, and procedural changes (e.g., changing song duration, requiring birds to land on perches rather than to peck keys, housing birds in vocal isolation) inhibited learning (ten Cate 1991). Nevertheless, ten Cate and Houx (1998) obtained learning by finches that simply heard audiotapes in isolation. The effects of these differences, particularly with respect to vocal interaction with conspecifics, remains unclear, and how birds used these songs afterward was not examined. A study using videotape input was inconclusive because of poor sound output from the TV monitor (Adret 1997). Clearly, much remains to be discovered about how video input affects a bird's ability to learn and engage in either conspecific or allospecific communication.

Parallels between humans and other animals with respect to learning from videotaped input are few, but some studies on children's and young adults' video instruction on, respectively, first and second language learning may provide such data. Normal children learn vocabulary even under conditions comparable to our basic (i.e., isolated) video sessions (Rice 1991; Pepperberg, Naughton, and Banta 1998).[11]

I suggest that children acquire and juvenile parrots fail to acquire referential vocalizations under such conditions because acquisition of referential English speech represents normal learning for human children, but is a case of *exceptional* learning for parrots—a form of communication characterized by vocal learning that, in the normal course of development, is unlikely to occur without intensive instruction (see Chapter 2). For children, too, video instruction is merely a part of their overall learning experience: By the time researchers begin studying a child's acquisition from video (generally at the age of 3–5 years; e.g., Reiser, Tessmer, and Phelps 1984; Rice and Woodsmall 1988), even the youngest child, unlike our parrots, has already been exposed to extensive social vocal interaction[12] and older children have considerable language skills. As noted above, 5-year-olds learn more from videotapes than 3-year-olds (Rice 1991), and children who have language impairments (and seem to lack concomitant social skills) learn significantly less than normal children of the same age (see Rice 1991).[13] Thus children who are already well on the way to language and who have more rather than less experience with videos can compensate for a lack of interaction and differential feedback; they may extrapolate their everyday experiences to what they see on tape (Pepperberg, Naughton, and Banta 1998). Younger and dysfunctional children do not (seemingly like our parrots) have this ability.[14] Thus I suggest that parallels between our birds and humans seem more appropriate with respect to second language learning in humans, which can also be considered exceptional (reviews in Pepperberg and Neapolitan 1988; Neapolitan et al. 1988; Pepperberg and Schinke-Llano 1991).

Interestingly, limited research on how video instruction affects second language learning in college students also emphasizes reference, functionality, and interaction and suggests parallels with our work. Students shown videos depicting the *use* of new vocabulary did better on comprehension tests but not on basic learning skills than students who learned through traditional interactive classroom drills (Secules, Herron, and Tomasello 1992); however, students given short summaries of the tape content by their teachers (presumably in an interactive setting) before viewing, surpassed, on several metrics, students who lacked such assistance (Herron 1994). Although Kyaaro and Alo neither acquired referential use of labels nor learned to associate their attempted sounds with corresponding objects presented via video, they did, like children (see, e.g., Watkins et al. 1980; Reiser et al. 1984; St. Peters et al. 1989), learn more when some social interaction occurred: They attempted targeted labels more and Kyaaro vocalized more when a trainer directed their attention and provided approbation, but they apparently learned not to vocalize in the reward situation (Pepperberg, Naughton, and Banta 1998). Of course, the presence of humans may, as proposed for children by Salomon (1977:1150), simply have been "a source of nonspecific general arousal, thus acting as a general energizer of all responses which were likely to be emitted in the learning situation (Zajonc 1965; Zajonc, Heingartner, and Herman 1969)." But, as noted above, data from the newest addition to the laboratory, Griffin, suggest that even use of on-line video in a fully interactive, referential situation fails to engender referential learning in Grey parrots (Pepperberg et al. 1999).

The Effects of Joint Attention on Learning

Overall, few researchers question the role of social interaction in language learning (i.e., conspecific communication) by children. Granted, several studies suggest that language development occurs in situations that lack such interaction, but proof is difficult to obtain. Research on deaf children demonstrates that some language develops even in the absence of a formal model (review in Goldin-Meadow 1997), but such children were raised with parents and caretakers who interacted with them extensively and who, in concert with the children, developed various means of decoding and responding to the children's codes for their various wants and needs. Children who were not deaf but who were raised in inhumane conditions that included lack of interaction failed to develop language (see Skuse 1988), but in such aberrant situations, the reasons for failure are many. In general, interaction plays some role in learning a communication code (see Chapters 12 and 13), and joint attention seems particularly important (see, e.g., Baldwin 1995). I thus compare our birds to children with respect to that aspect of social interaction (Pepperberg and McLaughlin 1996).

When joint attention is missing, input received by a subject (whether a child or a parrot) also often lacks a clear demonstration of reference, and parallels seem to exist with respect to the effects of such impoverished input on acquisition. Although I have not proven that Grey parrots specifically use joint attention to establish an association between an object and a label, my studies do show that Grey parrots fail to acquire referential human speech when input lacks such explicit associations; at most, birds will produce but not comprehend an allospecific code learned under such conditions (Pepperberg 1994b; Pepperberg and McLaughlin 1996). Studies on how children acquire language have also shown that the *way* a caregiver talks to a child may be more important than *what* he or she says (Tomasello and Todd 1983). Specifically, Baldwin (1991) demonstrated that if a caretaker intentionally produces a label while she looks away from the child and the object that engages the child's attention, the child fails to learn the label or the connection between label and object (see also Morales, Mundy, and Rojas 1998). Baldwin therefore suggests (1991, 1993) not only that simply hearing a label in the presence of an object is insufficient for label acquisition, but also that children use lack of joint attention as a specific nonvocal cue to avoid mapping the label to the object. Such a strategy is probably adaptive: A child who used temporal contiguity of label and object in the absence of joint attention to establish reference would learn many erroneous associations because 30–50% of mothers' labels fail to correspond to the object a child is viewing at the time (Collis 1977; Harris, Jones, and Grant 1983). A similar strategy may be involved in Grey parrot vocal learning in the laboratory. And although nothing is known about how Grey parrots learn to communicate in the wild, they do imitate other creatures (Cruickshank, Gautier, and Chappuis 1993), and probably employ a strategy that enables them to determine which aspects of input are worthy of learning (see Pepperberg 1998 for a detailed discussion).[15]

Joint attention affects not only reference but also the functionality of input. One reason for children, and birds in my laboratory, to learn to communicate is to get their needs met (Pepperberg 1981; Bruner 1983). An 8-month-old child points to a nearby object and vocalizes; the caretaker transfers the object to the child, and the child, having experienced how to obtain a desired object, learns both how to establish and how to use joint reference. If the caretaker consistently labels the object during this exchange, the child is more likely to acquire the label than if the child hears the label without the exchange (see, e.g., Snow 1984; Adamson, Bakeman, and Smith 1990), possibly because the label becomes associated with the object's transfer. Similarly, a Grey parrot that produces even a squawk in early stages of M/R training receives both the object in view and the object label. Baldwin (1989) suggests that infants learn that such an exchange not only satisfies an immediate need, but also is a way to establish word-object links. If joint attention is subsequently absent—a caretaker talks about and points to an object other than one that holds the child's immediate interest or simply talks while looking somewhere else—part of the script is missing: The child sees little reason to acquire the label because he or she knows that joint attention *during* the caretaker's speech is related to obtaining the object at hand. Our parrots' lack of speech during M/R-variant 2 sessions suggests a similar scenario (Pepperberg and McLaughlin 1996). Interestingly, Kyaaro and Alo may not only have failed to learn targeted labels because sessions lacked functionality, but may also have learned not to speak in these sessions for three reasons. First, they may have realized that utterances were not needed to obtain what they could simply reach out and grab. Second, because students continued to talk whether or not the birds responded, the birds learned that their behavior did not affect a trainer's actions. Finally, the total lack of *inter*action (as opposed to reaction) after they vocalized may, given their concurrent interactive experience in regular M/R sessions, have extinguished speech in the impoverished training situation. Quite likely, even though the same vocal rewards were used as in M/R training (e.g., "*Good* bird!! That's right, that's a key!"), the birds may not have discriminated these phrases from trainers' other speech acts and thus did not realize they were being rewarded.

Absence of joint attention most obviously results in lack of social interaction between teacher and student; for children and possibly Grey parrots this lack of interaction may have indirect ramifications in addition to the obvious effects (Pepperberg and McLaughlin 1996). Baldwin (1989, 1991), for example, reports that in the absence of joint attention, infants looked more frequently at the person who was talking than at the object that was present, possibly to pinpoint the direction of the speaker's gaze. Thus the infants had less experience with hearing a particular label while viewing a particular object. Interestingly, during some sessions Kyaaro got down from his perch, moved around the room, and once attempted to climb up the trainer's pant leg; conceivably his actions were attempts to engage her in some manner or determine her direction of gaze. We have no way of knowing if his behavior was intentional; however, the lack of joint attention may, just as for children, have diverted his attention, albeit unintentionally, from the task of learning the object label. Researchers have shown that input that intentionally redirects children's attention from an object they have chosen to one their caretaker selects negatively affects label acquisition (see, e.g., Tomasello and Farrar 1986; Harris, Kasari, and Sigman 1996). Possibly any diversion of a subject's attention from the task at hand (whether the subject is a child or a parrot) will have a similar negative effect. Interestingly, Kyaaro did not learn when he occasionally watched Alex request or identify and receive objects used in M/R-variant 2 training; such exposure was, in terms of reference, context, and interaction, no different from videotaped input, and often redirected Kyaaro's attention from whatever he was eating or manipulating (Pepperberg and McLaughlin 1996).[16]

Conceivably, our birds, like children, failed to learn without joint gaze and directional pointing because such training eliminated another potential attentional mechanism—that of monitoring facial movements coordinated with speech (Pepperberg and McLaughlin 1996). Although parrots in general are less likely to be attuned to movements of the human vocal apparatus than to physical attributes of conspecifics (e.g., beak movements, tracheal protraction; Patterson and Pepperberg 1994), my Greys may have become sensitized to human cues. I must eventually determine whether parrots attend to human facial features and the role such attention could play in learning.

The Need for Reference, Functionality, and Social Interaction If Exceptional Learning Is to Occur

Although the discussion above highlights the importance of joint attention for vocal learning in parrots and points out parallels with children who are learning their first language, joint attention is probably a necessary but not sufficient social condition for psittacine allospecific learning. As noted earlier, all my previous work suggests that acquisition of referential allospecific speech by Grey parrots—at least in the laboratory—is a form of exceptional learning. If such is the case, Grey parrots need additional aspects of social interaction for such learning. "Exceptional" refers not only to the extent to which a particular type of learning is unusual, but also to what might be considered the extreme richness of the input that such learning requires (Bandura 1977). Compared to the input necessary for normal learning, input for exceptional learning must (among other requirements, see Pepperberg 1993) include demonstrations that are even more explicit with respect to real-world use and the consequences of the targeted behavior. Specifically, effective input is a matter not of the number of available cues and modalities, but rather of the specific types of cues and modalities that must be available (Pepperberg and McLaughlin 1996; Pepperberg 1997, 1998).

I thus describe other situations in which Grey parrots will be unlikely to learn referential use of allospecific speech. First, I expect that Grey parrots will not learn if input includes referentiality and joint attention, but lacks a clear demonstration of functionality, as when only *one* trainer labels an object that is the focus of joint play (Pepperberg and Ellsworth, unpubl. data). I also expect failure if a bird receives multimodal stimulation of an object, a sound, attentional focusing (e.g., a point) and reinforcement (e.g., food for producing the sound): This exchange provides interaction but provides only incomplete reference and confounds functionality. The bird may learn to produce the sound, but will associate it with the food reward rather than with the targeted object (Pepperberg 1997, 1998). I also expect that a Grey parrot will fail to learn if joint attention is missing from the M/R procedure, for example, if two trainers chat about an object, but fail to point to it and either have their eyes occluded or maintain each other's gaze and avoid looking at the object or the bird: This situation would have reference, lack all but minimal functionality, and provide the same amount of interaction with the parrot as does our M/R-variant 2 procedure; the only difference would be a demonstration of a nonfunctional (for the bird) vocal interaction between two humans (Pepperberg, Naughton, and Banta 1998).

When taken in conjunction with previous studies on the acquisition of allospecific, human-based codes by nonhuman primates (review in Pepperberg 1997) and studies on normal language acquisition in human children, our findings suggest that many species require certain environmental conditions for learning. Children, for whom acquisition of referential English represents normal learning, do acquire some language competence when input is impoverished (see, e.g., Goldin-Meadow and Mylander 1990; Goldin-Meadow 1997); in contrast, although Grey parrots and other nonhumans will produce allospecific elements without comprehension when input lacks full social interaction, reference, or functionality, these subjects all learn either more readily, more extensively, or with full comprehension as well as production under the richer environmental conditions exemplified by our standard M/R procedure (see, e.g., Pepperberg 1998). Additional conditions, of course, must be tested to determine whether a subject being taught an allospecific code needs input that is fully referential, fully functional, and fully interactive to achieve full production and comprehension. My students and I have not, for example, yet examined the effects of varying other features of input (e.g., changing the number of exposures to the targeted behavior or altering the clarity of the input—by, for example, using accents or dialects differing from those routinely heard), nor have we examined the relative amount and speed of learning in M/R sessions conducted by two humans versus those in which Alex takes the role of a model for the juvenile Greys. Will Grey parrots learn referential labeling if exposed to interactive video with audio link-ups? Further, our birds' breeding and living conditions, which involve continual exposure to human and conspecific interactions, may conceivably have primed them to ignore all but interactive input; might a "Kaspar Hauser" Grey parrot learn from video or audio input?[17] I also wonder if Grey parrots might exhibit a sensitive phase specifically for learning from noninteractive (e.g., video or audio) input that can be extended by interactive input, similar to what has been shown for white-crowned sparrows (Petrinovich and Baptista 1987). My data and those of researchers studying child language acquisition (e.g., Bruner 1983) have shown that fully referential, functional, and socially interactive input is sufficient for complete learning of a communication code, but the extent to which even partial absence of some of these conditions affects learning is unclear (see, e.g., Pepperberg 1990d, 1994b).

Our results not only provide data on certain aspects of vocal learning in Grey parrots, but also suggest three intriguing parallels between psittacine and primate behavior with respect to learning to communicate. First, acquisition of *referential* labeling in Grey parrots, human children, and apes appears to be similarly affected by the absence of certain elements of input (see Pepperberg 1997). Conceivably, such behavior reflects comparable predispositions that can be examined further. Second, some level of interaction beyond joint attention should be present in a parrot's and probably an ape's training if referential use of a human-based communication code is to be learned. Third, my data may have implications for other forms of exceptional learning, such as human acquisition of a second language (see Neapolitan et al. 1988; Secules et al. 1992; Herron 1994) or learning by dysfunctional children (see, e.g., Kasari et al. 1990; Watkins et al. 1990; Harris, Kasari, and Sigman 1996).

Finally, because the extent to which a behavior is exceptional clearly varies across species (see Pepperberg 1985), the degrees of social interaction, referentiality, and functionality necessary and sufficient for communication to develop probably also vary across species; such differences must be examined in detail. For avian species with sizable repertoires, allospecific learning might be less exceptional than for species with limited repertoires: If the extensive learning capacity used to acquire large conspecific repertoires is correlated with an overall reduced selectivity toward what is learned, allospecific learning might then require less referentiality, functionality, or interaction (Pepperberg 1985). Marsh wrens (large repertoired birds for whom allospecific learning in nature is unknown), that were raised prior to singing with same-aged (i.e., not yet singing) sedge wrens *(Cistothorus platensis),* tape-tutored with a mixture of marsh wren, sedge wren, Bewick's wren, and swamp sparrow songs, and isolated before they could hear live sedge wren song, acquired mostly marsh wren songs but did incorporate some allospecific vocalizations (Kroodsma and Pickert 1984a). A correlation does, however, seem to exist between (1) the extent to which a behavior is indeed exceptional and (2) the degree of interaction, referentiality, and functionality of the input needed to facilitate learning. When a behavior is *not* exceptional, interaction, referentiality, and functionality may simply facilitate or modify the course of development (e.g., the extent to which a male cowbird shifts to potent songs while singing to a female; West and King 1988); when behavior *is* exceptional, the type of input can determine whether learning occurs at all (see, e.g., Baptista and Petrinovich 1984). Knowledge of these matters is likely to provide information applicable not only to matters touched upon in this book but also to the most general theories of learning.

15 How Similar to Human Speech Is that Produced by a Parrot?

> The acoustic signal that the parrot produces when it "talks" is not equivalent to that of human speech because birds cannot inherently produce the sounds of human speech (Greenewalt, 1968). Human listeners interpret the parrot's acoustic signal as speech although it is distorted.
>
> Philip Lieberman, *Uniquely Human: The Evolution of Speech, Thought, and Selfless Behavior* (1991:180)

As noted in Chapter 2, one reason for studying the Grey parrot was its ability to reproduce human speech. Birds are among the few non-humans to learn such sounds, and parrots are among a small subset (e.g., corvids, stringillids, cacatuids, psittacids) of 2 (Passeriformes, Psittaciformes) of 28 orders of birds that clearly mimic human speech (Kroodsma and Miller 1982). To the human ear, Grey parrot speech sounds almost indistinguishable from that of its trainers. Classic jokes about maintenance men mistaking a bird's utterances for those of a human have, indeed, been repeated in my laboratory. But is this similarity a matter of human perception, or do the physical characteristics of the bird's utterances match those of humans? If the latter, are these characteristics the same ones scientists use to describe human speech? My students and I were intrigued by these questions. A study of similarities and differences in human and parrot utterances would provide insights not only into parrot abilities, but also into how humans perceive speech.

Although analyzing qualities of psittacine speech might seem quite removed from my interest in cognition and learning, such analyses are indeed integral to my studies. Specifically, to reproduce human speech, Alex must discriminate among and appropriately categorize human speech sounds despite individual speaker's variations. Furthermore, to use such speech referentially, Alex must understand that minor differences in speech sounds are meaningful, that is, that "want corn" and "want cork" have different results. Such ability requires considerable cognitive processing, and might also provide insight into how birds characterize sounds in nature. Researchers have shown, for example, that humans and some songbirds categorize conspecific and allospecific songs and calls in similar ways (e.g., field sparrows, Nelson 1988; swamp sparrows, *Melospiza georgiana,* Nelson and Marler 1989; zebra finches, Cynx 1990; chickadees, Nowicki and Nelson 1990; cliff swallows, Loesche, Beecher, and Stoddard 1992). We also know that budgerigars perceive English vowels as do humans: Budgerigars discriminate between /i/, /E/, /a/, and /u/, and, for /a/ and /i/, find differences between human speakers less salient than differences between vowel categories (Dooling and Brown 1990). Budgerigars' categorizations are particularly striking because vowel sounds fall outside their range of enhanced spectral resolving power (2–4 kHz, Dooling and Saunders 1975). Budgerigars also parallel humans in their clustering of synthetic voice-onset and other speech stimuli (Dooling, Okanoya, and Brown 1989; Dooling, Best, and Brown 1995; Dent et al. 1997). Such data suggest that avian and human perception differ little,[1] and why humans and parrots might perceive the *same* signal in similar ways; but the data say nothing about why humans might perceive parrot speech (a *different* signal) as *comparable* to that of humans if, as some researchers suggest (e.g., Lieberman 1991), parrot speech is an acoustically distorted signal. Possibly relevant are data showing that female but not male red-winged blackbirds may react differently to conspecific male song versus mockingbird imitations; the reasons could be either perceptual or motivational (Searcy and Brenowitz 1988).[2] Extrapolated to humans, such data might suggest that humans perceive differences between human and imitated avian speech without acting upon these differences, but not explain why humans interpret avian speech accurately. My students and I thus decided to examine just how psittacine and human speech might differ.

How We Studied Avian Speech

Our goal was to determine why psittacine and human signaling are mutually comprehensible; clearly psittacine systems did not evolve for the production and perception of human speech (see, e.g., Homberger 1999). Of course, studies have shown that human speech signals are discriminated and classified by structures and perceptual mechanisms that are not uniquely human (e.g., Kuhl and Miller 1975; Morse and Snowdon 1975; Waters and Wilson 1976; Hienz, Sachs, and Sinnott 1981; Kuhl and Padden 1982; Kluender, Diehl, and Killeen 1987; Dooling et al. 1989, 1995; Dent et al. 1997), and we already knew that Alex and his trainers mutually perceived each other's signals. But what aspects of these signals facilitated such perception? To answer this question, we analyzed spectrograms and videotapes of numerous avian speech samples to search for and examine acoustic and articulatory parallels between human and psittacine vowels (Patterson and Pepperberg 1994) and stop consonants (Patterson and Pepperberg 1998). Our goal was feasible because of Alex's extensive repertoire (Pepperberg 1990d). We also compared some of his and my vowels to those of a New World psittacid, a Yellow-naped Amazon *(Amazonica ochrocephala auropalliata)*, Zaa.

We recorded Alex, Zaa, and me with a Sony TCM 5000 voice-activated recorder, AKG CK8 microphone, and Maxell XL-UDII tapes. Dianne Patterson categorized and transcribed utterances according to the International Phonetic Alphabet (IPA). For consonants, standard American English equivalents are /p/ = *p*at, /b/ = *b*at, /t/ = *t*ap, /d/ = *d*ab, /k/ = *c*ap, /g/ = *g*ap.[3] For vowels preceding or following our consonants, equivalents are /i/ = h*ee*d, /I/ = h*i*d, /E/ = h*ea*d, /æ/ = h*a*d, /a/ = h*o*t, /ə/ = h*u*t, /ɚ/ = h*ur*t, /o/ = h*oe*, /U/ = h*oo*d, /u/ = wh*o*'d. Various reliability checks ensured the accuracy of the transcription and categorization process (Patterson and Pepperberg 1994, 1998).

For acoustic analyses, we transferred signals from tape to a Kay 5500 DSP Sona-Graph, which converts sounds into a plot of time versus frequency, and measured features most often used to characterize human speech. For the vowel study, we measured fundamental frequency, F_0 (the source vibration at the syrinx), and the vocal tract resonances considered most critical for vowel perception in humans, that is, what are called the first and second formants, F_1 and F_2 (Borden and Harris 1984).[4] These formants are dark bands on the sonagraph output. They are not harmonics of F_0. Harmonics may also be present, but show up as multiples of F_0; formants are separate values that represent resonant frequencies in humans of the mouth opening (F_1) and oral cavity (F_2) and, as we will see (Chapter 16), very probably specific areas of the parrot's vocal tract as well. The third formant, F_3, indicates whether constrictions that produce speech occur toward the front or the back of the human vocal tract. Alex's F_3 for vowels is rarely observed or is so close to F_2 as to be nearly indistinguishable from it. We also used sonagraphic analyses to learn whether Alex used one or two sound sources (F_0s) to produce speech. For consonants, we attempted to measure VOT (voice-onset timing, the time between lip opening—in humans—to release air and the start of vibration of the sound source),[5] number of bursts (bursts are a result of the release of the air pressure built up by lip closure), and principal stop loci (high energy peaks associated with the voiceless aspirated portion of human stop consonants). Where possible, we collected three formant-related measures: (1) $F_x$onset—frequencies of F_1, F_2, and F_3 during the first 10 milliseconds of voicing following the consonant; (2) $F_x$target—average frequencies of F_1, F_2, and F_3 during the steady-state portion of the vowel following the consonant; and (3) duration (DUR)—time from the beginning of voicing to the first point where formants assume a steady-state pattern. We then calculated SLOPEx—the difference in Hz between $F_x$onset and $F_x$target divided by DUR in milliseconds.

We also videotaped Alex's speech acts with a Panasonic SVHS AG-450 camera and Maxell XR-S120 SVHS tape (30 frames/s). We analyzed the tape frame by frame and obtained stills. From stills, we traced externally visible articulatory correlates (e.g., beak opening, head angle) for various vocalizations (Patterson and Pepperberg 1994, 1998).

We then used many standard phonetic and statistical techniques to compare Alex's (and sometimes Zaa's) utterances with mine. I was the human subject because I was Alex's principal trainer and most observers claim that his vocalizations closely resemble mine. The actual methods we used for analyses are exceedingly complicated, and I will not give details here (see Patterson and Pepperberg 1994, 1998). What are of interest are the results, and I present a summary of our findings.

Comparisons of Vowel F_0s: Alex, Zaa, and Humans

F_0 analyses suggest that Alex, like humans and unlike most songbirds (Greenewalt 1968; Stein 1968; Miller 1977), uses one voice/fundamental frequency, that is, a single set of articulators, to produce voicing (Patterson and Pepperberg 1994). An oscine songbird, in contrast, can use the left and right halves of its syrinx (the avian sound source, located at the base of the trachea; see Chapter 16) independently to produce two different sounds simultaneously (Miller 1977; Robisson 1992). Specifically, some researchers suggest that avian articulatory apparatuses do not produce signals with the acoustic spectrum of human speech. In this view, human perception of avian mimicry as speech depends on a bird's using each half of the syrinx independently to produce a different sinusoidal, pure tone; the "tone is present at the formant frequencies of the original human speech sound that the bird is mimicking. The sinusoids are, in addition, interrupted at the rate of the fundamental frequency of the mimicked human speech . . . We perceive these nonspeech signals as speech because they have energy at the formant frequencies" (Lieberman 1984:156). Neither data from a mynah (an oscine that matches the F_0 of a woman's voice, Klatt and Stefanski 1974) nor data from Alex support such an interpretation.[6] We used four different methods to measure each parrot's F_0. Each method has advantages and disadvantages, although Cepstral analysis may be more appropriate than the other methods for parrots (Patterson and Pepperberg 1994). Numbers from narrowband spectrograms, two time wave analyses, and Cepstral analyses differ somewhat, but concur that Alex has a single F_0 source: one pair of vibrating membranes in the syrinx, the lateral tympaniform membranes (see Gaunt and Gaunt 1985; Scanlan 1988). Zaa's F_0 data did not differ significantly from Alex's.

We also analyzed F_0 because human vowel perception is somewhat sensitive to F_0 (Ryalls and Lieberman 1982; Traunmüller 1988; Kewley-Port et al. 1996), and human recognition of avian vowels would seemingly require similarities in avian and human F_0s. F_0 values may not, however, distinguish individual vowels: Average human F_0s, for a *given* sex and age range, differ little among vowels ($\sim$10%), and how much categorical information this difference provides is unclear (Traunmüller 1988). Also, for a given vowel, human F_0s differ greatly *between* sexes (male values can be $\sim$60% that of females; Peterson and Barney 1954; Whiteside 1998), and Zaa and Alex occasionally reproduce male voices. Nevertheless, humans apparently normalize input to account for large F_0 variations that reflect speakers' age and sex differences (Lieberman 1984), and the overall range for all parrot vowels should tell whether psittacine and human values are similar. Moreover, although F_0s may not distinguish avian vowels, F_0 data may reflect anatomical and control differences in avian species (see Scanlan 1988).

Psittacine and human F_0s do not match exactly, but Zaa and Alex both produce absolute F_0s in the general range of an adult human (124–276 Hz, Peterson and Barney 1954). Overall, F_0 ranges do not differ significantly between parrots, who can be considered representative of Old and New World psittacids. The range over the different techniques was 81–174 for Alex and 83–194 for Zaa; the mean Cepstral value over all vowels was 111 for Alex and 179 for Zaa. Physical differences also exist between these species: Grey parrots seemingly have greater control over syringeal aspects that may be critical for good reproduction of the human voice, which may explain why humans perceive a difference in speech quality between Amazon and Grey parrots (Scanlan 1988). These differences, whatever their source, are not obviously reflected in F_0 values (Patterson and Pepperberg 1994).

Comparisons of Alex's and My Vowels: F_1 and F_2

Although F_0 provides some linguistically salient cues, accurate perception of human speech generally requires processing F_1, F_2, and often F_3 (Lieberman 1984). Unlike human F_0s, human F_1s and F_2s vary across vowels for a given sex and age and thus provide characteristic data for a given vowel. We compared human and psittacine F_1s and F_2s, and tested whether Alex's formant values, like those of humans, could be used to predict his vowels. F_D, the value of $F_2 - F_1$, is often used to categorize human vowels (see, e.g., Peterson and Barney 1952); because Alex's vocal tract is fairly short and thus might shift absolute formant frequencies upward compared to those of humans, a *relative* measurement such as F_D might provide particularly valuable for comparing Alex with humans.

We found (Table 15.1) some striking differences and a few similarities between Alex's data and that of humans. For both F_1 and F_2, his range of values is smaller than the human range and especially lacks low frequency values compared to the range of humans. The range of Alex's means, particularly for F_1, is limited even compared to mine: His mean values for F_1 and F_2, respectively, cover only 64% and 83% of mine. For most vowels, Alex's F_1s differ considerably from mine but our F_2s are similar. However, /u/ is a special case: Alex apparently produces /u/ as two separate sounds, $/u_1/$ and $/u_2/$, that glide together sequentially; in humans such a sound is called a diphthong. I also produce a diphthong /u/, but my values differ considerably from Alex's. His F_2 for $/u_1/$ resembles that of my /E/ or /æ/ or a child's values for those vowels; his F_2 for $/u_2/$ resembles a woman's /u/ or my /U/. Figure 15.1 shows Alex's $/u_1/$ and $/u_2/$, /u/ for most humans, and my $/u_1/$ and $/u_2/$. Alex's mean formant values often differ from human values, but his F_Ds are similar to those of humans (Figure 15.1), particularly for /E/ and /æ/ (Zahorian and Jagharghi 1993).

Statistical tests reveal some additional similarities and differences in Alex's and my vowels (Patterson and Pepperberg 1994). Alex's vowels, like mine, are somewhat distinguishable by their formant values, particularly F_2. Figure 15.1, however, shows that his F_1 varies considerably less than mine with respect to vowel identity; his is almost invariant. Alex's F_2 differs more among vowels than F_1 and provides about the same amount of information as F_D. Another means of categorizing human vowels, according to tongue placement with respect to height and distance from the front and back of the oral cavity (Remez et al. 1987), is also important for Alex. Interestingly, the frontness/backness, and not height, of Alex's tongue relative to his oropharyngeal cavity

Table 15.1. Means of F_1 and F_2 for humans and Alex. Values on the first line for M (man), W (woman), and C (child) are taken from Peterson and Barney (1952); values on the second line are from Zahorian and Jagharghi (1993). A represents Alex and IMP represents me. (After Patterson and Pepperberg 1994.)

		/i/	/I/	/e/[a]	/E/	/ae/	/a/	/ə/	/o/	/U/	/u/[b]
F_1	M	270	390		530	660	730	640		440	300
		272	410		550	656	749	596	456	439	324
	W	310	430		610	860	850	760		470	370
		338	486		745	922	981	793	532	528	400
	C	370	530		690	1010	1030	850		560	430
		313	563		875	1116	1125	862	660	573	400
	Alex	932	805	821	812	848	872	806	684	837	844/829
	IMP	310	407	636	586	585	753	698	567	527	391/340
F_2	M	2290	1990		1840	1720	1090	1190		1020	870
		2209	1859		1740	1748	1192	1289	1176	1234	1396
	W	2790	2480		2330	2050	1220	1400		1160	950
		2837	2284		2123	2089	1440	1599	1419	1437	1617
	C	3200	2730		2610	2320	1370	1590		1410	1170
		2705	2615		2436	2345	1590	1627	1645	1558	1806
	Alex	2775	2330	2343	2117	2187	1433	1480	1360[c]	1604	2373/1637
	IMP	2988	2458	2520	2365	2382	1378	1505	1476	1626	2479/2148

a. We do not include /e/ for M, W, C because no published values are available.

b. Alex and IMP have two values for /u/ because they produce this vowel in two separate parts.

c. This entry comprises a single sample.

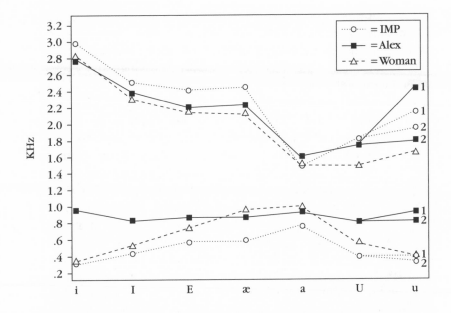

Figure 15.1. A comparative plot of F_1 (lower set of lines) and F_2 (upper set of lines) for Alex, me (IMP), and an "average" woman (data for the woman from Peterson and Barney 1952). (From Patterson and Pepperberg 1994.)

(the equivalent of our oral cavity) correlates with different vowels. This finding is consistent with his relatively flat F_1, which in humans roughly corresponds to tongue height, and his variable F_2, which in humans corresponds to front/back tongue position (see Chapter 16). Alex's /i,I,E,æ/ cluster together as "front" vowels and /U,ə,a/ cluster as "back" vowels. Although /u/ is a back vowel for most human speakers of Standard American English, its two parts are front vowels for me, and Alex's /u_1/ and /u_2/ appear to be front and back vowels, respectively. All his vowels, however, cluster midrange with respect to the high/low distinction.

Implications of Our Findings on Alex's Vowels

Although psittacine vowels do not vary predictably along a single dimension, for Alex they all somewhat resemble those of humans (Patterson and Pepperberg 1994). Despite the differences described above, we perceive his vowels as equivalent to ours. Our F_0 data, in particular, suggest that Alex, like humans, produces true resonances and not sinusoids at the appropriate formant frequencies. To perceive Alex's vowels, humans probably tune out information about his F_1, which is only weakly correlated with his vowels and is often unrelated to human values. Humans also probably switch the criteria for evaluating vowels from instance to instance (see Nearey 1989), or some of Alex's vowels would be unintelligible. Clearly, some mechanisms must exist for such perception to occur, either in the invariant feature(s) of human (note Borden and Harris 1984) or avian articulation that have not yet been discovered, or in the apparatuses that humans use to categorize vowels. Formant data are probably only some of many cues that humans use for vowel perception (Miller 1989). Humans may use dynamic (e.g., temporal) information (Strange 1989). Data showing that humans can match pure tones to vowels (Kuhl, Williams, and Meltzoff 1991), for example, are consistent with human listeners' apparent ability to rely primarily on Alex's F_2 to classify vowels.[7] Other proposed mechanisms for human perception involve phenomena such as nasalization (Fant 1970) that may have no direct avian correlates. Some researchers also suggest that human data can be expressed in alternative scales to account for natural perceptual mechanisms, but Miller (1989:2117) notes that "over most of the ranges of values of the center frequencies of F_1, F_2, F_3, the Koenig scale, the mel scale, the Bark scale, and frequency in hertz are nearly equivalent." Despite these issues, our research provides initial comparisons of psittacine and human vowels. The next step, of course, was to compare Alex's consonants to mine to see if phonetic and articulatory patterns in production and perception are generalized across other aspects of psittacine "speech."

Comparisons of Alex's and My Consonants (Stops)

As for vowels, significant differences and similarities existed between Alex's and my data in several acoustic parameters. Alex's stops, like those of humans, generally could be distinguished with respect to acoustic parameters. The particular parameters that were distinctive, however, sometimes differed for him and me (Patterson and Pepperberg 1998).

We first compared Alex's and my *range of* and *mean* VOT values for voiced /b,d,g/ and voiceless /p,k,t/; we also examined the data with respect to labial /p,b/, alveolar /t,d/, and velar /k,g/ subsets (Patterson and Pepperberg 1998). Such subsets are often used to describe human stops. (For the difference between voiced and voiceless stops, see note 5; labial, alveolar, and velar describe where in the human oral cavity the constriction occurs that produces the stop.) Alex's range of VOTs roughly matches mine and those for other humans (Lauter, Pearl, and

Baldwin 1986; Fant 1991). Ranges for /p,t,k/ for Alex and me are, respectively, 51.47–113.90 ms and 63.11–103.21 ms; our standard deviations (SD) are also similar. Alex's ranges for /b,d,g/ are −14.64–28.45 ms and mine are 5.88–24.71 ms; his SD is far greater than mine. Overall, his ranges are broader than mine and, for voiceless and voiced stops, respectively, are 150% and 200% of mine. Alex's mean values for voiced stops are ~75% of mine, but our values for voiceless stops are comparable. The differences in voiced values are the consequence of Alex's often negative value for /b/: For labials (both /p/ and /b/), his value is 53.4% of mine, but his /p/ is 81.5% of mine. For alveolars and velars, respectively, Alex's values are slightly larger and comparable to mine.

We measured number of bursts because these values are often used to distinguish human stops. Alex's mean numbers of bursts for /p,b,t/ are considerably smaller than mine, but for /d,k,g/ his means roughly match mine; his variability, unlike mine, often is equal or double that of the mean. Statistical tests reveal that for both Alex and me, /k/ is distinct from all other stops, but Alex's /d/ is also distinct from /t/, /p/, and /b/. Overall, Alex's data do not conform to human norms: Humans often have double bursts for /k/ (Olive, Greenwood, and Coleman 1993) and occasionally for /g/ (Fischer-Jorgensen 1954); Alex follows the human pattern for /k/, but has more double bursts for /d/ than /g/. For voiced stops, Alex's /d/ and my /g/ frequently displayed multiple bursts, but we also found occasional double bursts in all samples except my /d/. For both Alex and me, the number of bursts differs significantly for /p,t,k/ versus /b,d,g/. Alex's and my values also differ significantly from each other. The multiple bursts may, however, be an artifact of our difficulties in reading sonagrams, particularly Alex's: First, because 50% of the stops we evaluated occurred within a phrase,[8] a burst in one stop could be interpreted as belonging to the following one; thus for the /d/ in Alex's "go ea*t d*inner" we may have counted the preceding /t/ as an extra /d/ burst. In contrast, /m/ in Alex's "calm down" was probably not associated with a burst and thus the problem did not exist. Second, failure to count bursts correctly is more likely in a noisier signal, and Alex's signals were often noisier than mine. Such noise may have made it more difficult to identify the /t-d/ boundary in Alex's "go eat dinner."

We measured stop loci because these values may be used to predict the place of constriction (where stops occur) in humans (Kuhn 1975). In Chapter 16, I present a model that my students and I propose for Alex that is similar to the standard model for describing how humans produce speech (see, e.g., Ladefoged 1982); stop loci for Alex were measured for this model. For now, I will simply summarize the data.

The critical point is that stop loci values differ across stops and for Alex and me particularly for a given stop, and thus provide evidence that humans and Grey parrots differ considerably in how and where their vocal tracts produce stops. For example, humans produce /p,b/ at their lips, and parrot beaks do not function like human lips. We have seen Alex produce /b/ with his beak open and his tongue down (Patterson, unpubl. data). Calculating the place of constriction for his stops may help us understand his production mechanisms (see Chapter 16).

For humans, calculated values of formant onsets and targets, and transition durations are usually analyzed as SLOPE, where $SLOPEx = [F_x\text{onset} - F_x\text{target}]/DUR$, and are important because in perceptual experiments on humans, slopes provided almost as much information on stop recognition as other measurements (Dorman, Studdert-Kennedy, and Raphael 1977); furthermore, SLOPE2 and SLOPE3 were sufficient to cue place of constriction. The angle of the slope varies primarily with stop identity (Olive et al. 1993), but may be affected by context (e.g., the identity of following vowel). A slope is negative if it rises from a stop into a following vowel, and positive if it falls. Formants' slopes converge if they shift toward each other as they move from a stop into a following vowel (Stevens and House 1956). Slopes tend to rise out of labials, converge out of velars, and remain constant (horizontal and parallel) for alveolars (see, e.g., Liberman et al. 1954). We find correlations in our data that match those in humans, such as between SLOPE and stop identity, but fewer SLOPE and context correlations for Alex. We find that, consistent with our vowel data, Alex's SLOPE1 (based on F_1) does not vary significantly with either stop or context. For Alex and me, most correlations occurred for SLOPE2, which is where predictive power primarily resides for humans (Olive et al. 1993). SLOPEx is a poorer predictor of stop identity in our study than in Dorman et al.'s (1977), possibly because of Alex's invariance in F_1s.

Just as in the vowel study, we ran statistical tests to see how Alex's (and my) stops differ from each other with respect to various measurements. We clearly *perceive* differences in Alex's stops, but the tests identify what characteristics we might use for these perceptual judgments. We found that for both Alex and me, all 15 pairings of stops are distinguishable with respect to some measure. Interestingly, all of Alex's pairs, but only 12 of mine, are distinguishable with respect to VOT; my remaining 3 pairs (b-d, b-g, d-g) are distinguishable with respect to stop loci.

We had found that vowels within the front/back categories are less distinguishable from one another than vowels across such categories;

was the same true for stops within the voiced/voiceless and place of constriction subsets? Interestingly, within these subsets, stops for both Alex and me are generally as distinguishable as or more distinguishable than they are across such categories. Thus the stop data differ from the vowel data (Patterson and Pepperberg 1994, 1998).

Other statistical tests confirmed much of what we had observed (Patterson and Pepperberg 1998). Stop identity was more closely connected to F_2 for Alex than for me, and for Alex any F_1-related measurement correlated little with stop identity. Alex's F_3 was not always measurable or always produced, and tests confirmed that F_3, when present, was less closely related to stop identity for Alex than for me. In humans, F_1 and F_3 as well as F_2 are used for perception; for Alex, however, F_2 appears to carry most information about vowel and stop identity—just as for his vowels, F_2 seems to *compensate* for lack of information in F_1 and F_3. If redundancy in humans is critical for disambiguating speech signals in noisy environments (Repp 1983), then Alex's intelligibility should drop off more sharply than that of humans amid noise. Such a possibility needs study.

Our data suggested and calculations were based on the premise that Alex's stops fall into the same natural categories as those of humans, but we had not tested the *extent* to which this premise was true. Remember, Alex's vowels, like those of humans, separated into front/back categories, but unlike humans', not into high/low categories (Patterson and Pepperberg 1994). Statistical tests showed how well human divisions based on voicing (/p,t,k/ and /b,d,g/) and place of constriction (/p,b/, /t,d/, and /k,g/) characterized Alex's data. Interestingly, his stops, like mine, separate into voicing and place subsets, but my subsets are more coherent. To see how Alex's stops were divided and how his divisions compared with mine, we performed a cluster analysis based on VOT data.[9] Alex's and my patterns differed somewhat (Figure 15.2). For me, voiced /b,d,g/ clustered closely and were separated from the less closely clustered voiceless /p,t,k/. Overall Alex's voiced/voiceless distinction was less coherent than mine. His /p/ clustered somewhat more closely to /d,g/ than to /t,k/ and his /b/ clustered somewhat less closely to /d,g/ than did /p/.

We next examined statistical correlations between Alex's and my data to see if human perception of similarities in our speech was based on similarities in acoustic structure. Positive correlations reveal how closely Alex's speech resembles mine. The same nine measures for Alex and me are significantly correlated (Table 15.2) for the full set and voiceless subset; only five of these nine are correlated for the voiced subset. Correlations are somewhat stronger for voiceless than voiced

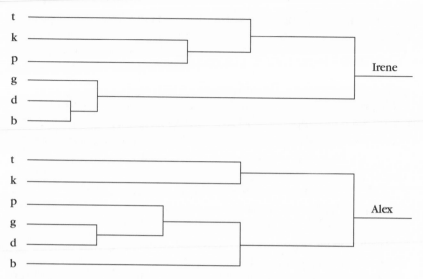

Figure 15.2. Cluster diagram of my (top) and Alex's (bottom) stops. Note the differences in placement of our /b/ and /p/. (Adapted from Patterson and Pepperberg 1998.)

stops. As in the vowel study, F_1 correlations either do not exist or are the weakest significant correlations (Patterson and Pepperberg 1994, 1998).

The Implications of Our Data on Alex's Stops

Just as for vowels, similarities and differences exist for Alex's and my stops (Patterson and Pepperberg 1994, 1998). Similarities involved acoustic characteristics and separation of stops into subsets based on voicing and place of constriction. Alex's stops do have statistically distinct acoustic characteristics. Traditional acoustic measurements used to characterize human stops were generally applicable to Alex's speech. Differences involved specific ways stops separated into subsets and probably the mechanisms Alex and I used to produce speech. Although Alex's stops were, like humans', divisible on the basis of voicing and place distinctions, his subsets were not identical to mine: His /b/ was an outlier, and his /p/ clustered nearer /d,g/ than /t,k/ (Patterson and Pepperberg 1998). Remember, Alex must produce /b,p/ without lips; he may compensate by using mechanisms not available to humans. I discuss such possibilities in Chapter 16.

Alex's data suggest another phenomenon considered uniquely human, anticipatory coarticulation. Anticipatory coarticulation implies preparatory strategies: It occurs when a phoneme is produced so as

Table 15.2. Correlations between measures for consonants for Alex and his principal trainer (me) in order of descending value. (After Patterson and Pepperberg 1998.)

	Measure	Pearson's r	Signficance (*p*)	No. of observations
All	F_2target	0.85	.0001	210
	VOT	0.85	.0001	307
	F_2onset	0.82	.0001	205
	loci	0.68	.0001	228
	SLOPE2	0.58	.0001	168
	F_3target	0.55	.0026	28
	F_3onset	0.50	.0050	30
	BURSTS	0.32	.0001	306
	F_1target	0.15	.0294	219
[b,d,g]	F_2target	0.79	.0001	111
	F_2onset	0.77	.0001	109
	VOT	0.59	.0001	125
	loci	0.55	.0001	65
	SLOPE2	0.46	.0001	87
[p,t,k]	F_2target	0.90	.0001	99
	F_2onset	0.88	.0001	96
	loci	0.73	.0001	163
	F_3onset	0.73	.0076	12
	F_3target	0.71	.0065	13
	VOT	0.65	.0001	182
	SLOPE2	0.60	.0001	81
	BURSTS	0.40	.0001	185
	F_1target	0.22	.0228	105

to configure the vocal tract for a subsequent sound. Anticipatory co-articulation may thus imply top-down processing (Ladefoged 1982). A student and I examined two possible instances of psittacine anticipatory coarticulation and a third related observation; such data enabled us to discuss, if not decide, whether Alex's behavior gave evidence of top-down processing.

Tracings of SVHS video stills of Alex's utterances reveal different degrees of beak openness during the burst of /k/ depending upon following context: "*k*ey" /ki/ versus "cork" /kork/ (Patterson and Pepperberg 1998). The burst for any /k/ is associated with a rapid retraction of his slightly protracted trachea and thus is easily tracked. For "key," Alex opens his beak as much as 300 ms before the /k/ burst; in contrast, his beak is completely closed before he produces "cork," and opens only slightly during production. These data suggest that Alex may prepare his vocal tract for the production of particular sounds, and thus are consistent with anticipatory coarticulation. Such data are not available for Alex's voiced stops: Voiced VOTs are short, and insufficient time exists to identify the effect of a following vowel before it is actually produced. Moreover, voiced stop bursts are weak and the tracheal protraction associated with pressure buildup is not visible (Patterson and Pepperberg 1998).

Other observations of Alex and our juvenile Grey, Griffin, and X-ray videos of Alex (Warren, unpubl. data) also suggest that our birds (whether consciously or unconsciously) probably position their vocal tracts for the second sound in an utterance. Thus they both protract their trachea before an /f/ followed by /or/, as in "four." We see this anticipatory positioning in Alex even if he is interrupted and does not immediately produce "four" (Patterson and Pepperberg 1998). Analyzing his behavior is difficult because /f/ can be too weak in amplitude to be identified sonagraphically. Nevertheless, the behavior suggests a form of anticipatory positioning, if not top-down anticipatory coarticulation.

Statistical patterns in the acoustic data are also consistent with articulatory coarticulation for voiced stops, despite a lack of physical evidence for this phenomenon (Patterson and Pepperberg 1998). For Alex and me, VOT is significantly affected by the identity of the following phoneme, which implies that VOT is adjusted in preparation for this phoneme. The effect is strongest for voiceless stops, but the presence of a following /ɚ/ strikingly shortens Alex's but not my VOT for /b/: This VOT becomes negative for Alex. Although likely caused by a combination of goal-directed imitation and restrictions imposed by Alex's anatomy, his acoustic patterns are nevertheless consistent with a mechanism involving anticipatory coarticulation for voiced stops.

Whether these data provide proof for anticipatory coarticulation in a parrot is, however, unclear (Patterson and Pepperberg 1998). Even researchers who examine human data debate the implications of anticipatory coarticulation. Ladefoged (1982) treats this behavior as evidence of top-down processing because the phenomenon seems to presuppose knowledge of sounds before they are produced (see Wood 1996), but Repp (1986:1618) suggests that "phenomena commonly lumped together under the heading of 'coarticulation' may have diverse origins and hence different roles in speech development. Some forms of coarticulation are an indication of advanced speech production skills whereas others may be a sign of articulatory immaturity, and yet others are neither because they simply cannot be avoided." Alex's vocal tract is not immature, but his anatomy might make anticipatory coarticulation unavoidable. Thus whether Alex's behavior reflects top-down processing or anatomical constraints cannot be determined until we know more about the exact mechanisms he uses for such productions.

The Overall Implications of Our Findings

Our data have implications not only for human perception of Grey parrot speech, but also for the uniqueness of human speech. Although basic aspects of Alex's speech resemble mine, he does not generally mimic either my articulatory motions or my acoustic idiosyncrasies; his apparent need to manipulate his F_2 to do more work than mine and his possible involvement of different anatomical structures (e.g., glottis and esophagus; see Chapter 16) to compensate for his lack of lips suggest that he constructs his own articulatory and acoustic solutions to the task of producing speech. That he generally maintains the voiced/voiceless distinction and, to some extent, distinctions between labial, alveolar, and velar stops, and the front/back distinction for vowels, however, implies that such distinctions may be basic to vertebrates rather than to mammals.

Furthermore, in addition to observed similarities in Alex's and my acoustic measurements, Alex's recombination of human phonemes to produce new human speech patterns from existent ones (as in solitary sound play, Chapter 13) just as humans do, suggests that he may maintain a complex acoustic map between the human system and his own. To understand Alex, humans must likewise maintain some kind of mapping between the two systems. These observations not only provide additional evidence for flexibility in human speech perception, but also suggest that other species share this flexibility and that nonhumans develop phonetic categories of considerable generality. Moreover, evidence that Alex may exploit anticipatory coarticulation and data on his solitary sound play are consistent with the possibility (but do not prove) that he engages in top-down processing. Still, Alex not only separates specific phonemes from the flow of speech, but also may produce these phonemes to facilitate production of upcoming phonemes. Our data thus suggest, contra Lieberman (1996:57), that voluntary control of portions of the vocal tract for linguistic ends is not a unique human characteristic, and that at least Grey parrots, like humans, can use phonetic distinctions to produce meaningful sounds that are "largely independent of affect and psychological state." We now needed to determine what *mechanisms* Alex uses to produce human speech sounds.

16 How Does a

Grey Parrot Produce

Human Speech Sounds?

The voluntary control of the supralaryngeal vocal tract for linguistic ends appears to be a unique human characteristic. The neural structures that are implicated in the regulation of the human SVT essentially store and rapidly access *automatized* patterns of muscular activity that have a linguistic rather than an emotive function. Automatized muscular commands are, in essence, "overlearned" responses that are performed without conscious thought or effort. The automatized SVT patterns of human speech are largely independent of affect and psychological state. Humans, therefore, can produce sounds that have an arbitrary relationship to their emotional state. The sounds that make up the word *help,* for example, in themselves do not have any inherent emotional quality. The sound [h] can be used in the word *hello.* In contrast, the vocal signals of other mammals appear to be tied to their affective state. A chimpanzee will produce a food-bark on seeing food. However, he cannot produce an arbitrary sequence of sounds to signal food.

Philip Lieberman, *Signal to Syntax: Bootstrapping from Speech to Grammar in Early Acquisition* (1996:57)

Given the similarities and differences described in the previous chapter between psittacine and human utterances, my students and I continued to be fascinated by how parrots, and Alex in particular, learn to communicate with humans in the vocal mode. My birds do not broadly mimic; rather, like humans, they appear to have flexible, voluntary control of portions of their vocal tracts: Specifically, Alex (as well as, to a lesser extent, the juveniles still in training) uses different combinations of human phonemes to produce meaningful sounds that are referential, rather than a consequence of his emotional state (see Griffin 1984). Thus Alex not only produces "tea" and "pea," but also recognizes that the slight difference in phonemes in these utterances (/t/ versus /p/) allows him to request a sip of herbal drink versus a bit of vegetable. My students and I wondered how he manages this feat.

Clearly, brain mechanisms are involved (see, e.g., Brauth et al. 1994; Striedter 1994; Durand et al. 1997), but great apes and dolphins have impressive brains and yet only minimal competence in human sound reproduction (Lilly 1967; Hopkins and Savage-Rumbaugh 1991). Such data suggest that Alex's competence might lie, at least in part, in the anatomy of, and mechanisms controlling, his vocal tract.

Unfortunately, little is known about how parrots use their vocal tracts to produce conspecific sounds, and even less about how they reproduce allospecific utterances (review in Warren, Patterson, and Pepperberg 1996). We do not know the extent to which mechanisms differ among mimetic species (Nottebohm 1976; Brackenbury 1982, 1989; Gaunt and Gaunt 1985; Brittan-Powell et al. 1997; Banta 1998), and few studies compare acoustic and articulatory aspects of avian and human speech (e.g., Klatt and Stefanski 1974; Nottebohm 1976; Scanlan 1988; Patterson and Pepperberg 1994, 1998; Silaeva 1995; Chapter 15). Although Homberger (1986) and Nottebohm (1976), respectively, published detailed descriptions of the Grey parrot lingual apparatus and Orange-winged Amazon syringeal anatomy, researchers have not provided correlations between vocal behavior and anatomical data that permit detailed comparisons between psittacine and human speech mechanisms. In some cases, the limiting factor was the parrot subject's small lexicon (see Warren et al. 1996). Scanlan (1988), for example, obtained cineradiographic production data for only four instances of an isolated /a/ from a hybrid Amazon parrot (species not provided) and three instances of "Coco" from a Grey parrot.

Although studying Grey parrot vocal mechanisms, much like analyzing features of speech (Chapter 15), seemed tangential to my interest in Grey parrot cognition and vocal learning, I thought that such research might provide insight into both areas. Data on how Alex (and my other parrots) developed "speech" (Chapter 12) clearly suggested that they *learned* to control their vocal tracts and that cognitive processes (such as matching output to input) were probably involved in such learning.[1] My students and I thus began studies to determine the physical structures and mechanisms used by Grey parrots to produce recognizable speech. Even without the rationale of tying our studies to cognitive and learning processes, we thought we owed it to the research community to pursue such studies because we had one of the few subjects whose repertoire was adequate for analysis. Not only was Alex's repertoire large, but he also produced consonants and vowels in a variety of phonological contexts. Thus our analyses could be correlated with articulatory data from these contexts. We first used three noninvasive video imaging methods—Super VHS video (SVHS), infrared, and x-ray radiography—to obtain data for a preliminary model of Alex's vowel production (Warren et al. 1996). Subsequently, we used magnetic-resonance imaging (MRI) on one of our juveniles and dissections and electron beam computed tomography (EBCT or CT) scans on cadaver birds to learn more about the sizes of various vocal tract cavities in order to develop a theoretical model of Grey parrot speech (Patterson et al. 1997, in prep.). Because few people are familiar with the basics of avian sound production, I describe the anatomical structures involved in parrot vocalization before introducing our model.

Anatomical Structures Used for Speech Production

Just as in humans, speech production in parrots is a complex process involving many structures that are configured differently for each sound (Warren et al. 1996; Patterson et al. 1997). Human speech is produced by vibratory elements of the larynx upon exhalation or, occasionally, inhalation and is modified by the supralaryngeal resonating chambers, that is, the pharyngeal walls, tongue, tongue root, velum, sinuses, teeth, and lips (Fant 1970; Olive et al. 1993). Sound in the avian vocal system, in contrast, is produced in the syrinx (Greenewalt 1968). The extent to which psittacine suprasyringeal structures (those physically above the syrinx) such as the trachea, larynx, tongue, and both upper and lower mandibles (Figure 16.1) modify resonant properties of the vocal tract is still under debate, though some modification of syringeal sound appears necessary (see Westneat et al. 1993; Brittan-Powell et al. 1997; Patterson et al. 1997; Banta 1998).

Although parrots lack teeth and lips, and their morphology differs dramatically in other ways from human structures, Alex's data show that acoustic characteristics of avian speech resemble those of humans (Chapter 15); such similarities suggest that production mechanisms are in some ways analogous to those of humans. Spectrograms of "eat" produced by Alex and me, for example (Figure 16.2), show striking similarities, although Alex's formants are less distinct than mine (Warren et al. 1996). Remember, these are true formants, not harmonics: Harmonics are integer multiples of the source vibration (i.e., of the fundamental frequency, F_0, of vocal folds of the human larynx or parrot's syringeal membranes); formants (e.g., the first and second formant frequencies, F_1 and F_2) primarily are a function of natural resonances of particular configurations of human supralaryngeal or psittacine suprasyringeal tracts (Warren et al. 1996). How, then, do different physical apparatuses produce comparable utterances? Let's examine specific structures involved in parrot vocalizations.

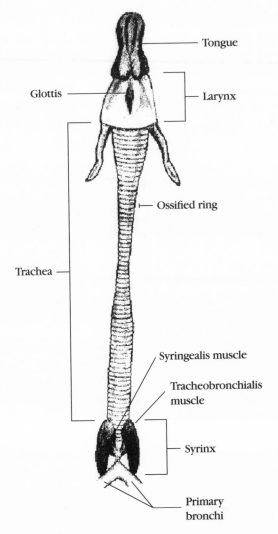

Figure 16.1. Dorsal view of the Grey parrot vocal tract. (Adapted from Warren et al. 1996.)

The Psittacine Syrinx

Avian syringes differ across species and are classified in different ways. A syrinx that comprises the posterior end of the trachea at its bifurcation and cranial portions of the bronchi is termed *tracheobronchial* (King 1989); a syrinx is *tracheal* if the medial tympaniform membrane is missing or nonfunctional (Gaunt and Gaunt 1985). Unlike many songbirds' syringes (including that of the mimetic mynah), which prob-

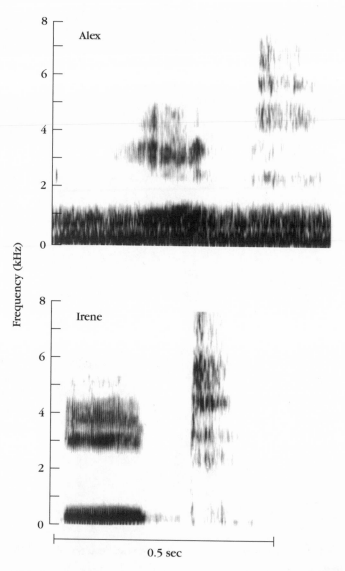

Figure 16.2. Spectrograms of "eat" (/it/) by Alex and me. (From Warren et al. 1996.)

ably have two sites for sound production, parrot species examined so far appear to have only one site (Nottebohm 1976; Gaunt and Gaunt 1985; Patterson and Pepperberg 1994). Grey parrot syringes may have other special properties. According to Scanlan (1988:140), "The shape and position of the syringeal cartilages in the Grey parrot *(Psittacus erithacus)* differ from those of other psittacine species . . . the dorsoventrally oriented cranial edges of the syringeal cartilages are straight, not semi-oval in outline . . . Also, the syringeal cartilages are positioned further craniad than in all other parrot species studied" (Figure 16.3). Scanlan consequently postulated that these modifications may "facilitate control of the intra-syringeal aperture," perhaps precisely controlling contact between opposing lateral tympaniform membranes and tightly coupling movements of cartilages and membranes (Warren et al. 1996). The acoustic effects of these technical specifications are that Grey parrots may be able to control the periodicity and frequency of their vocalizations more carefully than do other psittacids.

The role of the syrinx in psittacine sound production is still under examination, and probably differs from that of songbirds. Some researchers propose (on the basis of songbird studies) that all or most frequency modulation (i.e., specification of formant values) is performed by the syrinx, and that resonant properties of the rest of the avian vocal tract play little or no part in the modification of sound to produce specific utterances (Greenewalt 1968). Of greater relevance is the work of Scanlan (1988), who discusses the roles of several anatomical structures in psittacine speech. Scanlan clearly notes the role of the syrinx—he states that a syringeal constriction mechanism in parrots (that involving the lateral tympaniform membranes) functionally resembles that of the human vocal folds in phonation (note Gaunt and Gaunt 1985)—but also discusses in detail the roles of suprasyringeal structures. Interestingly, Gaunt (1983) has shown that the degree of syringeal complexity across avian species is not directly correlated with complexity in vocal production, with some syringeal complexity being necessary but not sufficient for vocal plasticity. For species with relatively simple syringes but complex vocal behavior (such as Grey parrots), these data imply that other structures must modify syringeal output. Gaunt (1983) and Stein (1968) suggest that vocal plasticity arises in part from neurological adaptations. Although parrots have a complex neurological vocal control system (Striedter 1994), my students and I, on the basis of our analyses of speech production, believe that suprasyringeal structures also play a major role (Chapter 15).

Suprasyringeal Structures and Their Role in Psittacine Vocal Production

Parrot vocal abilities probably require that the syrinx and suprasyringeal structures work in concert (Warren et al. 1996; Patterson et al. 1997; Homberger 1999). Many researchers suggest that suprasyringeal resonating chambers play a significant role in vocalizations, although most data are for nonpsittacids and even nonoscines. Nowicki (1987),

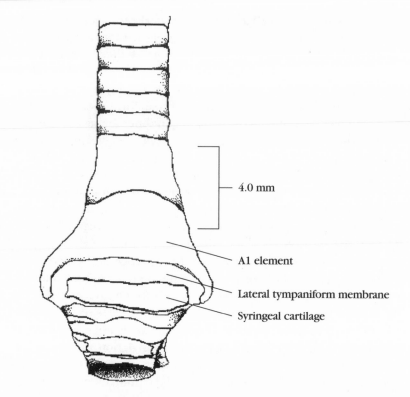

Figure 16.3. Cross-sectional view of the syrinx of a Grey parrot. (Adapted from Warren et al. 1996.)

for example, argues that suprasyringeal resonance effects are demonstrated by the frequency shift observed when song sparrows and black-capped chickadees produce whistled songs in a helium atmosphere. Correlational evidence implies that suprasyringeal structures also modulate frequencies in other species.[2] Our data suggest that such structures are even more important for Grey parrots (Patterson et al. 1997). I thus describe the major suprasyringeal structures and review correlational evidence for their role in sound modification.

The first structure above the syrinx is the trachea. Its role in sound modification is unclear. Alteration of tracheal length or configuration could modify the effective vocal tract length (Stein 1968); the trachea would then be a resonant chamber modulating frequencies emanating from the syrinx. The extent of such alteration probably varies across species. Greenewalt (1968:87), on the basis of research on song sparrows, argued that "at least for whistled song, the trachea does not significantly modulate the sounds produced at the source." Other investigators, however, noted apparent changes in tracheal length during avian vocalizations, although mostly for nonpsittacids and nonpasserines. Artificially shortening the trachea of hens raised the pitch of their calls (Myers 1917). Harris, Gross, and Robeson (1968:112) determined that the harmonics of artificially produced sound in domestic fowl varied with tracheal length; they hypothesized that a chicken's trachea—but only in concert *with* the beak and oral cavity—may "tune the sound of vocalization . . . to a resonant frequency, which causes the pitch to be more sharply defined," and concluded that the trachea and primary bronchi combine to form a single resonant tube. One function of extrinsic syringeal muscles in ducks *(Anas, Aythya)* may be to vary the length of the trachea and thus its resonant characteristics (Warner 1971). The calls of Ross' and snow geese may be modulated by tracheal resonance (Sutherland and McChesney 1965). Some data also exist for passerines: Elongated tracheal morphology in birds-of-paradise (Paradisaeidae) may "lower the pitch of, and perhaps amplify, their vocalizations" (Clench 1978:428; see also Frith 1994). Brackenbury (1978), through an acoustic analysis, surmised that sound pulses in grasshopper warblers *(Locusta naeiva)* exhibit features characteristic of a pulsed tracheal resonator.

Tracheal effects on Grey parrot vocalizations may, however, be relatively small. The Grey parrot trachea, which when excised measures approximately 11 cm, consists of a series of ossified, complete rings, with minimal intervals between them (Pepperberg, Howell, et al. 1998). These rings can overlap, allowing changes in length or configuration, but only by about 10%; the range of length in a talking bird, however, is not known (see Hersch 1966 and Pepperberg, Howell, et al. 1998 for critiques of measurements taken from formaldehyde-fixed tissue samples and by other techniques). Scanlan (1988) suggested that the tracheal protraction observed in an x-ray of a parrot's production of the vowel /o/ (in "Coco") was due to lengthening; our data suggest that although protraction (i.e., bending) occurs for certain human vocalizations, lengthening is not significant (Patterson and Pepperberg 1994, 1998).

Unlike the human larynx, that of birds apparently lacks vibratory elements capable of producing sound (McClelland 1989). Homberger's (1979) study of the Grey parrot larynx found anatomical differences from that of *Corvus* (Bock 1978) and *Gallus* (White 1975), but did not clarify how such differences might affect vocal production. She showed that "the extrinsic musculature moves the larynx with respect to the hyoid skeleton" and that "laryngeal action is influenced by the position of the hyoid skeleton with respect to the lower jaw" (p. 988). The avian larynx does not produce sound, but it probably modifies sound produced by the syrinx. The larynx could, for example, constrict the glottis to varying degrees, thus changing vocal tract resonance to achieve a targeted vocalization (Warren et al. 1996; Homberger 1999). White (1968) noted laryngeal descent and changes in laryngeal configuration during chickens' crowing. Nottebohm (1976) suggested that rostrocaudal movements of the Orange-winged Amazon larynx could alter resonating properties of the nasopharyngeal and oral cavities and the tracheal tube. Scanlan (1988) noted that a vocalizing Grey parrot engaged in preparatory movements that transport the larynx to a vocalizing position, and synchronic movements that occur during vocalization.

Among birds, the psittacine tongue has a unique skeleton and musculature (Burton 1974), and, although it is likely to have evolved for eating (Homberger 1986), probably plays a considerable role in these birds' vocal abilities. For the Grey parrot, three joints and six extrinsic and seven intrinsic pairs of lingual muscles can affect tongue motion, and the glottal opening in the larynx is just dorsal to the tongue (Homberger 1986). This unique structure not only allows particularly flexible laryngeal movements (Homberger 1986; Scanlan 1988), but also may enable certain tongue movements to affect the larynx and trachea and, possibly, tracheal movement to affect the tongue. Nottebohm (1976) suggested that parrots may use their tongues to modify the shape and consequently the resonant properties of the vocal tract. (The tongue, its extrinsic musculature, and the hyoid apparatus usually are called the "lingual apparatus.")

Beak movement, although studied primarily in nonpsittacids (see, e.g., Hausberger et al. 1991; Westneat et al. 1993), may serve not only as a visual display, but also to modify amplitude and frequency during vocalization. For geese, beak opening or gape increase is positively correlated with higher call frequencies (Hausberger et al. 1991). White-throated sparrows and swamp sparrows may use beak gape to change the effective length of the vocal tract not to modify amplitude independent of frequency (i.e., by projection), but rather to track (and change with) the fundamental frequency produced by the syrinx so as to maximize amplitude for particular frequencies (Nowicki et al. unpubl. data; Westneat et al. 1993). A parrot's lower mandible has a wide range of motion, and thus mechanisms similar to those in sparrows must be considered in the study of Grey parrot phonation. Research in progress suggests that by producing some vowels with a mostly closed beak, Alex forms a slotted tube to extend the length of his oropharyngeal cavity and thus lowers the value of his second formant (see below; Patterson et al. in prep.).

We know little about the morphology of parrot nasal cavities, but these structures also may affect the quality of psittacine vocalization (Nottebohm 1976), just as they do for humans. Air moves between the pharyngeal and nasal cavities via the choana, which is framed by paired fleshy choanal folds (Homberger 1980, in press). The elasticity of these folds probably affects how sound energy is transferred from pharyngeal to nasal cavities (Warren et al. 1996).

Early Models of Avian Vocal Production

Because speech production, whether avian or human, is based upon the configuration and interaction of the sound source and the vocal tract (Sulter et al. 1992; Banta 1998), modeling strategies used for humans should apply to parrots. For humans, researchers propose physical and mathematical models for structures involved in speech, obtain physical and acoustic measurements, test the models against these measurements, and then revise the models as needed (see, e.g., Baer et al. 1991; Ferguson, Menn, and Stoel-Gammon 1992; Sorokin 1992). Researchers similarly design models for birds (e.g., Fletcher 1988; Fee et al. 1998), usually based on analogies either with wind instruments (e.g., Myers 1917) or the human voice (e.g., Nowicki and Marler 1988). The wind-instrument model assumes that the avian vocal tract is a single tube that can be open at both ends (syrinx and glottis) or only one end (glottis; see Brackenbury 1982), with vocal resonances tightly coupled to the source (syrinx). The human voice model assumes that the avian vocal tract has at least two resonating chambers, and that vocal resonances need not be coupled to the source (Olive et al. 1993).

Early studies and models concentrated on nonpsittacine calls and song; researchers then attempted to adapt these models to avian speech. Thus Thorpe (1959, 1961) proposed, on the basis of prior studies (e.g., Rüppell 1933), that application of wind-instrument or human models depended upon whether a species did or did not, respectively, have intrinsic syringeal muscles. Greenewalt (1968), in contrast, favored the wind-instrument model for all production of avian speech. The most complicated model, based on an electrical circuit (Harris et al. 1968), treated the trachea as a cylindrical tube, the oropharyngeal area as an expandable chamber, the syrinx as an acoustic oscillator, and the glottis and beak as variable slits; this model corresponds to the more sophisticated multichamber models proposed for humans.

The few researchers who proposed psittacid speech models generally favored the single-tube approach; given their lack of anatomical and kinematic information, this simple model made sense. Thus Nottebohm (1976:1633) accounted for tracheal resonance effects by assuming that the tube could be either opened or closed at one or both ends, or that "the glottal aperture of the larynx could vary from (a) fully open to (b) nearly closed," such that one end of the tube could assume a variable opening. Another possibility is that both ends approach closure (Westneat et al. 1993; Brittan-Powell et al. 1997). Data from my laboratory, however, suggest that, for Grey parrots, a multiple-tube (component) model, related to that of Harris et al. (1968), is most realistic. I describe the generally accepted human model, explain how psittacine and human structures might be functionally analogous and, on the basis of these analogies, propose a model similar to, but slightly simpler than, that of Harris et al. (1968).

A Model for Human Speech Production and Parallels with Parrots

The earliest acceptable "first-order" human speech model differed from the wind-instrument model in that it involved two chambers, the oral cavity and pharynx, divided by the tongue; this model excluded the nasal cavity (Stevens and House 1961). The model was judged by how well it predicted vowel formant frequencies, because vowel formants were easily obtained from sound spectrograms. More complicated models, interestingly, provide little additional accuracy based on this criterion (Maeda 1991). The two-chamber model is consistent with the idea that each vowel is uniquely defined by its two lowest formant frequencies, F_1 and F_2; these formants result from resonance in, respectively, the pharyngeal and oral cavities. Tongue placement changes for different vowels, and these changes modify the relative and absolute attributes of both chambers and create unique resonant characteristics for each vowel (Nearey 1978). Some researchers argue that the abrupt area-function discontinuities created by the tongue in the two-chamber model are required for vowel production and cannot be generated in a single-tube system (e.g., Fant 1970; Lieberman 1984).

Given the tongue's purported role in a human two-chamber model, researchers examined how tongue placement correlates with vowel formant frequencies (Lieberman 1984). For humans, F_1 roughly corresponds to tongue height and F_2 to tongue position (front versus back) within the oral cavity (Remez et al. 1987); vowels are characterized by where they lie on a tongue-placement chart in which F_1 is plotted against F_2-F_1 (see Figure 16.4). Human vowels tend to fall into four categories: high front, high back, low front, and low back (Borden and Harris 1984). Alex's vowels, in contrast, cluster in only two statistically distinct categories, front versus back (Chapter 15).

Given that some acoustic similarities exist between human and psittacine vowels, a two-chamber model, similar to that of Harris et al. (1968), may also be appropriate for Alex. F_1, the lower formant, which varies little across Alex's vowels, should thus result from resonance in and be correlated with a long tube that changes little in length. F_2, the higher formant, which differs significantly across Alex's vowels, should result from resonance in and be correlated with a short tube that can quickly and substantially change in length (Warren et al. 1996; Patterson et al. 1997).

Two anatomical candidates exist for resonant chambers in the psittacine vocal tract: the trachea and the oropharyngeal cavity (Warren et al. 1996). The obvious candidate for F_1 production is the trachea (from syrinx to larynx) because of its length. The trachea might be a candidate for F_2 production because this length (and thus resonant frequency modulation) can be changed by the alteration of tracheal ring overlap. Fletcher (1988), however, suggests that a change in length would be slight, and anatomical data confirm this suggestion (Pepperberg, Howell, et al. 1998); thus tracheal changes are unlikely to create the frequency shifts observed in F_2 across vowels. In contrast, the oropharyngeal cavity is ideal for producing frequency shifts seen in F_2 (see below). Note that beak opening could also effectively shorten this chamber and thus increase its resonant frequency.

This model may make more sense if I describe how sound travels in a parrot. As outlined by Harris et al. (1968), sound produced by the syrinx travels up the trachea, then encounters a point of constriction— where the trachea (which has a small diameter) meets the oropharyngeal cavity (with a large diameter). The constriction changes the impedance, which causes resonance in the trachea. When sound exits the oropharyngeal cavity and enters an essentially anechoic arena, it once again encounters an impedance change, causing resonance in the oropharyngeal cavity.

The anatomical feature corresponding to the first constriction point and thus the divider between the two chambers is probably the glottis (Warren et al. 1996; Patterson et al. 1997): a slit that exits from the larynx. Back-front laryngeal movement could quickly alter oropharyngeal cavity resonant frequencies by changing the position of the glottal opening. Moreover, glottal size can be altered by intrinsic muscles (Homberger 1979, 1980); thus the glottis can be a site of maximum constriction, acting like the neck of a Helmholtz resonator (Rossing

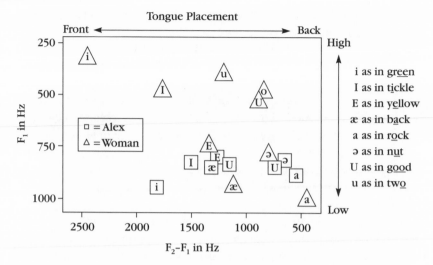

Figure 16.4. Predicted placement of Alex's and a woman's tongue based on acoustic parameters. Symbols are from the International Phonetic Alphabet. (From Warren et al. 1996.)

1982), and significantly affecting sounds produced by the Grey parrot vocal tract (Patterson and Pepperberg 1998).

The parrot's big fleshy tongue may also separate chambers; remember, such is the case for humans, who constrict various points along the vocal tract with structures such as the tongue, lips, and glottis separately or in combination (Ladefoged 1982). Figure 16.4 suggests how Alex's tongue placement might be predicted from acoustic data and thus be involved in vowel production. As noted earlier, Alex's vowels split into front/back categories of tongue placement (i.e., his F_2 varies significantly across vowels). Adapting a scheme specific to human vocal anatomy to that of a parrot may not be optimal, but the presence of front/back categories in Alex's speech may help determine how he produces these sounds. "Fronted" vowels, for example, have shorter F_2 resonating chambers and hence higher F_2s than "back" vowels.

Recent work by my students supports a two-chamber model of psittacine speech production. My students and I analyzed, via angular and spatial measurements, Alex's vocal tract configurations during vowel production via three noninvasive imaging methods: superVHS video, infrared, and x-ray radiography (Warren et al. 1996). We obtained specific tracheal measurements via (1) MRI of a different Grey parrot,

(2) caliper measurement of four preserved specimens, and (3) EBCT (or CT) of three of these specimens (Pepperberg, Howell, et al. 1998). We scanned (via EBCT) two Grey parrot cadaver heads with intact vocal tracts to obtain nasal, oropharyngeal, esophageal, and tracheal area functions; we measured how tongue and larynx positions correlated with beak opening (Patterson et al. 1997). Our goal was not only to test the two-chamber model, but also to describe various factors contributing to psittacine vowel production. Specifically, we wished to discover the role, if any, of suprasyringeal structures in vowel production and to correlate anatomical data with data on acoustic output. I review our experimental designs (details are in the cited journal articles), then concentrate on our results.

The Choice of Vowels to Study via Noninvasive Imaging of Alex's Speech

Analysis of video images of speech, whether human or parrot, is extremely time-consuming and we thus initially limited our study to human "point vowels," /i/ and /a/ (Warren et al. 1996): vowels that, for humans, differ most from one another both in acoustic characteristics and in tongue placement (Borden and Harris 1984). We generally analyzed /i/, as in *ea*t, and /a/, as in r*o*ck, in the context of a word but occasionally in isolation. (Our notation, as in Chapter 15, is that of the International Phonetic Alphabet, IPA, which is the system used by phoneticians and linguists to represent human speech; see Pullum and Ladusaw 1968.) We obtained six samples of each vowel (two for each word or isolated instance). Words or isolated instances analyzed for /i/ were "green," "ee," and "eat," and for /a/ were "rock," "want," and "pasta." Alex closely imitates my New York–Boston accent, and for me vowels in "rock," "want," and "pasta" are phonetically equivalent.[3]

When possible, we examined vowels in context rather than in isolation (see, e.g., Scanlan 1988) for two reasons. First, Alex's targeted vowels are more easily identified in the context of a referential term we specifically elicit. Vowels taken at random or from nonreferential speech may vary because of sound play (Chapter 12)—blends, rearrangements, or substitutions (Todt 1975b; Pepperberg et al. 1991). Second, data extracted from examination of entire words are more easily compared with the data from human speech studies, which rely on vowels in context because human formants, "to be identified with certainty, must often be perceived in relation to the frequencies of some other bit of speech uttered from the same vocal tract" (Borden and Harris 1984:194).[4]

Experiment 1: Noninvasive Imaging Techniques
to Analyze Alex's Speech Production

We used three noninvasive imaging techniques: x-ray radiography, SVHS video, and infrared video (Warren et al. 1996); the last two procedures provided qualitative data and context for the first. For all three methods, we attempted to keep Alex lateral to the camera to obtain optimal views of his vocal tract (Figure 16.5). To obtain data comparable to that for humans on movement of internal structures during speech production, we x-ray videotaped Alex with a Toshiba X-ray Machine and an Altronics Medical System HRV 3000 EM High Resolution Multiscan Video Recorder. Merlin Engineering Works downscanned the radiograph from 1026-line-rate to 512-line-rate format so that we could view the tape on conventional monitors. We used SVHS to observe external movements associated with speech production. We filmed with a Panasonic SVHS AG-450 camera and Maxell XR-S120 SVHS tape at 30 frames/s. We obtained external contexts for internal movements observed in the x-ray video, but could not record clear views of the tongue, although its movement might be important for Alex's speech production. We therefore videotaped Alex with an infrared camera with sensitivity better than 0.1° C (ImagIR, Santa Barbara Focalplane, Goleta, California) at 30 frames/s; his warm tongue was easily visible when not obstructed by his beak.

The analysis process was complicated and I present only a brief summary here. First, we identified which sounds were produced (and the order of their occurrence) in the series of vocal movements we saw on tape (Warren et al. 1996). As in human studies, we assumed that directional change in vocal structure movement correlated with production of a targeted sound (see, e.g., Subtelny, Whitehead, and Subtelny 1989). Frames containing vowel configurations were extracted for enhancement and analysis. Second, we digitized data captured on video and analyzed it using NIH Image.[5] Digital image processing enabled us to enhance specific anatomical structures. The radiopacity of such structures, which are buried in layers of hard and soft tissue, varies depending on their position in the bird. Consequently, such structures cannot be easily examined on unprocessed images. We used operations such as sharpening and smoothing, histogram equalization, density slicing, contrast manipulation, image magnification, and image subtraction to enhance each area of interest. Third, using procedures outlined by Subtelny et al. (1989), we identified visible structures in the x-ray video—the axis of the vertebral column, the hyobranchial junction (intersection of the larynx and the

Production of /i/   Production of /a/

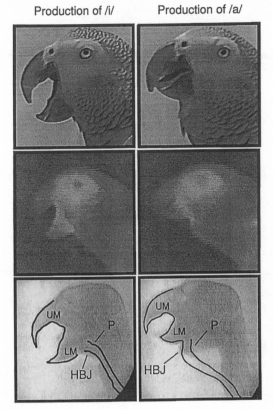

Figure 16.5. Images of Alex during production of /i/ and /a/ from (top) SVHS, (middle) infrared video, and (bottom) x-ray video. UM is the upper mandible, LM is the lower mandible, HBJ is the hyobranchial junction, and P is the procricoid cartilage of the larynx. (From Warren et al. 1996.)

hyoid bone of the tongue), the procricoid of the larynx, the cranofacial joint (hinge on upper mandible), and both bill tips—and placed small dots on predetermined landmarks on each structure. These landmarks were chosen because they could be identified consistently and their placement allowed us to perform necessary measurements. Points on the mandibles, for example, provided measurements of head tilt and beak gape. Given these tagged structures, we drew horizontal and vertical reference lines to measure angular and spatial variation in vocal tract configurations during vowel production (see Figure 16.6). Even with these techniques, we could measure the position only of the mandibles, the hyobranchial junction, and the procricoid of the larynx, and only the relative position and configuration of the trachea. We inferred tongue position, but could rarely view the tongue during vowel production, even in the infrared. Examination of the respiratory system, syrinx, trachea, larynx, glottis, nasal cavity, and tongue, which all play direct or indirect roles in psittacine speech production, thus had to await future experiments.

We used several statistical tests for data analysis (Warren et al. 1996). We determined which measurements differed significantly between /i/ and /a/, how much variation in each measurement was attributed to vowel versus word, and whether suprasyringeal tract motions were predictable for these vowels. We also identified which measurements covaried significantly for each vowel. We wanted to see which measurements, if any, provided distinct information and whether measurements we expected to be related were indeed correlated. Also, structural movements that were correlated but were not visible in the x-ray would suggest additional functional relationships during Alex's speech acts.

Experiment 1: The Results of Noninvasive Imaging

We found striking differences in articulatory configurations for /i/ and /a/ (Figure 16.5). Beak gape and the protracted area below the lower mandible differ during production of these vowels. The tongue is never in a high front position relative to the beak; in such a position, the tongue would be a light grey in the infrared. For each vowel, x-ray images show different positions for the trachea, the hyobranchial junction, part of the larynx, the vertebral column, the beak, and the skull.

Given that we had proposed that the trachea was responsible for the relatively invariant F_1s of Alex's vowels (Chapter 15), we searched for possible differences in tracheal configuration for /a/ and /i/. We could not obtain exact tracheal length and protraction measurements (see below), but the images suggest that, despite obvious protraction for /a/, little change in length occurred. The trachea, however, lies to one side of Alex's medial axis; thus, depending upon our view, a protracted trachea could be foreshortened or otherwise altered in perspective (Warren et al. 1996).

Statistical tests confirmed the significance of many observed differences. Significant differences ($p < 0.05$, df = 6) existed between vowels for the angular position of the lower mandible and hyobranchial junction, overall beak gape, head tilt, hyobranchial junction protraction, and position of the lower beak tip. Some measurements were more strongly affected than others by vowel identity. The relationship between vowel and angle of beak gape (E in Figure 16.6) is stronger than that between vowel and angle of head tilt (C in Figure 16.6), for example. Statistical tests also showed how vowel context (i.e., the word containing the vowel) affected physical measurements: When we

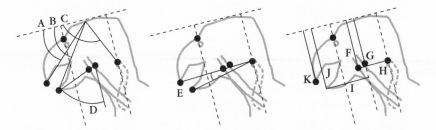

Figure 16.6. Depiction of the angular (left and center) and spatial (right) measurements used to characterize Alex's vocal tract configurations during his production of /a/ and /i/. A is the angle of the upper mandible to the horizontal; B is the angle of the lower mandible to the horizontal; C is the angle of head tilting; D is the angle of the hyobranchial junction to the procricoid with respect to the vertical; E is the angle of beak gape, which can be a negative value if beak tips overlap; F is the vertical position of the hyobranchial junction; G is the vertical position of the procricoid of the larynx; H is the horizontal position of the hyobranchial junction; I is the horizontal position of the lower mandible; J is the vertical position of the lower beak tip. (E is depicted separately from A–D for clarity.) (After Warren et al. 1996.)

examined context, we found that head tilt did not differ significantly between words containing /a/ versus /i/, but that differences for the five remaining measurements were stronger (see Warren et al. 1996).

A correlation matrix clearly depicted relationships among the physical measurements, A-K (Figure 16.6; see Moore 1992). We saw the expected close correlations between some physically related measurements, such as beak opening size and the angle the lower mandible makes with the horizontal, but not others, such as beak opening size and the upper mandible's vertical position. Other significant correlations were not as robust (Warren et al. 1996). Physical measurements could be divided into two groups based on correlational evidence. In group 1, the angle of the lower mandible to the horizontal and the beak gape angle were closely correlated, as was the angle of the lower mandible to the horizontal and the vertical position of the lower beak tip (B-E) and, consequently, beak gape angle and vertical position of the lower beak tip (B-J). In group 2, the angle of the hyobranchial junction to the procricoid was closely correlated to the horizontal position of the hyobranchial junction (E-J). Head tilt angle (C), which decreases as the beak comes up, was significantly negatively correlated with group 1 and positively correlated with group 2. The head thus tilts back as the beak opens. Most intriguing, however, were the consistent negative correlations between group 1 and 2 measurements, which suggest how external (B,E,J: related to beak opening) and internal (C,D,H: related to internal changes) structures might work in concert to produce vowels. In general, /i/ is characterized by an open beak, a nonprotracted trachea, and probably a frontal tongue position with an anterior orientation of the glottal opening; in contrast, /a/ is characterized by a closed beak, a protracted trachea, and probably a back tongue position, with a superior orientation of the glottal opening. We also found an unexpected statistical correlation between structures whose physical relationship was not obvious from the x-ray: As the beak opens, the hyobranchial junction (from which we infer laryngeal position, see below) moves forward (for /i/) and, as the beak closes, the hyobranchial junction retreats toward the spinal cord (for /a/). Anatomical connections, although indirect (i.e., through the influence of other anatomical parts) may exist between these structures. Whether movement of these two structures is correlated for other vowels is still unknown.

What was exciting was that these findings not only helped us see how Alex produced vowels, but also allowed us to compare his mechanisms to those of humans. We found striking differences and similarities between Alex's and human vowel production. I will try to describe these data without too much morphological detail (details in Warren et al. 1996).

Not surprisingly, differences we observed between vocal strategies used by humans and Alex are probably a consequence of the differing constraints and flexibilities of their vocal structures: Alex's production of point vowels /a/ and /i/ is, like that of humans, correlated with the functional morphology of his vocal tract, but one quite unlike that of humans. We observed considerable flexibility in Alex's trachea, which protracts during production of /a/ but not /i/. His tracheal protraction also correlates with changes in other vocal tract structures, and may be facilitated by movements such as the backward head tilt also associated with /a/ (see Warren et al. 1996). A human trachea, in contrast, is stationary and plays little role in sound modification, probably because it is below the laryngeal sound source (Dickson and Maue-Dickson 1982). We also found a striking difference in beak versus mouth opening: An open mouth is characteristic of a human /a/ (Fromkin and Rodman 1983), but Alex almost closes his beak to produce /a/ (Figure 16.5); in contrast, Alex's /i/ is associated with a very open beak (Figure 16.5), compared to the relatively closed mouth of humans (Fromkin and Rodman 1983). Although both of Alex's mandibles are hinged, statistical tests show that the angle of the lower mandible to the horizontal contributes most to beak gape. Lower mandible horizontal position, however, does not change significantly between /a/ and /i/. Clearly, we need to determine the roles of tracheal configuration and beak gape in the production of other vowels.

Despite the anatomical differences between Alex and humans, certain speech production mechanisms appear to be similar. As noted earlier, acoustic data on Alex's vowels suggest frontal and back tongue positions for /i/ and /a/, respectively, as is seen for humans, and our x-ray data support this finding. Given that "gross movements of the larynx in birds are necessarily associated with movements of the tongue" (Scanlan 1988:155), and given what we can infer during /a/ and /i/ production about laryngeal movement from Alex's tracheal protraction and position of the hyobranchial junction (which is on the ventral surface of the larynx, where the larynx and tongue's hyoid bone intersect), Alex's tongue, like that of humans, is probably retracted for /a/ and protracted for /i/ (Warren et al. 1996). We cannot tell, however, whether tongue position in the oropharyngeal cavity is incidental to tracheal position or beak opening, or whether the tongue itself actively modifies sound.

Our main goal was to see how well the observed mechanisms corresponded to predictions from our two-tube model. We thus used our vowel formant values to calculate cavity lengths for each measured formant frequency. Using the formula $X = 35,400/4\,F_n$, where X is the length of a uniform tube in cm, F_n is the nth formant frequency in Hz, and the constant is the speed of sound in moist, warm air (comparable to that inside a bird), we found cavity length values for /a/, /i/, and a mean for all vowels (Table 16.1). We compared these values with anatomical measures (from a dissected *Psittacus erithacus;* see below) to see how closely the two-tube model approaches physical reality. If the two sets of values are roughly comparable, the model has some physical validity; if not, the model needs revision.

The data suggest that a two-tube model is a useful starting point for understanding how Alex produces human speech. Certain aspects of his speech are consistent with our model; others are not. Looking at the contributions of the two tubes shows these consistencies and inconsistencies in detail.

Our data are consistent with an association between the oropharyngeal cavity and F_2; statistical tests show that both this formant and anatomical movements vary significantly and are correlated for /i/ and /a/ (Patterson and Pepperberg 1994; Warren et al. 1996). Oropharyngeal cavity length, for example, is influenced by the degree of beak gape and probably by the tongue and glottis. When the beak is open,

Table 16.1. Cavity length calculations associated with measured formant frequencies. (After Warren et al. 1996.)

Formant (F) value	Calculated cavity length (cm)	Acoustic correlate of F value
932	9.50	Value for /i/ for F_1
872	10.15	Value for /a/ for F_1
841	10.52	Mean F_1[a]
684	12.94	Value for /o/ [b]
2775	3.19	Value for /i/ for F_2
1433	6.18	Value for /a/ for F_2
2028	4.36	Mean F_2[a]

a. /o/ is excluded from calculations because of its outlier status. In 29 of 30 cases, /o/ was characterized by a single, low, broad formant (Patterson and Pepperberg 1994).

b. Value for /o/ presented for completeness.

the tongue forward, and the glottis relatively far to the front, the oropharyngeal cavity length effectively decreases, and its resonant frequency increases, which we observe for /i/ (Table 16.1). With the beak closed, the tongue back, and the glottis less far to the front, the oropharyngeal cavity is effectively enlarged, lowering its resonant frequency, which we observe for /a/ (Table 16.1). Although our calculations suggest that Alex's oropharyngeal cavity must vary between 3.19 and 6.18 cm to produce the highest and lowest F_2s (/i/ and /a/, respectively), the cavity need be only 3.68 cm to produce the mean F_2 across vowels. The actual combined length of oral and pharyngeal cavities is 3.69 cm. Increasing oropharyngeal cavity length by 67% through laryngeal retraction is unlikely; additional structures must thus be involved in the production of /a/. Alex might, for example, constrict his glottis to produce particular resonant frequencies through mechanisms other than changes in oropharyngeal or tracheal length (see below; see Gay et al. 1991 for human data). A 14% decrease from the measured value for /i/ production is, however, within reason (Warren et al. 1996).

Our data are also consistent with an association between the trachea and F_1, which does not vary significantly between /a/ and /i/. Calculations (Table 16.1) suggest that, to produce F_1s for /i/ and /a/, respectively, Alex must vary his trachea only between 9.50 and 10.15 cm; to produce the mean F_1 across vowels, tracheal length should be 10.52 cm (10.77 including /o/). Our x-ray data show that tracheal configuration but not length differs considerably between /a/ and /i/ (Warren et al. 1996). A trachea from a dissected Grey parrot (see below) was

11 cm (almost three times the oropharyngeal cavity length) and could be stretched by only ~10%. A protracted trachea may be a source of low frequencies; before making definite claims, however, we must compare x-ray and other, more sophisticated data for tracheal configurations (see below) for vowels such as /i/ and /o/, between which F_1 does vary significantly (Patterson and Pepperberg 1994).

The vowel /o/ presents a challenging case for any model, because in almost all instances Alex's /o/ is characterized by only a single formant (Patterson and Pepperberg 1994). Our calculations also yield a tube length of 12.94 cm (Table 16.1). A model would therefore require coupling of the tracheal and oropharyngeal cavities to form a single tube. Whether and how that is possible are as yet unknown.

Our model and x-ray data suggest that beak opening is important in Alex's vowel production (Warren et al. 1996). Little, however, is known about the role of beak gape in avian productions except oscine song (Fletcher 1988). Until the results of an immobilization study were evaluated (Nowicki et al. unpubl. data), beak gape in sparrows was assumed to be correlated with frequency rather than amplitude at a given frequency. Conceivably, beak gape in the Grey parrot, like that in sparrows, tracks movement of other structure(s) responsible for F_1 and F_2 of vowels; alternatively, according to Harris et al. (1968), the beak might act as a slotted tube and the bird could vary the size of the slots to modify production (Patterson et al. in prep.).

Overall, our results are a first step in learning how one nonhuman, nonprimate, nonmammal produces human speech sounds. Our study also revealed that we needed more dimensional data on the Grey parrot vocal tract to design an appropriate model and determine exactly how the vocal tract functions. We thus embarked upon a study to learn how best such measurements might be obtained.

Experiment 2: Comparative Measurements of the Grey Parrot Trachea

Several procedures provide shape and dimension data for static vocal tracts, but no single technique is absolutely accurate, and measurements of static structures do not allow us to correlate movements and speech production (Pepperberg, Howell, et al. 1998). Nevertheless, static vocal tracts provide information about anatomical areas such as the trachea that do not appear to change much during speech acts; by comparing tracheal measurements obtained by different procedures, we can determine which technique provides the best measurement of other structures. Methods currently in use and available when we performed our study were dissection, MRI, and EBCT (or CT).

Each of these procedures has advantages and disadvantages. Traditionally, researchers dissect preserved specimens, but the specimen and thus the resulting data may be affected by tissue shrinkage and loss of flexibility from rigor mortis, decay, and formalin storage (Hersch 1966). Baer et al. (1991) suggested that MRI, which poses no known danger to subjects, affords an advantage over invasive methods by allowing observation of an unaltered living system (see Filipek et al. 1989), although in many cases *functional* MRI—that of the living system in action—is not possible, and the system must be studied at rest. Nevertheless, image analysis techniques purportedly are more precise than caliper measurements on dissected specimens. MRI, however, despite its accuracy (Sulter et al. 1992), still involves three inherent limitations. First, constraints (such as exposure time) exist on scanning; second, quantitative data analyses rely on subjective decisions and thus are as liable to human error as caliper measurements; third, MRI images blur somewhat at air-tissue boundaries (Baer et al. 1991; Stone 1991), and this blurring could affect exact tracheal area calculations. In contrast, CT provides extremely sharp contrast at air-tissue interfaces and finer resolution scans than MRI (Story, Hoffman, and Titze 1996; Story, Titze, and Hoffman 1996), but data are also subject to scanning constraints and subjective measurements. Also, CT, which uses ionizing radiation and poses nontrivial hazards to live subjects, is preferably performed on preserved specimens and thus the data probably reflect some of the same inaccuracies as do dissection studies. We thus compared tracheal data from an MRI scan of a live parrot with that obtained from traditional dissections, and compared both MRI and dissection data to CT scans of dissected tracheas we had measured with calipers.

We used different subjects for some studies and the same ones for others (Pepperberg, Howell, et al. 1998). We did MRIs on Alo, who was similar in size and general physique to Alex, who could not be imaged for several medical reasons, including the presence of a metal pin in his chest cavity. Alo was anesthetized before being placed in the imaging coil to prevent movement during scans. We dissected a formalin-preserved Grey parrot specimen donated by the Smithsonian Institution (#54099) and three extracted tracheas similarly preserved and donated by G. Dorrestein, D.V.M. (Veterinary Pathology Department, University of Utrecht, The Netherlands). We removed the vocal tract from the Smithsonian specimen and stained and cleared it according to procedures in Dingerkus and Uhler (1977); we then preserved it in glycerin. Netherlands specimens were neither stained nor cleared. For the CTs, we imaged the three excised Netherlands vocal tracts.

I'll provide only basic details of the equipment and procedures used for each experiment (see Pepperberg, Howell, et al. 1998 for specifics). MRI scans used a GE Sigma Clinical Magnetic Resonance Imaging Scanner (Department of Radiology, University Medical Center, University of Arizona) and a standard human "knee" coil; we obtained three series of images—two coronal (images perpendicular to the spinal column) and one sagittal (images parallel to the spinal column). For dissections, all length and diameter measurements were taken with Starrett 120 mm dial calipers. We measured tracheal length between the tracheobronchialis muscle and larynx, and represented the trachea by ring diameters. Cross-sectional tracheal shape is mostly ellipsoid; thus we measured two diameters: (a) those that bisect the trachea longitudinally into right and left halves and (b) those that also bisect the trachea longitudinally, but perpendicular to those in (a). We were requested not to dissect the Smithsonian trachea; thus all its diameter measurements included thicknesses of the ossified tracheal rings. One Netherlands trachea was sliced open to obtain thickness measurements of the rings. To calculate inner areas, we subtracted corresponding values from each tracheal radius before calculating areas. We also subtracted these ring thicknesses from outer measurements of the Smithsonian trachea to estimate its inner areas. We obtained CT images of the three excised Netherlands vocal tracts using the Imatron C-100XL at the University of Iowa.

We analyzed MRI and CT scans with digital imaging processing programs. For MRI scans, we used NIH Image. For CT scans, we used VIDA (volumetric image display and analysis), a program written specifically for such analyses (Hoffman et al. 1992).

Results from Tracheal Measurements via MRI, Dissection, and CT

I compare the results of these techniques in Figure 16.7. We plotted MRI and CT areas against distance along the trachea, and dissection data against ring number. Inner cross-sectional areas of CT and dissection data cluster, whereas MRI data cluster instead with data from outer cross-sectional areas from CT scans (Figure 16.8) and dissection results (Figure 16.9). Thus MRI scans apparently included tracheal ring diameters. We knew MRI blurred air-tissue boundaries, but did not expect measurements to match the outer tracheal areas so closely. Overall, a striking similarity exists between the various data sets for changes in cross-sectional area across tracheal length, even though dissection/CT and MRI data came from different individuals: MRI and dissection/CT outer measurements differ (except at the ends of the tracheas) by ~10% (range 2–24%), and where MRI and dissection data diverge in the last slices, the standard deviation of the MRI data was particularly high. A discussion of these results not only provides some reasons for these differences, but also suggests what technique might be most useful for future studies.

Differences may simply have been due to individual subject variation. Grey parrots, like all creatures, vary in size, and Greys particularly vary in how far they can extend their necks (Banta, Patterson, and Pepperberg, pers. observ.). Also, we did not know the ages, sexes, or subspecies of all subjects. None were *Psittacus erithacus timneh* (a distinctly smaller subspecies, Forshaw 1989), but some may have originated in different areas of Africa and possibly represent different genetic populations (Forshaw 1989); however, birds from different areas vary in size by < 10%. The Netherlands tracheas were all consistently slightly shorter than the Smithsonian trachea, although not to a striking degree (e.g., the shortest Netherlands trachea stretched 10% was comparable to the Smithsonian compressed 10%). Age and sex differences may also translate into tracheal length and area variations (e.g., differences in tracheal ring thickness and ossification).

Furthermore, each of our three techniques has distinct possible confounds (Pepperberg, Howell, et al. 1998). Even initial dissection procedures might have influenced the data: The Smithsonian trachea was stained, cleared, and stored in glycerine, but Netherlands tracheas did not undergo these procedures. Clearing may remove residual tissue and allowed additional tracheal stretching; staining may add thickness; glycerine storage may have kept the Smithsonian trachea from drying (and slightly shrinking) during measurement. These changes may well have caused divergence of CT and dissection data from data obtained from a live subject during MRI scans. Also, caliper measurements during dissection are subject to human error. In MRI and CT calculations, we obtain area measurements by counting pixels in a digitized picture of the selected area, but dissection areas are calculated from diametric measurements—because the trachea is not perfectly ellipsoid, the dissection calculation has a built-in imprecision. MRI has comparable problems with quantitative data analysis; the measurement error is ~0.5–2% (see Plante and Turkstra 1991). Moreover, although pixel counting would seem straightforward, pixel area is defined and limited by scanning technique; in our MRI study, pixel dimensions were about the same size as the tracheal wall thickness and thus probably often strad-

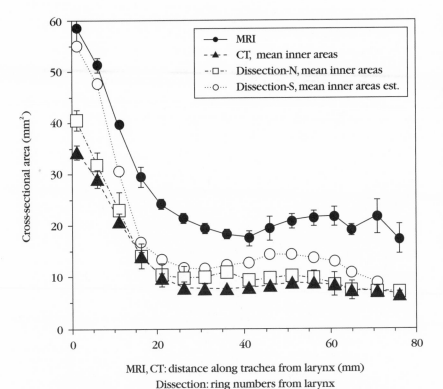

Figure 16.7. Cross-sectional area comparison of dissection, CT, and MRI methods. Dissection values are reported after subtraction of the estimated tracheal thickness, and plotted against ring number along the trachea; CT (inner areas) and MRI values are plotted against distance in mm along the trachea. (From Pepperberg, Howell, Banta, Patterson, and Meister 1998.)

dled a boundary. Given the irregular nature of a parrot trachea, errors can occur in determining the air-tissue boundary (Stone 1991; Russ 1992). Specifically, we could not estimate tracheal ring thickness from the MRI images because we could not distinguish tracheal tissue from the air it surrounded.

Other measurement differences could have been a consequence of measuring an *in vivo* versus an excised trachea (Pepperberg, Howell, et al. 1998). An obvious factor might have been how we positioned our subjects. We may have inadvertently stretched or compressed excised tracheas with respect to their normal positions; also, although we attempted to place Alo in a straight position in the MRI, her trachea may also have been stretched, compressed, or otherwise deformed.

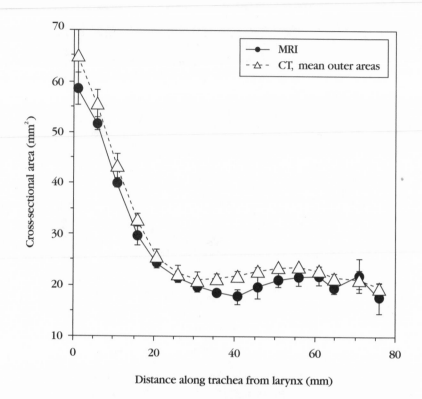

Figure 16.8. Cross-sectional area comparison of MRI and CT data, using mean outer areas for the CT data (that is, including tracheal wall thickness). (Adapted from Pepperberg, Howell, Banta, Patterson, and Meister 1998.)

We did find that Alo's trachea was bent at about a 4° angle at one place where MRI and Netherlands dissection data deviate somewhat. Moreover, other structures in a live bird may exert pressure and cause normal deformations that are absent in excised tracheas. Where the interclavicular air sac surrounds the trachea (which thus may have been torsionally affected by air sac pressure), for example, MRI and dissection data again diverge; little divergence exists, however, between MRI and CT outer data in this region. MRI scanning can take up to 30 min and any motion (even a subject's heartbeat and respiration) during this period may affect the clarity of the scan (Stone 1991). Such motion may have caused somewhat increased fuzziness and thus increased standard deviation in certain measurements. MRI may also have measured some inter-ring areas; because both positive and negative air

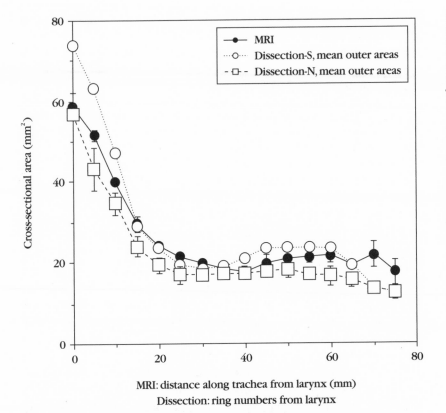

MRI: distance along trachea from larynx (mm)
Dissection: ring numbers from larynx

Figure 16.9. Cross-sectional area comparison of MRI and dissection data, using mean outer areas for the dissection data (that is, including tracheal wall thickness). (Adapted from Pepperberg, Howell, Banta, Patterson, and Meister 1998.)

pressure in a live trachea could expand and contract tracheal cross-sectional area relative to a preserved specimen, all MRI areas may not directly correspond to outer areas of the dissected specimens.

Given the problems described above, CT scans of a live subject would seem optimal for determining Grey parrot vocal tract dimensions. CT is far better than MRI for human vocal tract measurements for different vowels (Story, Hoffman, and Titze 1996). CT, however, requires a significant dose of radiation, and the small size of a parrot thus precludes such an *in vivo* study. The correspondence between MRI data on a live subject and outer CT cross-sections of preserved specimens and that between inner CT and inner dissection data nevertheless implies that inner CT cross-sections on preserved specimens provide accurate information, and suggests that we could use other information from CT scans of cadaver birds to learn more about the Grey parrot vocal tract.

Experiment 3: CT Scans of the Entire Grey Parrot Vocal Tract

Recent papers (e.g., Story, Hoffman, and Titze 1996) have correlated human CT data with vocal tract shape for several vowels; we therefore performed a related study on Grey parrots (Patterson et al. 1997). Using the same CT equipment as in Experiment 2, we scanned two Grey parrot cadaver heads with intact vocal tracts (i.e., attached tracheas); cadavers were donated by a Utah veterinary practice. We scanned a mature bird with its beak propped open to simulate production of /i/, and a juvenile with a closed beak to simulate production of /a/. These CT scans provided data unobtainable from the x-ray and MRI studies, particularly for tiny structures such as the procricoid and hyobranchial junction. As before, we analyzed CT scans with the VIDA program (Hoffman et al. 1992). We also measured the resting and maximum size of the glottal aperture in the excised tracheas with calipers and used a millimeter ruler to estimate minimum, maximum, and resting lengths of each trachea.

To predict vowel formant frequencies, we used a one-dimensional wave propagation model originally designed for humans (Story, Hoffman, and Titze 1996). The model simultaneously takes into account area functions of *all* tubes involved in vocal production and estimates formant frequencies and relative intensities for each tube configuration. (Note that, in contrast, Warren et al. 1996 calculated values for each tube independently.) We substituted area functions of parrot tubes for human values. By modeling changes in tracheal and esophageal length and area functions of the glottis, mouth, and pharynx, we simulated changes in the living bird and estimated formants associated with various vocal tract shapes (Patterson et al. 1997). We could then see which tube configurations produced our measured formants (Chapter 15).

Observations and Results of CT Scans

Interestingly, we found that beak position may affect both the esophagus and the choana (Patterson et al. 1997), structures we mostly ignored in our other experiments. We were not entirely surprised because some consonant data might be explained by esophageal involvement (see Chapter 15). Closing the beak enlarges the laryngeal box and pushes it back against the pharyngeal wall, occluding the esophagus. The choana is more elliptical when the beak is closed rather than open and appears larger from a lateral view, but its area is not significantly dependent upon the size of the beak opening. Opening the beak depresses, flattens, and moves the larynx forward; as the larynx drops down and away from the pharyngeal wall (e.g., for /i/), the esophagus may be added to the vocal tract as a side branch resonator, and possibly the more rounded choana also is involved (Patterson et al. 1997).

Opening the beak also appears to affect area functions and the length of the pharynx and the mouth. With the beak closed and tongue up, these cavities are quite small. Opening the beak enlarges and slightly lengthens the oropharyngeal cavity (Patterson et al. 1997).

By plugging our anatomical data into the one-dimensional model, we created over a thousand possible combinations for the various structures. We used statistical tests to compare each anatomical value to its neighbors and identify significant changes in frequency and intensity that result from changing tracheal length, glottal aperture, the area of the pharynx and mouth, and, if appropriate, esophageal involvement (Patterson et al. 1997). These data, combined with information on means and ranges, identify the effects of anatomical variables and their interactions on formant frequencies and intensities. Table 16.2 shows the relative contributions of each anatomical structure variation to mean formant value. These percentages do not reflect the extent of structural change, but rather the extent to which any variation, whatever its extent, affects formants. As we predicted earlier, tracheal length was more closely correlated with F_1 than F_2, and area functions of the pharynx and mouth and the amount of esophageal involvement were more closely correlated with F_2 than F_1. Glottal area correlated more closely with F_1 than F_2, but affected both formants more than any other structure.

In general, models predict F_1s and F_2s consistent with mean measured values for /i/ and /a/, though the predicted F_1 for /i/ is ~10% too high (but within a standard deviation; Table 16.3); the data are also consistent with Alex's observed anatomy for these vowels. For /i/, we find an open beak and glottis, no tracheal protraction, and little or no esophageal resonance. For /a/, we find a closed beak, an open glottis, a trachea slightly longer than for /i/, and clear esophageal involvement.

Evaluating the Results of Experiment 3

We treated each anatomical component of the Grey parrot vocal tract as a simple tube (albeit conjoined) and used a one-dimensional wave

Table 16.2.   Relative contribution to the mean formant value of variation in each anatomical structure. (After Patterson et al. 1997.)

	Tracheal length	Glottal aperture	Mouth/ pharynx	Esophagus
F_1	36%	57%	6%	1%
F_2	20%	36%	26%	18%
$F_2 - F_1$	22%	17%	33%	3%
Amplitude of F_1	16%	42%	25%	3%
Amplitude of F_2	6%	56%	12%	31%

propagation model (Patterson et al. 1997). Do these simplifying assumptions reflect reality? Let's look at our assumptions in order: First, we assumed that the trachea stretches uniformly along its length, that is, that inter-ring distances increase for all rings of the trachea proportionately. This approximation may not be entirely valid. At the cranial end where the diameter increases, several rings overlap almost completely when the trachea is at rest, but all evidence of tracheal "lengthening"—actually protraction—occurs at exactly this point: Might only this area lengthen? Initial testing suggested that such localized protraction increases F_1 but decreases F_2 compared to more uniform stretching; such adjustments would make the model correlate less well with measured formants. Second, we treated the glottis as a 5 mm constriction. This approximation is probably correct, but laryngeal musculature is complex, and the constriction might be shorter. Initial testing suggested that formant frequencies are raised and F_2-F_1 decreases as glottal length decreases, a result that would be more in line with measured values. Third, we simplified relationships among values in the pharynx and mouth and considered their shapes to be uniform. Constrictions created at different oropharyngeal points by tongue movement, however, might affect sound production (Hersch 1966; Homberger 1986). Fourth, the model does not examine acoustic effects of the upper and lower beak beyond the end of the mouth. Treating the beak as a slotted tube for some productions of /a/ lowers formants (Patterson et al. in prep.) Finally, we assumed that our specimens were appropriate models for Alex, and the results of Experiment 2 suggest that this assumption was valid. We nevertheless know that individual subject differences exist, and could cause variations of about 10%.

Other speech production models may also be consistent with our data. Sophisticated models for human speech, for example, take into

Table 16.3. Actual and calculated formant values for Grey parrot vowels /i/ and /a/. (After Patterson et al. 1997.)

	/i/	SD	/a/	SD
F_1 actual	932	109	872	100
F_1 calculated	1033		869	
F_2 actual	2775	123	1433	90
F_2 calculated	2706		1377	

account tongue movement (e.g., Stone 1991), nonlinear energy exchange in the vocal tract to learn what aspects of the speech spectrum should be attributed to the source versus the filter (McGowan 1991), and the limitations of one- versus three-dimensional models (Kagawa et al. 1992). Researchers studying humans, however, believe that models are constrained by the quality of empirical knowledge available even with sophisticated scanning techniques (Crelin 1987); the same is probably true for parrots. Fant (1991:490) states, "There is an apparent lack of anatomical and physiological data, a lack of insight into dynamic variations of control parameters and lack of flexibility to continuously adapt to configurational variations such as overall tract length, lateral dimensions and essential cavity structures of consonants. We need more insight into voice and noise source interaction, mutually and with respect to the tract system function." Complex three-dimensional models might provide more detailed form and function correlations, but may not improve correlations between formant values and physical structures: Even in humans, one-dimensional linear wave propagation models adequately calculate the formants of a given vocal tract (Maeda 1991; Story, Titze, and Hoffman 1996).

Finally, our consonant data also support a two-tube model. Very briefly, because such work is still in progress, I note that we can use stop loci to calculate where Alex's point of constriction may occur in his vocal tract for stops. These calculations place points of constriction for /b/ at 4.15 cm and /p/ at 5.45 cm within his oral cavity; however, the cavity is at most 3.7 cm. Thus Alex must produce these stops by another mechanism (Patterson and Pepperberg 1998): Possibly, he lengthens the front cavity to produce the observed low frequencies by involving his esophagus. This mechanism is consistent with models of a tube closed at one end or approaching closure at both ends (see, e.g., Nottebohm 1976; Westneat et al. 1993; Brittan-Powell et al. 1997). The end of a tube that is not entirely closed but approaches closure resembles the neck of a Helmholtz resonator (Rossing 1982); the effect is similar to that achieved by lengthening the tube. The data are also consistent with a perturbation theory model (Chiba and Kajiyama 1941, discussed in Johnson 1997:100), which predicts that "constriction of the vocal tract near a point of maximum velocity . . . *lowers* the formant frequency." (Actual calculations for either model, however, are beyond the scope of this chapter.) Finally, Alex may lower resonances by involving his nasal cavities or limiting his glottal aperture.

Overall Conclusions

Despite our simplifying assumptions, difficulties in comparing live and cadaver structures, and our use of different measurement methods, I believe our findings do provide important insights with respect to the study of Grey parrot speech. Clearly, we have only begun to describe the complex alterations that a living Grey parrot can make to its vocal tract. Nevertheless, we can model formants of Alex's vowels. Our models also fit expectations about anatomical configurations of the vocal tract during production of /i/ and /a/ and provide strong evidence that, at least for the Grey parrot, vocal tract resonances play a primary role in production of human speech patterns. Furthermore, consonant data support our two-tube model (Chapter 15). Whether our model and claims can be extended to other birds, or even other talking parrots, is unclear. Other mechanisms are likely for budgerigars (Brittan-Powell et al. 1997; Banta 1998) and mynahs (Klatt and Stefanski 1974). Quite likely the absolute size of a bird and its vocal tract, the presence or absence of complex syringeal mechanisms, and the type of vocalization being produced (i.e., song versus speech) dictate the specific mechanisms that are used. Our findings nevertheless raise questions about the degree to which human structures and mechanisms are necessary for the production of English speech. Possibly, by comparing and contrasting production mechanisms in "talking" birds and humans, we can more precisely define what is required, and propose numerous alternative mechanisms, for producing "speech."

17 CONCLUSION

What Are the

Implications of Alex's Data?

In humans, intelligence certainly means more than flexible learning: terms like "thinking clearly," "solving difficult problems," and "reasoning well" recur in attempts to define the ability. The scope of intelligence is quite wide, including learning an unrestricted range of information; applying this knowledge in other and perhaps novel situations; profiting from the skills of others; and thinking, reasoning, or planning novel tactics. (Which should remind us not to expect that intelligence is a single "thing," but a bag of devices and processes, endowments and aptitudes, that together produce behaviour we see as "intelligent.")

What all these (possible) components of intelligence have in common is that they contribute to general purpose skills, not highly specialized ones. This may help understanding what intelligence is *for*. All the abilities above are what one should expect in generalist animals: intelligence should most benefit extreme generalists, species adapted to exploit continually changing environments since they must daily cope with novelty in order to survive.

Richard Byrne, *The Thinking Ape* (1995:38)

In the preceding chapters I have described over two decades of my research on the cognitive and communicative abilities of Grey parrots, and on the type of input needed for referential, allospecific vocal learning to occur. I have, however, avoided even trying to answer a question that arises at many of my public talks: *Why do these birds have such abilities?* I have suggested that a parrot's capacity to learn what I teach in the laboratory must be based on an existent cognitive architecture, but have not intimated why this cognitive architecture should exist— or, from an evolutionary standpoint, what selection pressures may have shaped such an architecture and hence Grey parrot behavior. Such hesitancy comes from my propensity to propose testable hypotheses rather than exciting theories and, at present, I see few ways of designing appropriately rigorous tests of any possible theory. The only currently viable approach is to draw parallels to other creatures and other theories. The resulting correlations are, however, untestable; furthermore, correlations merely suggest causation. With those strictures in place, however, a bit of speculation might be in order—if only to try to place theories about avian intelligence into some perspective with current theories about other nonhuman animals.

The most obvious parallels are those with primates. The cognitive and social capacities of primates have been studied extensively (e.g., by de Waal 1989; Byrne 1995; Tomasello and Call 1997), and researchers have proposed numerous theories to explain the origins of these capacities. Humphrey (1976), for example, proposed that intelligence (and presumably the need for cognitive processing) is a correlate of having a complex social system and a long life: that intelligence is the outcome of a selection process favoring animals that can remember and act upon knowledge of detailed social relations among group members. More generally, Rozin (1976) defined intelligence as flexibility in transferring skills acquired in one domain to another. How the patterns suggested by Humphrey and Rozin might drive parrot cognitive skills and vocal behavior seems obvious: Parrots, long-lived birds existing in complex social systems, not unlike those of some primates, use abilities honed for social gains to direct other forms of information processing and vocal learning. If we add parrots' need to discern categorical classes (e.g., to distinguish neutral stimuli from predators, poisonous from healthful foods, etc.), their abilities both to recognize and to remember environmental regularities and adapt to unpredictable environmental changes over an extensive lifetime, and a communication system that is primarily vocal, the capacities of parrots are not surprising. In fact, Marler (1996) has already proposed some similar parallels between birds and primates, although not specifically for parrots. I suggest, however, a scenario with one additional contingency, but first let's look at psittacine cognitive abilities that are used in nature for both social and other skills.

Individual Recognition

Individual recognition can occur between members of mated pairs, among members of a flock, and between parents and offspring. Such recognition not only is critical for mediating interactions, but also requires cognitive processing: the ability to learn, memorize, and, most important, for birds that move between flocks, the flexibility to update information and transfer it among situations. Aspects of individual recognition have been observed in a number of parrot species. Bahama Amazons *(Amazona leucocephala bahamensis)* may use calls for individual recognition of mated pairs within groups (Gnam 1988), and spectacled parrotlets *(Forpus conspicillatus)* appear to recognize siblings' as well as mates' contact calls (Wanker et al. 1998). Individual short-billed white-tailed black cockatoos *(Calyptorhynchus funerus latirostris)* can be identified by differences in the lengths of parts of their calls (Saunders 1983). Calls that can be used to identify individuals may be part of the repertoire of Glossy Black cockatoos *(Calyptorhynchus lathami;* Pepper, pers. comm.). Rowley (1980) proposes parent-offspring vocal recognition in galahs. Wright (1996) has demonstrated the existence of dialects in Yellow-naped Amazons *(Amazona auropalliata):* Several roosts can share a dialect, and some birds at roosts bordering two dialects use the calls of both neighboring dialects interchangeably; the dialects may possibly be used for flock recognition (note Schindlinger 1995). Bradbury and Wright (pers. comm.) are now studying whether vocalizations are used for individual recognition in these birds.

Sentinel Behavior

During foraging, some psittacids engage in sentinel behavior: One member of the flock perches in an exposed position and calls at the approach of danger. Like vervet monkeys (Cheney and Seyfarth 1990), parrots probably respond to these calls based to some extent on the reliability of the caller, that is, on some cognitive evaluation of the caller's identity and status. Any proposal suggesting how sentinel behavior derived from social skills would be a "just so" story at present; but Curio (1988; Curio, Ernst, and Vieth 1978) has shown cultural (social) transmission of enemy recognition in passerine birds, and McGowan and Woolfenden (1989) showed that Florida scrub jays coordinate group sentinel behavior (note Wickler 1985). Parrot behavior is probably similar. The critical point is that both cognitive and social skills are involved: Smith (1997) suggests that use of signals with external referents evolved when advantage accrued to a group that could be alerted by *specific* information from a distant signaler. Furthermore, even if the alarm calls do not refer to specific predators, the *nature* of each call must be specific. False positives or negatives would each have serious consequences for survival; thus sentinel behavior reflects some cognitive capacity for categorization and for using information flexibly. A predator observed routinely in one location, for example, must be recognized even when the location varies (Pepperberg 1998). Sentinel behavior has been observed in indigo macaws (*Anodorhynchus leari,* Yamashita 1987), Puerto Rican parrots (*Amazona vittata,* Snyder, Wiley, and Kepler 1987), maroon-fronted parrots (*Rhynchopsittica terrisi,* Lawson and Lanning 1980), and white-fronted Amazons (*Amazona albifrons,* Levinson 1980). A study of the short-billed white-tailed black cockatoo does not mention the presence of birds specifically acting as sentinels, but does report that any individual in a flock will emit alert screams upon noting anything that is unusual (Saunders 1983). Interestingly, some psychologists (e.g., Hearst 1984) argue that detecting something "unusual" would require a bird to have some form of representation of what is "usual." Given the extensive areas over which psittacids forage (Saunders 1983), such representations would require extensive memory for what was "usual" in a particular spot; a bird would also need considerable flexibility to process "usual" versus "unusual" changes with respect to each spot in the area.

Duetting

Some birds, particularly ones that form long-term monogamous pairs, learn duets. When dispersed in a flock, birds may use duets different from those they use when in close contact with their mates. Such behavior might mediate interactions among flock members (Gwinner and Kneutgen 1962; Mebes 1978; Wickler 1980), and the appropriate set of vocalizations must be chosen for a given situation. Although only rarely recorded in the wild (e.g., by Nottebohm 1972), parrot duets seem more complex and seem to take longer to learn than those of shorter-lived species (e.g., wrens, Farabaugh 1982). In general, duetting appears to require considerable plasticity and learning capacity. Although duetting appears unrelated to pair bonding in aviary-dwelling canary-winged parakeets (*Brotogeris v. veriscolurus,* Arrowood 1988), the exact role of duetting in the behavioral repertoire of wild, longer-lived psittacids remains unknown. Antiphonal duetting between mated pairs has been reported in the wild only for the Orange-winged Amazon (*Amazona amazonica,* Nottebohm 1972) and the Grey parrot; in the latter case, such vocalizations were observed at nest holes and such contiguity suggested (although did not prove) that these utterances are involved in precopulation behavior or territory defense (May, unpubl. observ.).

Categorical Classes

The biological basis for categorical classes was discussed in Chapter 4, but a slightly different approach may now be useful. We should not be surprised by an animal's ability to form categorical classes at some level. In fact, the absence of such abilities is what would prove notable: The need to extract regularity from the world seems to be a basic capacity for survival. Of course, not all creatures might need (or possess) all the various levels of categorization, and determining the level at which an animal can form categorical classes is not a trivial matter. Studies on the formation of categorical classes must be designed to take into account the ecological and ethological framework of the creature that is being studied—a framework consisting of the niche in which the subject lives and its species-specific behavior patterns (Pepperberg 1996). Results must be interpreted within this same framework (Pepperberg 1998). Of particular importance in this regard is that many studies to determine the abilities of animals to form various classes, my own included, require that the subjects form classes that are outside their natural domain. Alex, for example, can form some abstract classes related to human labels that refer to ecologically relevant aspects of his world (e.g., color), but he also forms some classes whose ecological relevance is not clear (e.g., "picture toy" for photographs). Interestingly, some researchers have demonstrated how learning can be greatly facilitated under ecologically and ethologically relevant conditions (Wright and Delius 1994) and how learning that will not occur under traditional paradigms will occur under ethologically relevant ones (reviews in Pepperberg 1985, 1992b). In this regard, data demonstrating that birds—and parrots in particular—might learn categories through social transmission would be particularly interesting. As noted above, enemy recognition may be transmitted socially in passerines, and in some birds food choices may also be transmitted, but the data are unclear (see Marler 1996 for a review). Possibly what is of basic importance in some species of birds is not merely the capacity to perform a particular task, but rather the capacity to learn a general type of task in a particular manner—for a parrot, for example, not only to use its general capacity to form ecologically relevant classes, but also to use an ethologically based learning situation to acquire other, less relevant classes.

Communication and Intelligence

Smith (1997:31) suggests that communication is best understood as being "forged in the linking of different individuals," and that if we know both the information a recipient receives from a signal and the recipient's response to that information, we have data leading to an understanding of the recipient's processing abilities. Of critical importance is Smith's thesis that both the content and the contextual information—the source and the circumstances surrounding the emitting of a signal—are part and parcel of what a recipient must process. And that the result of this processing is a *weighting* of input before a reaction is emitted; this weighting not only tell us that processing has occurred, but also provides some details of how an information hierarchy is formed.

This argument leads directly to that additional contingency concerning intelligence at which I hinted earlier: Possibly intelligence was an evolutionary outcome of the need not only for memory and flexibility, but also *for choosing what to ignore as well as what to process.* Social primates must actively engage in such choice; an animal that could not form a hierarchy of information would be unable to act. Only in parrots, however, can we clearly see, by what is vocally reproduced and what is not, the outcome of such choices in the vocal domain. Specifically, wild Grey parrots have the capacity to reproduce environmental noises (e.g., sounds of a nearby stream) as well as conspecific and allospecific vocalizations—but they reproduce only the last two forms (see, e.g., Cruickshank et al. 1993; May, unpubl. observ.). As noted in Chapter 14, reproduction of environmental noises would be maladaptive, and some evolutionary pressure probably selected against such behavior, that is, directed a hierarchy of what was important to learn. Such selection probably occurred in primates, but determining the results of such selection for primates is much more complicated. When, for example, a vervet monkey ignores a leopard kill strategically placed by an experimenter, one argument is that the monkey does not recognize the kill as evidence for the likely presence of the predator, that is, lacks a certain form of intelligence (Cheney and Seyfarth 1990). An alternative explanation, however, is that the monkey uses the kill as evidence that the leopard is satiated and is unlikely to hunt again soon; the monkey thus "ignores" the information as an indicator to increase its vigilance. Choosing between these (and possibly other alternatives) is not a simple matter for researchers using such an experimental design.

All I propose is that, at least with respect to *vocal learning,* parrots provide a clearer observational case for what they treat as viable or nonviable input. Clearly, birds attend and react to a large variety of input. Nevertheless, although they may locate a stream by its sound, they rank the sound at a low level with respect to their need for its reproduction, and we can use this lack of reproduction as evidence for hierarchical learning and cognitive processing. Parrots thus demonstrate a biologically relevant cognitive ability that can be measured relatively easily in nature (Balda, Kamil, and Bednekoff 1996). I do not wish to belabor this point; as noted above, such observational and correlational material does not lend itself to rigorous hypothesis testing. Moreover, evolutionary pressures may affect different cognitive abilities differently in different species in different habitats; and of course intelligence is not a unitary "thing" (Byrne 1995). The data merely suggest that the combination of intelligence and advanced communication skills may have arisen not only in primate or even mammalian lines, but also in avian lines, and that it directs not just learning but also what is appropriate to learn.

Why We Need King Solomon's Ring

In sum, the overall point of this book (and complementary studies by large numbers of my colleagues who could not be cited in a work of this length) is not to overwhelm readers with facts and figures, or to be a compendium of the past two decades of comparative cognitive research, but rather to encourage an awareness of, and a sensitivity to, the abilities of nonhumans, particularly nonprimate and nonmammalian subjects. Little reason exists to summarize the field of animal cognition or review arguments for synthesizing field and laboratory studies to uncover the origins of cognitive capacities. Several reviews of animal cognition and its importance have recently been published: *Species of Mind,* edited by Colin Allen and Marc Bekoff, addresses intentionality in social carnivores and the cognitive underpinnings of antipredator behavior in passerine birds. Sara Shettleworth's 1998 book *Cognition, Evolution, and Behavior* emphasizes the adaptive value of cognitive mechanisms across taxa. These researchers, and contributors to the recently published compendium *Animal Cognition in Nature* (Balda, Pepperberg, and Kamil 1998), argue that the application of cognitive theories to field studies and the synthesis of laboratory and field experiments can uncover unexpected cognitive abilities in a myriad of species.

My goal is not to reiterate the material in these books or even to discuss the evolution of avian intelligence (a subject I address in a volume edited by Robert Sternberg and James Kaufman, *The Evolution of Intelligence*), but rather to present my own data to encourage a reevaluation of animal capacities. For far too long, animals in general, and birds in particular, have been denigrated and treated merely as creatures of instinct rather than as sentient beings (see, e.g., Welty 1962). The data in this book are meant to provoke awareness in humans that animals have capacities that are far greater than we were once led to expect, and to remind us that all we need to examine these capacities are some enlightened research tools. How these data are to be used—and interpreted—are, of course, up to the reader. But if the data help us respect the processing abilities of brains that are structured differently from those of humans or are used to better the life of even a single captive parrot, prevent habitat destruction and capture of birds in the wild, or enable researchers to develop better animal models for various human dysfunctions, my work and the work of my students, my colleagues and, not in the least, my parrots, will not have been in vain—we will have found the best possible use for King Solomon's ring.

1. Introduction

1. Conceivably, the animal initially may have learned nothing about sameness and difference, but rather only the association between white-blue and black-triangle; the savings in learning with novel objects might simply be that the animal has clued in to the need to make such associations.
2. These ideas were not published until the 1980s, but were presented at conferences and seminars in the 1970s.
3. Many more recent studies exist that could be cited; my goal here, however, is to provide a snapshot of the field in the 1970s and 1980s, when I began my research.
4. By flexible, I refer to the range of responses available to the subject for any given question. In studies such as Premack's, only a limited number of plastic chips relevant to the question were available at any given time, rather than the subject's entire repertoire.
5. Of interest here is a recent article by Gannon, Holloway, Broadfield, and Braun (1998), showing that an anatomic correlate of the planum temporale, a key component of Wernicke's receptive language area in the human brain, is also present in chimpanzees *(Pan troglodytes)*. The authors thus suggest that the evolutionary origin of human language may have existed in the common ancestor of chimpanzees and humans.
6. We also now know a fifth and a sixth parallel. The fifth is that birds, like humans, learn not just what to ''say,'' but the appropriate context in which to produce their utterances (e.g., Kroodsma 1988; Chapman 1986). The sixth is that for songbirds, humans, and parrots, social interaction is an extremely important factor in determining what and how much is learned (for reviews, see Pepperberg 1994, 1997). Such information, however, was not available in the early 1970s. Note also that in the past few years, studies on avian vocal control systems have been extended to budgerigars; see Striedter 1994; Brenowitz 1997; Jarvis 1998.
7. Again, many more studies have been undertaken and much more knowledge has been gathered about both the common and pygmy chimpanzee since the

1970s; the point here was to describe what was currently known at the time I began my studies in the mid-1970s.

2. Can We Really Communicate with a Bird?

1. Specifically, Mowrer's birds did learn to use labels instrumentally (e.g., to request foods; Mowrer 1950), but they confused instrumental and conditional use: Failure to learn appropriate use of sounds such as "Hello" conditionally (i.e., at the appearance of the trainer) probably occurred because the bird had learned to use the term instrumentally, to obtain food from the trainer. Also, the birds' instrumental use of specific food labels was not well documented: No data were given as to whether a bird actually intended to request a *particular* food (i.e., would reject something else) or simply knew that random use of a set of sounds made something edible appear.

2. Recently, Wright (1996) and Wright and Dorin (under review) have found dialects in another parrot species, the Yellow-naped Amazon *(Amazona auropalliata)*, and Cruickshank, Gautier, and Chappius (1993) have shown that Grey parrots in Zaire mimic nonpsittacine species.

3. Birds may also learn from neighbors not necessarily because the neighbors are of high status, but because knowing a neighbors' repertoire may help the bird maintain its own "status"—its territory. See the recent paper by Beecher, Stoddard, Campbell, and Horning (1996).

4. Many studies of course exist on the effects of social tutors on nonvocal behavior (e.g., Lefebvre and Giraldeau 1994), but I am concentrating on vocal learning.

5. I have proposed (Pepperberg 1997) a scenario in which different levels of teaching are provided without the tutor or the student acting intentionally or demonstrating conscious awareness of the other's state (see Caro and Hauser 1992): Imagine, for example, that something about the behavior, presence, or appearance of the juvenile unconsciously elicits singing and other behavior from the adult that differ slightly from what would be elicited by the presence of other adults, and that these behavior patterns alter in concert with the maturation of the juvenile. For example, post–"critical period" hatching-year white-crowned sparrows (those more than 50 days old) are brown-crowned and often "streaky" (Baptista, pers. comm.); such birds also act differently from considerably younger birds. Whether a brown-crown elicits slightly different singing and other behavior in adults and whether such behavior patterns would help the bird extend its learning period is unknown; the fine-grained analyses required for such a study have not been attempted to date.

6. Note that reference and functionality cannot accurately be subsumed under the operant term "reinforcement." Reinforcement simply implies that some, not necessarily specific, positive outcome is associated with an action. Reference and functionality specify which particular positive outcome out of several is intimately associated with an object or action.

7. This point is particularly important, because I am often asked why, given the importance of social interaction in training, I use any reinforcements other than positive vocalizations. With children, for example, parental approbation ("yes," "good," even "hmm") can sometimes act as an effective nonreferential reward

(see Bowerman 1978 for a review). My students and I do, in fact, find such approbation useful in maintaining an established behavior and in directing a bird's attention during training. Approbation alone, however, does not provide the reference that appears necessary for a bird to learn to communicate with humans.

8. At the beginning of the project, my students and I used a slightly different procedure (Pepperberg 1981, 1990a). During that period, each object was assigned a number, and the test order was determined by a random number table. When the number of objects to be tested exceeded 10, the trainer not involved in testing picked the objects out of the toy box and wrote down the order of her choice. Once we began testing concepts such as same/different and numbers, the procedure described in the text was used.

9. When the number of items in a test reached 20, we found that the bird's attention span was too short to conclude a test in a single session. Thus the test is divided into sections, and the sections of the test are administered on subsequent days. Not only does this method increase the chances that the bird will attend to the test, but it also helps prevent cuing.

10. Similar "boredom" behavior has been observed in other animal subjects (meerkats, *Suricata suricata,* Moran, Joch, and Sorenson 1983; raccoons, *Procyon lotor,* Davis 1984; Long-Evans hooded rats, Davis and Bradford 1986; chimpanzees, Putney 1985).

11. When my students and I repeat a question (the correction procedure), a bird's reply is probably influenced by his erroneous first response and also by our reaction to this error. We therefore do not use the overall scores to judge the bird's competence, nor do we calculate their statistical significance.

12. See Bateson (1979) for an excellent discussion of the difficulties inherent in comparing results from different laboratories because of differences in experimental design.

3. Can a Parrot Learn Referential Use of English Speech?

1. For example, we trained Alex to respond to "climb," by saying the word and pressing one hand against his chest just above his feet. Parrots that are not afraid of hands will step up because they prefer to be at the highest level possible. Such training is common for all pet birds.

2. A question arises as to how to interpret the actions of, for example, Herrnstein, Loveland, and Cable's pigeons (1976), who signaled if some form of a tree was present or absent from a slide. I argue that these birds were still in the associative, rather than the contextual category. First, their actions had no communicative intent: They pecked a button at an appropriate instance of a tree to get seed, not necessarily to inform the researchers about the slide. Second, although they had learned to recognize some collection of properties, the appropriate identification of which led to a food reward, they were not tested on their understanding of what this collection represented; that is, they had no way to demonstrate any flexibility in their response and could simply denote presence or absence of this collection of properties. For a detailed discussion, see Zentall (1996). Subsequent work on pigeons such as that of Bhatt

et al. (1988) demonstrates more advanced capacities that approach conceptual use: Birds in these studies could categorize novel items into each of four different classes of previously learned natural and artificial stimuli. Thus they were not responding merely to the presence or absence of one particular collection of properties, but were forming some concept of categories.

3. The actual training paradigm, which involved subsequent transfer of the calls to shapes associated with the colors, additionally argues against the referential nature of the birds' calls.

4. I prefer the term peri-referential because I think that the prefix "proto" is so overused as to have become meaningless.

5. Our inabilities may be physical, such as being unable to see in the ultraviolet and thus be unaware of an animal's use of its coloration to signal intent; more likely, however, we are simply unaware of the subtleties of animal communication. See, for example, Beer's (1976) description of how he learned to detect the details inherent in the long calls of gulls.

6. We initially used food colors rather than nontoxic paints because the former allowed the grain of the object to show through. We now occasionally use nontoxic paints, but the birds will not chew objects that are coated with these paints, and therefore lose interest in the objects more quickly.

7. At the time Alex reacted negatively to all red lumpish objects; a rose woolen ball 1 inch in diameter also elicited distress calls. Rose pieces of wood, clothes pins, and keys did not bother him, and he subsequently habituated to all other rose items.

8. The first error also provided weak evidence for overextensions similar to those made by children.

9. Bowmaker (1986; also Bowmaker et al.1994, 1996) showed that parrots probably see in the ultraviolet; thus an item, depending upon the pigment used in its coloring, might appear quite different to a parrot and a human (see Bennet and Cuthill 1994). The food colors we used to dye objects were unlikely to cause such problems, but commercially colored items, such as metal keys, might indeed present some unexpected variety to Alex. His errors, however, occurred across all items.

10. We found similar behavior in our other juvenile parrots (Neal 1996), which suggests that such practice was not unique to Alex.

11. We ranked the desirability of objects by noting how long Alex interacted with them during "free play sessions": periods when neither testing nor training were in progress. Alex would chew on the corks and clothes pins, would tear off pieces of paper and ball them in his beak; he would scratch the area under his beak with the key.

4. Does a Parrot Have Categorical Concepts?

1. Note that part of the problem given the animal subject involves learning to agree with humans on how to divide the world into appropriate parcels. Lock (1980) provides a detailed description of the inherent difficulties and how they relate to the Sapir-Whorf hypothesis. Basically, he reminds us that although greenness is an inherent quality of an item, the decision to call some-

thing "green" is as much based on social convention as on human vision in the visible spectrum, because different cultures divide the color spectrum in somewhat different ways.

2. Note that Waxman and Markow (1995) suggest that repetition of category labels in appropriate situations may assist a young child in forming categories.

3. Conceivably, Alex might have learned the meaning of the phrases as a serendipitous side effect of hearing the relevant vocal exchanges, or possibly even through a process described by Premack (1976) as "no-fault testing," in which his chimpanzee was asked questions such as "What color apple?" and given only a symbol for "red." The most conservative interpretation, however, was that Alex had not engaged in such learning.

4. Although erroneously responding with the wrong color label would demonstrate some hierarchical concept of what labels fit into the category color (see Gardner, Van Cantfort, and Gardner 1992), I insisted on the more stringent criterion.

5. By this time Alex would chew on paper dyed with food colors for such an extended time that the dyes would be transferred to his beak, feet, feathers, perches, and ultimately to his trainers. We therefore chose not to test him on paper objects. Subsequent work with construction paper, even though the colors were not identical to the training dyes, showed that he had no trouble transferring to paper items (Pepperberg, unpublished data).

6. Note that Alex's task was also different from, and more complex than, subsequent categorizations required in studies on pigeons (e.g., Bhatt et al. 1988), in which birds indicated, by pecking four different buttons, that they could simultaneously classify four different types of objects. The pigeons, for example, did not have to decode a question to determine which category of object or which aspect of the object was being targeted, nor did they have to reclassify the same object in different ways. Remember that Alex had to categorize objects with respect to several different color and object labels, and to determine, for a given trial, whether to use the category of color *or* shape.

5. Can a Parrot Learn the Concept of Same/Different?

1. Here I am again reporting the state of affairs when I began my study. Contemporary (Shyan et al. 1987) and subsequent (Herman et al. 1989; Wright, Shyan, and Jitsumori 1990) studies further demonstrated how appropriate experimental design was critical if scientists were to show the extent to which animals could acquire abstract concepts. Even studies on monkeys had problems ruling out alternative mechanisms (see Wright, Shyan, and Jitsumori 1990). More recently, however, some studies have shown what does seem to be some level of same/different comprehension in pigeons (e.g., Cook, Cavoto, and Cavoto 1995; Wasserman, Hugart, and Kirkpatrick-Steger 1995; Young, Wasserman, and Dalrymple 1997), although not necessarily at a level that Premack (1983) would find acceptable: In some studies, birds might have responded to homogeneity versus nonhomogeneity; also, in the Cook et al. study, birds might have responded to the presence or absence of a "target" (a dissimilar spot in a field of similar patterns). With respect to the Young et al. study, Premack (1983)

would argue that the birds responded to familiar (from multiple presentations) versus nonfamiliar items.

2. Note that the purpose of the studies on songbirds was not to examine concepts of same/different, but rather to determine the sensitivity of birds' perception of differences in song. Thus these studies cannot be critiqued as failing to demonstrate same/difference at the level Premack requires. I present results of the birdsong studies merely to suggest that understanding the concept of same/difference at some level may indeed be important for birds in their daily life. The performance/perception distinction may also be involved: A subject may perceive a difference, but the experimental design may not force the subject to perform in a manner that demonstrates such perception.

3. Approximately 70 different possible objects (food items were rarely used), three possible correct responses, and two different questions were involved in the same/different task alone. For example, if Alex was asked, "What's same?" and I desired the response to be "color" and I chose a round green key as one of the objects on the task, this key could be paired with two-, three-, four-, five-, or six-cornered objects of paper, wood, or rawhide; three-, four- or six-cornered objects of plastic; plus objects such as green clothes pins, wooden or plastic cubes or spheres, plastic boxes and cups, etc. A similar set of permutations existed for responses of "shape," "matter," and the question "What's different?" Also, Alex was concurrently being tested on numerical concepts ("How many?") and additional labels ("What's this?", "What color?", etc.).

4. Alex had acquired these vocalizations incidentally from their use by trainers in the previous study (Pepperberg 1983a). Whether he was producing the phrases "What color?" and "What shape?" intentionally as requests for information was unclear (see Chapter 11; Pepperberg 1990c). Nevertheless, all trainers were instructed to respond *as if* these were intentional requests, and provide the color or shape of the object in question. Alex acquired the color labels "orange" and "grey" in this manner by querying us, respectively, as to the previously unlabeled colors of a carrot and his reflection (Pepperberg 1983a, 1990c).

5. Sessions ended when Alex insisted on preening rather than working, or when he began to make numerous requests for alternative objects.

6. Interestingly, even several years later, Alex will say phrases such as "What's the matter?" but will continue to produce "mah-mah" for the same/different task.

7. Some of these items were previously used as novel objects in a study on recognition of quantity (Pepperberg 1987b); in those trials, Alex was required to produce only the numerical label representing quantity, not the object label. Although labels for these particular objects were not trained, Alex occasionally learned their labels from students' replies to his queries of "What's that?"

8. The argument extends to children, too. Some researchers claim that language development reflects concept development (e.g., Bloom 1973); others claim that language development shapes concept development (Vygotsky 1962). Current theories (see, e.g., Bates 1993:233–234) suggest that "specific linguistic skills are associated with specific abilities outside language, in a many-to-many relation that is quite different from the one-to-one stage shifts predicted by orthodox Piagetian theory."

6. Can a Parrot Respond to the Absence of Information?

1. Although most research has been conducted with English-speaking children, the order of stages seems to be similar across at least some different languages (Bloom 1991). Note that Bloom and Lahey (1978) and Vaidyanathan (1991) suggest that prohibition may be a different category from rejection, and that it may follow nonexistence, but this interpretation is based on syntactic use: Use of the single word "no," which precedes any syntactic combination of words, generally expresses rejection and prohibition as the first evidence for negation. The order of stages of acquisition changes, however, if using syntactic combination to express negation is the criterion; then nonexistence precedes rejection (Bloom 1991). See Anderson and Reilly (1997) for development of negation in children using American Sign Language, and Litowitz (1998) for arguments for separating rejection and refusal into separate stages.

2. Many researchers have, as noted in Chapter 2, examined the extent to which such warning calls may be referential, but such investigations do not provide information about the extent to which such calls involve rejection in terms of territorial defense. In contrast, such investigations emphasize the animals' reactions to the presence, rather than the absence, of information (detection of a predator, for example, rather than scaring off or "rejecting" a competitor).

3. Note that such behavior is qualitatively different from learning what type of stimulus leads to the absence of reward (see, e.g., Astley and Wasserman 1992): In these cases the subjects may simply be learning what to avoid.

4. The extent to which the sea lion was actually "commenting" upon absence was unclear. Conceivably, the animal simply did not know what to do when the object was missing, and hence just remained stationary.

5. Zentall and Hogan (1978) found that if tests on match-to-sample for shape included trials in which the correct response was absent (i.e., the correct answer was actually to withhold responding), pigeons did better when transferred to hues than birds without "absence" training. A similar result was found for tests involving oddity-from-sample.

6. As in the same/different study, the first transfer series involved items that had not been used in training but that were familiar from other studies; we thus avoided novelty effects (see Zentall and Hogan 1974; Zentall et al. 1981). For this series, there were about 70 different possible objects that could be paired (food items were rarely used), four possible correct responses, and two different questions (Pepperberg 1988b). Thus if we asked "What's same?" desired the response to be "color," and chose a round blue key as one of the items, this key could be paired with two-, three-, four-, five-, or six-cornered objects of paper, wood, or rawhide, three-, four- or six-cornered objects of plastic, plus objects such as blue clothes pins, wooden or plastic cubes or spheres, plastic boxes and cups, etc. A similar set of permutations existed for responses of "shape," "matter," "none," and responses to the question "What's different?" The second transfer series involved a similar number of objects of novel combinations of colors, shapes, and materials that were obtained especially for these tests (Pepperberg 1988b). Again, as in the earlier study, these objects were shelved and handled by human trainers (as part of laboratory cleaning) in Alex's

view for several days prior to their use to avoid a possible fear response, but the colors, shapes, or materials of these items were never labeled.

7. Most novel sounds are learned in 4 to 5 weeks (Pepperberg, unpubl. data).

8. For example, although we rarely colored corks and basically used bottle-stoppers for this item, we did purchase thin corkboard, color and shape it, and ask Alex to tell us about the colors, shapes, and the material; he was correct on first trials.

7. To What Extent Can a Parrot Understand and Use Numerical Concepts?

1. Requiring that animals symbolically encode "more" versus "less" does, of course, require a higher level of processing than merely choosing the greater or lesser quantity, but such processing does not involve specific quantities. For a discussion, see Menzel and Draper 1965; Menzel 1969; Rensch and Dücker 1973; Hulse 1978; and Mitchell, Yao, and Sherman 1985. More recently, Zorina et al. (1991) have shown that pigeons and crows can choose the feeder that represents the greater of two quantities, but the English abstract of the Russian paper does not provide experimental details. Note that the subjects in the Emmerton et al. (1997) experiment transferred the concept of more versus less from training to testing quantities, but still did not label the individual values.

2. Premack (1971) used this technique to train chimpanzees on labels for modifiers such as color.

3. We wondered if such errors might have been related to Alex's disinterest in obtaining multiples of a single object (see Davis 1984); unfortunately, we could not test this possibility.

4. Identifications of novel shaped objects, however, demonstrate an important ability unrelated to quantity: The data show that Alex competently linked separate, appropriate vocalizations into novel combinations (e.g., "two-corner wood") to describe novel objects. Whatever the mechanism, this segmentation-like behavior provides further evidence for a lexical sophistication not expected in a parrot (see, e.g., Fromkin and Rodman 1974).

5. Note that Alex consistently rejected all food items in play sessions; thus although objects were edible, he did not consider them so, and therefore they did not provide any particular incentive.

6. We did not collect latency data on this task, even though Mandler and Shebo (1982) suggest that response times in humans on this task can provide information about the mechanisms used (e.g., counting versus some other strategy). For Alex, response time to the targeted question varied with his interest in the objects used and the number of his requests for nontargeted objects, and would have provided no information with respect to the task itself.

7. Two studies also show related abilities in animals. Beran, Rumbaugh, and Savage-Rumbaugh (1998) present data suggesting counting ability in a chimpanzee, but have not completely ruled out the possibility of advanced sequential subitizing (Burns 1988) and serial pattern recognition. Olthof, Iden, and Roberts (1997) purport to demonstrate that squirrel monkeys show ordinality with

respect to Arabic numbers, but, because of the form of training, the monkeys may merely have shown transitive inference (e.g., Wynne, von Fersen, and Staddon 1992). Moreover, first trial responses of the monkeys during testing were not reported, and, because animals improved over sessions, training was likely to have occurred during the test situation. The monkeys, however, were generally able to chose the greater of two *sums* of Arabic numbers, on first trials, suggesting that the Arabic numbers represented different quantities (even if the numbers were only represented perceptually) that could be summed. Other recent work (Brannon and Terrace 1998) also indicates that rhesus monkeys have abilities that are clear precursors to counting.

8. In the interest of having each chapter describe a particular type of competence, I discuss studies that occurred several years apart. Thus the work in this section postdates material in some subsequent chapters.

9. Gallistel and Gelman (1992) would interpret the data with respect to whether Alex was using preverbal or verbal mechanisms.

10. No correlations occurred in either the linear regression ($F[1,96] = .55$, $p =$ ns) or in a test for nonlinearity ($F[4,96] = .58, p =$ ns).

8. How Can We Be Sure That Alex Understands the Labels in His Repertoire?

1. Note that Skinnerians refer to requesting as "manding" and labeling as "tacting" (Skinner 1957).

2. A subject searching for items defined by the combination (known in the psychological literature as "conjunction") of two features may, under some conditions, perform the search as a parallel rather than as a sequential (recursive) task (Langley and Riley 1988). That is, the subject may combine information about "triangular" and "rawhide" and perform one search, rather than first finding triangular items and then eliminating all that are not rawhide (or vice versa). Including conjunction in a task, therefore, may not affect its *recursive* complexity.

3. According to Crain (1991), children can comprehend sentences with relative clauses before their third birthday.

4. These objects were "boring" because all combinations of attributes for these items (colors, shapes, materials) had been used previously. Training labels for new attributes (e.g., pink) would have confounded other concurrent studies (e.g., Pepperberg 1987a,b) for which we needed attributes Alex could not label.

5. This discussion does not address the obvious problem of intra-trial learning, which would occur if the same question was repeated sequentially for a single set of objects. Intra-trial learning may have confounded some studies of comprehension (see the critique by Gardner and Gardner 1978).

6. In a study published after mine, however, Savage-Rumbaugh et al. (1993) demonstrated that a bonobo, Kanzi, responded correctly on a recursive task involving a large number of items when asked to distinguish between a nearby and distant instance of a particular object (object A in the room versus else-

where), but not when asked to distinguish between two instances of the same object that were nearby (e.g., object A that is wet versus object A that is dry).

7. Note, however, that humans find searches for three conjunctions (color, size, *and* form) easier than searches for two (Wolfe, Cave, and Franzel 1989); the suggestion is that three parallel processes guide attention more effectively than two. How birds might react is not yet known.

8. Studies by Cook, Cavato, and Cavato (1996) on conjunction fall into this category, because these researchers tested how figure/ground discriminations (in this case, patterns) were affected when attributes common to the background were conjoined in the figure.

9. I could not completely avoid these confounds given the constraints of forming the collections.

10. A sea lion (Schusterman and Gisiner 1988), but not a dolphin (Herman et al. 1984), did make more errors when relational tasks included conjunction (e.g., [modifier] pipe ball fetch vs. pipe ball fetch, where the subject moves one object in relation to another). Such tasks, however, involve additional concerns not relevant to the present study: Not only did the marine mammals' tasks involve a relational condition, but also the modifiers in the sea lion's task referred to object qualities (e.g., brightness) whereas the dolphin's modifiers involved positional cues (e.g., an object's place in the tank) that could actually have simplified the task (see Schusterman and Gisiner 1988).

9. Can a Parrot Understand Relative Concepts?

1. Note that response to a relative class concept is different from response to relational same/different. In relational same/different, the subject designates whether the relation between A and A′ is the same as that between B and B′ or different from that between C and D . . . *whatever* that relation is. Conceivably, that relation can be relative (e.g., "Is A bigger than A′?" and, if so, "Is B bigger than B′?"), but the relation need not be relative, and the organism need not be able to designate the relation specifically. In contrast, a relative concept generally involves a very specific relation between two items (A and A′), and the organism must respond with respect to—and thus comprehend— this very specific relation.

2. Pasnak and Kurtz (1987) have shown that rhesus monkeys *(Macaca mulatta)* also tend to discriminate on the basis of absolute, rather than relative, brightness and size.

3. To ensure that trials were random, we first formed a 7×7 matrix of the possible colors labels and numbered the unique entries. We then formed a matrix of these numbers and the objects that could be colored, and numbered those entries. These latter numbers were placed in a jar and drawn at random. Drawing a pair with the same color labels indicated a trial in which Alex would have to tell us which object (i.e., material such as wool or wood) was bigger or smaller. We flipped a coin to determine whether the question was about the bigger or smaller object.

4. Remember, too, that Rice (1980) suggests that linguistic input is not sufficient to teach a nonlinguistic concept for which a subject is unready.

10. What Is the Extent of a Parrot's Concept of Object Permanence?

1. For example, according to Baillargeon (e.g., Baillargeon et al. 1985; Baillargeon 1986, 1987; Baillargeon and DeVos 1991), a subject ostensibly in Stage 1 can demonstrate some understanding that objects exist when occluded *if* tests do not require active search for a hidden item. The habituation paradigm used by these researchers, however, is difficult to apply with appropriate controls in animal studies. Most researchers thus make comparisons to standard Piagetian stages.

2. One might argue that the experimenter unconsciously looks at the place she or he expects the subject to choose (the "logical" hiding place), even though body and head position are being controlled. Note, however, that in most studies the covers used to hide the targeted object are separated by small distances (usually < 3.5 cm), and that the experimenter's eyes are usually at relatively far distances from the objects (e.g., 80–90 cm); thus such distances make eye orientation an unlikely cue (Miller and Murphy 1964).

3. Wise et al. 1974 reported Stage 6 competence for rhesus monkeys, but Dumas and Doré (1987) argue that the procedures that were used may invalidate conclusions on the rate and level of object permanence attained; see also de Blois and Novak (1994). Orangutans appear to have some understanding of invisible displacements, but memory problems interfere with the most complex tasks (de Blois et al. 1998). Dogs achieve Stage 6 according to standard invisible displacement tests (Gagnon and Doré 1995), but fail (Doré et al. 1996) when the task is a variant of the "shell game" (Sophian 1985).

4. Subsequent studies are even more compelling; see Clayton and Dickinson 1998; for reviews, see chapters in Balda, Pepperberg, and Kamil (1998).

5. Butterworth et al. (1982) suggest that if the object that is hidden is changed, the subject may not recognize that the set of trials in the second location is related to the first set of trials.

6. We recorded mistrials when a bird was clearly fixated by the covers. Even if an object was uncovered in such a trial, birds sometimes appeared oblivious to the item and focused their entire attention on the covers. Occasionally, a bird continued to manipulate the covers until we removed them from its beak or feet. Such behavior contrasted with occasions in which, for example, a bird interacted with an object that it uncovered by accident.

7. Although we do not intend to anthropomorphize Alex's behavior, we have learned that the "slit-eye" look is usually a prelude to our being bitten if we continue the actions that have caused Alex to adopt this pose. We are in similar danger of being bitten if we ignore his beak-banging behavior.

8. Chapman's (1987) research with children also provides weak evidence that increased task exposure does not improve performance: He divided subjects in his longitudinal study into two groups; one had roughly twice the number of tests as did the other. The former group scored slightly better than the latter during the course of the experiment, but the difference in scores was not significant.

9. Although mirror exposure might aid the development of general spatial awareness, Griffin did not have an opportunity to practice any specific spatial skills: Unlike subjects in a previous mirror study (Pepperberg, Garcia, Jackson, and

Marconi 1995), he was not asked to find an object on the basis of its reflection; thus he did not experience any additional search-related trials. Also, according to Chapman (1987), mirror recognition tasks require a higher level of cognitive ability than object permanence tasks (the former require the coordination of four sensorimotor schemes whereas the latter require combining two schemes with only a doubling of a third); thus it would seem that mirror tasks are unlikely to provide training for object permanence tasks. Of course mirror trials might, simply because they also provided a break in the general routine, have somehow acted as a general energizer of all responses that might be emitted in other situations.

11. Can Any Part of a Parrot's Vocal Behavior Be Classified as "Intentional"?

1. Such behavior is not surprising, given psittacine anatomy: Balancing on one foot in order to point with a claw would be challenging, and for similar reasons stretching to obtain an object with the beak seems to occur only when the object actually is within reach. Alex did, however, often stretch his body in the direction of a particular perch (e.g., toward a favorite chair back), and if placed at that position would sit quietly.
2. Down Syndrome children in intervention programs whose caretakers responded *to* communicative acts learned a balanced distribution of comments and requests; those in programs whose caretakers used explicit prompts for production responded more, but predominantly with comments (Salmon, Rowan, and Mitchell 1998).
3. Independent transcribers agreed on 97.5% of the human words and 98% of Alex's vocalizations (Pepperberg 1988a).
4. Note that such gestures were iconic and that researchers had always previously trained these particular chimpanzees by using a computer system based on non-iconic symbols (Rumbaugh 1977).

12. Can a Parrot's Sound Play Assist Its Learning?

1. Examples include the developmentally disabled (Fey 1986), learners of a second language (Krashen 1976), and even normal preverbal children who are being hurried to attain recognizable competence (Rogoff 1990).
2. Recent studies (e.g., Baddeley, Gathercole, and Papagno 1998) provide evidence for specific mechanisms to explain such findings: Supposedly, when learning unfamiliar material, children use a phonological loop system to store sounds temporarily while long-term phonological representations are being constructed; learning is thus slower than it is when existing representations can be accessed. Moreover, the short-term system appears to be located in the perisylvian region of the left hemisphere (see Smith and Jonides 1995). Avian analogues of this system and anatomical area are unknown, but our data suggest that a search for such analogues might be fruitful.

3. Optimally, we would have used a person not connected with the study to transcribe the tapes so that the transcriber could not be influenced by what she or he expected to hear. For three reasons, however, we believe that expectation cuing did not affect the results. First, we had good interobserver agreement between the two transcribers of the targeted vocalizations even though the secondary transcriber did not know the order of the tapes. Second, because Alex did utter "one" (although not frequently) in the presence of the primary transcriber (Katherine Brese) prior to training, she was both familiar with the utterance *and* actively searching for it during both the baseline and training periods. Third, because Alex had, on occasion, learned to produce a novel vocalization perfectly after a single session (e.g., "carrot," Pepperberg 1990c), the transcriber had no reason to expect any delay in acquisition and was always actively searching for the targeted vocalizations.

4. Although the percentages of "one/un" use were small, a significant difference existed between 0 (observed) and 5 (expected) vocalizations of "one/un" before training ($p = 0.023$, two-tailed z-test); the difference between 18 (expected) and 20 (observed) vocalizations after training began was not significant ($p \sim 0.73$). Given that "one" was in Alex's repertoire prior to training on "none," we wondered why it was noticeably absent in pre-training monologues. The reason appeared to be that Alex's use of a vocalization in monologue speech in a given session often correlated with the familiarity of the vocalization and its use during the previous day by the trainers. "One," unlike the other number labels, was at the time used only for questions on sequential quantities (metronome clicks), and few such trials were then in progress.

5. The difference between 2 (observed) and 11 (expected) vocalizations of "green" before training was significant ($p = 0.0052$, two-tailed z-test); the difference between 7 (observed) and 11 (expected) vocalizations of "bok" was not significant ($p = 0.19$).

6. Difference between 5 (observed) and 5 (expected) productions of "green" during training was not significant, but 1 (observed) versus 11 (expected) productions of "cracker" was significant ($p = 0.003$).

7. At that time, Alex had referential use of 80 different vocalizations for objects, colors, shapes, quantity, categories, and requests. The remaining 150 utterances tabulated during monologue speech were (1) phonological variations of these 80 utterances (e.g., "four-orner" for "four-corner," "eee" for "key" or "three"); (2) single vocalizations that Alex uses in a contextually applicable manner but that we have never formally tested (e.g., "sorry," "climb," "gentle") and their variations (e.g., "sor-"); and (3) parts of phrases that Alex uses contextually but that have not been tested (e.g., "tell," which comes from "You tell me!") and their variations.

8. These studies may, however, have been affected by the age and sex of the subjects. Subjects in the nonstructured study were juvenile males, whereas subjects in the structured studies were adult females.

9. Petitto and Marentette (1991) describe manual babbling for deaf children exposed to ASL; these authors find a correlation between elements used in manual babbling and the childrens' first signs.

13. Can a Parrot's Sound Play Be Transformed into Meaningful Vocalizations?

1. For a recent review comparing animals with normal and dysfunctional children with respect to fast mapping, see Wilkinson, Dube, and McIlvane (1998).
2. Alex has a number of utterances that linguists call "minimal pairs"—e.g., "key" /ki/ and "tea"/ti/—that suggest he processes information on a phonemic level, but we cannot yet prove this assumption.
3. Bloom et al. (1996) argue that children take a major role in language development by learning labels for expression and interpretation that are used to engage adults and share the contents of their minds.
4. At the time of this study, we were concentrating on examining concepts such as same/different (Pepperberg 1987a) rather than on training that would increase the size of Alex's repertoire; thus almost all additions to the lexicon during that period came about through referential mapping.
5. For health reasons, we stopped giving Alex seeds in November 1990.
6. Errors on subsequent identifications of the color orange were often "rose" (red) or "yellow" (Table 13.1). As noted earlier, the color of these hand-dyed objects may have been perceived differently by Alex than by his human trainers because of differences in psittacine and human visual systems (Bowmaker 1986; Bowmaker et al. 1994, 1996; note Bennet and Cuthill 1994).
7. For various reasons (e.g., I found that training or testing is not possible when he is on a trainer's shoulder), Alex often must be placed in locations other than those he requests. Additional data are thus difficult to collect.

14. What Input Is Needed to Teach a Parrot a Human-based Communication Code?

1. Tomasello, Strosberg, and Akhtar (1996) suggest that children can use joint attention to, and labeling of, a hidden object to infer a label-referent association; we did not attempt such a study with our birds.
2. Parrots, of course, might simply not *see* video. Studies have argued that flicker fusion effects (e.g., Adret 1997), distance, size of image, or lack of ultraviolet input might make learning from videotaped input impossible. My students and I have discussed these problems in some detail (Pepperberg, Gardiner, and Luttrell 1999), and although these factors may affect the extent of learning, we do not believe they completely prevent learning.
3. He insisted, however, on producing *nuk* instead of *yuk*.
4. Interestingly, many Grey parrots (including our newest addition, Griffin) seem unable to produce referential English speech until sometime in their second year, although they babble English sounds much earlier. I term this babbling "Greylish," because the vocalizations include some English prosody, but no clear phonemes, and the babbling occurs along with parrot whistles, calls, and begging noises. The delay may involve maturation of the vocal tract: Many parts of the psittacine vocal tract must work in concert to produce human phonemes (see Chapters 15, 16; Patterson and Pepperberg 1994, 1998; Warren, Patterson, and Pepperberg 1996). Interestingly, in the first 11 months, Kyaaro acquired some extremely clear vocalizations from informal interactions with trainers: "Hi

Kyo," "Want tickle," "Kiss" (Pepperberg 1994b). These utterances were generally contextually appropriate; they were used, respectively, when we entered his room but ignored him (e.g., during cleaning or a time-out), while he bowed his head and stretched toward our hands, and when he stretched his beak toward our faces. Because Kyaaro always accepts tickles or beak rubs, such utterances cannot be tested and no claims can be made for their referentiality.

5. These interpretations might explain why birds may discriminate video images and exhibit behavior an experimenter wishes to elicit (see, e.g., Evans and Marler 1991; Herman, Morrel-Samuels, and Pack 1990; Redhead and Pierce 1995; Dawkins and Woodington 1997; Patterson-Kane et al. 1997), but fail to acquire vocalizations or certain types of social behavior (see, e.g., Ryan and Lea 1994).

6. Treatments involving virtual reality have had some success in alleviating physical phobias (e.g., fear of heights); however, such treatments differ considerably from mere exposure to two-dimensional video. Also, whether such work can be extended to social situations is unclear (Glantz et al. 1996). For successful use of modeling in treating animal phobias, see Goetestam and Berntzen (1997).

7. A similar situation is likely for pet birds trained with only slightly more reference and functionality, who appropriately produce, for example, "Hello" or "Bye, bye" routines—"Good night dear," "Good-bye, and thank you" (Amsler 1947:68). These birds may have a more general sense of situations in which their vocalizations can be used than do birds taught without any functionality, but they neither comprehend the individual words in these routines nor fully comprehend that part of the allospecific code they have acquired (Pepperberg 1997).

8. Itakura and Anderson (1996) have, however, shown that capuchin monkeys can be trained to use eye gaze to guide object-choice responses.

9. In this study, Griffin (our youngest bird) sat in his room with an interactive trainer and watched a video of a live, but broadcast M/R session being conducted with Alex.

10. Note that although wild song sparrows respond preferentially to normal songs versus those to which harmonics have been added (Stote and Nowicki 1996), juveniles do not preferentially chose to learn unaltered songs but may delete harmonics from learned material when singing (Nowicki et al. 1992). Such data suggests that songs may be altered after acquisition on the basis of how the songs are received.

11. Even normal preschool children appear to learn less from videotaped input than from live interaction: In a comparison of exposure to novel words via computer video display (still story pictures in color with an accompanying soundtrack), via videotape (a fully animated story), and via live adult reading (of a book whose pictures and narrative matched those of the video), children exposed to live reading showed a small but significant advantage in word recognition (Terrell and Daniloff 1996). Because the adult did not provide additional explication, differences were probably a consequence of other attentional effects.

12. Our birds do receive considerable vocal social interaction (e.g., students label foods, drinks, and certain actions such as "climb"), but their experiences are considerably more limited than those of human children or even of encultur-ated chimpanzees (see, e.g., Tomasello et al. 1993).

13. Note, too, that successful attempts to teach autistic children conversational speech with video modeling (Charlop and Milstein 1989) involved presenta-tion of videotapes with immediate interaction (testing) by therapists who rep-licated the video modeling and rewarded initially echoic (mimetic) speech. See also Watkins et al. (1990), whose results were described earlier in this chapter.

14. Specifically, older children recognize that two-dimensional input can repre-sent three-dimensional form, and use two-dimensional information appropri-ately; younger children cannot (DeLoache and Burns 1994).

15. For example, a subject can, by assessing whether input comes from a source that is also attending to the same object or action, determine the reference of the input and possibly whether learning would be useful. In contrast, non-social background sounds (e.g., the rustling of leaves, mother talking while facing away) might be processed for their information content (the approach of a predator, the presence of the parent in the room), and one might imagine a functional use for reproducing such sounds (deceiving a sympatric bird into taking cover and missing out on a food source, or possibly gaining the parent's attention). Nevertheless, the lack of social interaction, including joint atten-tion, should inhibit acquisition of such sounds by signaling that learning in such a condition would be maladaptive (because, e.g., rustling may have many implications and the context and use of such a learned vocalization would be unclear; note that a parent that incorrectly assumes that the child's vocaliza-tion implied joint focus would remain functionally nonengaged). See Pepper-berg (1998).

16. Kyaaro and now Griffin have learned to produce labels after informally watch-ing Alex identify objects, but such learning appears to center on items that are particularly "desirable," such as walnuts, and also often requires some mapping (see Chapter 13) before the label is fully referential.

17. The term "Kaspar Hauser" refers to an individual who is brought up in iso-lation from normal social contacts and who thus fails to learn social and com-municative skills. The original Kaspar Hauser was a German youth who ap-peared to have been incarcerated until late adolescence.

15. How Similar to Human Speech Is That Produced by a Parrot?

1. At least with respect to general vocal behavior. Zebra finches and budgerigars, for example, are, however, extraordinarily sensitive to mistunings of single harmonics (i.e., in discriminating inharmonicity); see Lohr and Dooling (1998).

2. Or, in this case, be based on the quality of the single mockingbird song used as the stimulus.

3. Other consonants (such as fricatives, e.g., /f/, and nasals, e.g., /n/) were not examined.

4. We measured F_{0}s for both Alex and Zaa, but because Zaa does not produce a

complete set of vowels, we measured F_1s and F_2s only for Alex's and my vocalizations.

5. For "voiced" consonants, /p,t,k/, the lips separate *after* the vocal folds of the larynx begin to vibrate; for "voiceless" consonants, /b,d,g/, the vocal folds are vibrating when the lips part (Borden and Harris 1984). Thus VOT is extremely short for voiceless consonants and somewhat longer for voiced consonants in human speech. What Alex uses in place of lips is still under study (Chapter 16; Patterson et al. 1997; Patterson and Pepperberg 1998).

6. Although mynahs do not independently use two halves of their syrinx to produce formants from F_0s as Lieberman (1984) suggests, they probably use a different strategy for formant production than do Grey parrots. I do not compare mynah and parrot formant production; for details, see Klatt and Stefanski (1974). Recent studies (Banta 1998) show that budgerigars do produce human vowels by amplitude modulation (as Lieberman suggests), but, because a budgerigar's syrinx cannot produce two independent sounds, production occurs via a somewhat different, as yet undetected, mechanism.

7. Also, preliminary data (Patterson, unpubl. observ.) suggest that humans' abilities to identify Alex's vowels improve with exposure; the same "normalization" phenomenon is observed for people who learn to produce and comprehend English as a second language (Flege, Bohn, and Jang 1997).

8. Even though Alex produces many labels, some of his vocalizations occur only in phrases.

9. Although VOT might not be considered the best variable to chose for this analysis, it is the one used most commonly and is also less affected than other variables by the context of the consonant.

16. How Does a Grey Parrot Produce Human Speech Sounds?

1. Given that a recent study (Fee et al. 1998) suggests that intrinsic mechanical properties of the syrinx can affect temporal and acoustic patterns in songbirds, some aspects of vocal behavior thus appear independent of complex central control. Nevertheless, some form of learning is probably still involved, because different birds must somehow account for individual differences in morphology in order to generate similar acoustic patterns (see Goller 1998).

2. Such data exist for domestic chickens (*Gallus domesticus,* Myers 1917; Harris, Gross, and Robeson 1968; White 1968), Ross' geese (*Chen rossii,* Sutherland and McChesney 1965), snow geese (*Chen hyperborea hyperborea,* Sutherland and McChesney 1965), barnacle geese (*Branta leucopsis,* Hausberger, Black, and Richard 1991), white-throated sparrows (*Zonotrichia albicollis,* Westneat et al. 1993), and swamp sparrows (Westneat et al. 1993).

3. Interobserver reliability of identification of Alex's vowels in context was 96% for transcribers (see Patterson and Pepperberg 1994 for details).

4. Any observed differences between /a/ and /i/, moreover, would not be significantly affected by context—that is, use of whole words—because only ~10% of variation in Alex's F_2 across all his vowels was due to phonological context, and his F_1 was relatively unchanging overall (Patterson and Pepperberg 1994).

5. NIH Image was written by Wayne Rasband, U.S. National Institutes of Health; it is available from the Internet from zippy.nimh.nih.gov.

REFERENCES

Adamson, L. B., R. Bakeman, and C. B. Smith. 1990. "Gestures, words, and early object sharing." In *From Gesture to Language in Hearing and Deaf Children,* ed. V. Volterra and C. Erting, pp. 31–41. New York: Springer-Verlag.

Adret, P. 1993. "Operant conditioning, song learning and imprinting to taped song in the zebra finch." *Animal Behaviour,* 46: 149–159.

———1997. "Discrimination of video images by zebra finches *(Taeniopygia guttata):* Direct evidence from song performance." *Journal of Comparative Psychology,* 111: 115–125.

Adrien, J. L., J. Martineau, C. Barthelemy, N. Bruneau, B. Garreau, and D. Sauvage. 1995. "Disorders of regulation of cognitive activity in autistic children." *Journal of Autism and Developmental Disorders,* 25: 249–63.

Amsler, M. 1947. "An almost human Grey parrot." *Aviculture Magazine,* 53: 68–69.

Anderson, D. E., and J. S. Reilly. 1997. "The puzzle of negation: How children move from communicative to grammatical negation in ASL." *Applied Psycholinguistics,* 18: 411–429.

Anderson, D. R., and P. A. Collins. 1988. *The Impact on Children's Education: Television's Influence on Cognitive Development.* Office of Research, Working Paper no. 2. Washington, D.C.: U.S. Department of Education, Office of Educational Research and Improvement.

Anderson, J. R., P. Sallaberry, and H. Barbier. 1995. "Use of experimenter-given cues during object-choice tasks by capuchin monkeys." *Animal Behaviour,* 49: 201–208.

Antinucci, F. 1990. "The theoretical framework." In *Cognitive Structure and Development in Nonhuman Primates,* ed. F. Antinucci, pp. 11–17. Hillsdale, N.J.: Erlbaum.

Antinucci, F., G. Spinozzi, E. Visalberghi, and V. Volterra. 1982. "Cognitive development in a Japanese macaque *(Macaca fuscata).*" *Istituto Superiore di Sanità Annali,* 2: 177–184.

Aoki, T. 1977. "On the counting process of patterned dots." *Tohoku Psychologica Folia,* 36: 15–22.

Arrowood, P. C. 1988. "Duetting, pair bonding and agonistic display in parakeet pairs." *Behaviour,* 106: 129–157.

Astley, S. L., and E. A. Wasserman. 1992. "Categorical discrimination and generalization in pigeons: All negative stimuli are not created equal." *Journal of Experimental Psychology: Animal Behavior Processes*, 18: 193-207.

Atkinson, J., F. W. Campbell, and M. R. Francis. 1976. "The magic number 4 ± 0: A new look at visual numerosity judgments." *Perception*, 5: 327-334.

Averbach, E. 1963. "The span of apprehension as a function of exposure duration." *Journal of Verbal Learning and Verbal Behavior*, 2: 60-64.

Baddeley, A., S. Gathercole, and C. Papagno. 1998. "The phonological loop as a language learning device." *Psychological Review*, 105: 158-173.

Baer, T., J. C. Gore, L. C. Gracco, and P. W. Nye. 1991. "Analysis of vocal tract shape and dimensions using magnetic resonance imaging: Vowels." *Journal of the Acoustical Society of America*, 90: 799-828.

Baillargeon, R. 1986. "Representing the existence and the location of hidden objects: Object permanence in 6- and 8-month old infants." *Cognition*, 23: 21-41.

———1987. "Object permanence in 3½ and 4½ month-old-infants." *Developmental Psychology*, 23: 655-664.

Baillargeon, R., and J. DeVos. 1991. "Object permanence in young infants: Further evidence." *Child Development*, 62: 1226-1246.

Baillargeon, R., E. S. Spelke, and S. Wasserman. 1985. "Object permanence in five-month-old infants." *Cognition*, 20: 191-208.

Balda, R. P. 1980. "Recovery of cached seeds by a captive *Nucifraga caryocatastes.*" *Zeitschrift für Tierpsychologie*, 52: 331-346.

Balda, R. P., A. C. Kamil, and P. A. Bednekoff. 1996. "Predicting cognitive capacity from natural history." In *Current Ornithology*, vol. 13, ed. V. Nolan, Jr., and E. D. Ketterson, pp. 33-66. New York: Plenum Press.

Balda, R. P., I. M. Pepperberg, and A. C. Kamil, eds. 1998. *Animal Cognition in Nature.* London: Academic Press.

Baldwin, D. A. 1989. "Infants' contributions to the achievement of joint reference." Ph.D. thesis, Stanford University.

———1991. "Infants' contributions to the achievement of joint reference." *Child Development*, 62: 875-890.

———1993. "Early referential understanding: Infants' ability to recognize referential acts for what they are." *Developmental Psychology*, 29: 832-843.

———1995. "Understanding the link between joint attention and language." In *Joint Attention: Its Origin and Role in Development*, ed. C. Moore and P. J. Dunham, pp. 131-158. Hillsdale, N.J.: Erlbaum.

Baldwin, D. A., E. M. Markman, B. Bill, R. N. Desjardins, J. M. Irwin, and G. Tidball. 1996. "Infant's reliance on a social criterion for establishing word-object relations." *Child Development*, 67: 3135-3153.

Baldwin, J. M. 1914. "Deferred imitation in West African Grey parrots." *Ninth International Congress of Zoology*, 536.

Ball, S., and G. A. Bogatz. 1970. *The First Year of Sesame Street: An Evaluation.* Princeton: Educational Testing Service.

Bandura, A. 1963. "The role of imitation." In *Social Learning and Personality Development*, ed. A. Bandura and R. H. Walters, pp. 47-108. New York: Holt, Rinehart & Winston.

———1971a. "Analysis of modeling processes." In *Psychological Modeling,* ed. A. Bandura, pp. 1–62. Chicago: Aldine-Atherton.

———1971b. "Psychotherapy based upon modeling principles." In *Handbook of Psychotherapy and Behavior Change: An Empirical Analysis,* ed. A. E. Bergin and S. L. Garfield, pp. 653–708. New York: John Wiley.

———1976. "Effecting change through participant modeling." In *Counseling Methods,* ed. J. D. Krumboltz and C. E. Thoresen, pp. 248–265. New York: Holt, Rinehart & Winston.

———1977. *Social Modeling Theory.* Chicago: Aldine-Atherton.

Banta, P. A. 1998. "Neuroethology of acquired English and conspecific vocalizations in the budgerigar, *Melopsittacus undulatus.*" Ph.D. thesis, University of Arizona.

Baptista, L. F. 1983. "Song learning." In *Perspectives in Ornithology,* ed. A. H. Brush and G. A. Clark, Jr., pp. 500–506. Cambridge: Cambridge University Press.

———1988. "Imitations of White-crowned Sparrow songs by a Song Sparrow." *Condor,* 90: 486–489.

Baptista, L. F., and S. L. L. Gaunt. 1994. "Advances in studies of avian sound communication." *Condor,* 96: 817–830.

Baptista, L. F., and M. L. Morton. 1988. "Song learning in montane white-crowned sparrows: From whom and when." *Animal Behaviour,* 36: 1753–1764.

Baptista, L. F., and L. Petrinovich. 1984. "Social interaction, sensitive phases and the song template hypothesis in the white-crowned sparrows." *Animal Behaviour,* 32: 172–181.

———1986. "Song development in the white-crowned sparrow: Social factors and sex differences." *Animal Behaviour,* 34: 1359–1371.

Baptista, L. F., M. L. Morton, and M. E. Pereyra. 1981. "Interspecific song mimesis by a Lincoln Sparrow." *Wilson Bulletin,* 93: 265–267.

Baron-Cohen, S. 1995. "The eye direction detector (EDD) and the shared attention mechanism (SAM): Two cases for evolutionary psychology." In *Joint Attention: Its Origin and Role in Development,* ed. C. Moore and P. J. Dunham, pp. 41–59. Hillsdale, N.J.: Erlbaum.

Bates, E. A. 1979. *The Emergence of Symbols: Cognition and Communication in Infancy.* New York: Academic Press.

———1993. "Comprehension and production in early language development." *Monographs of the Society for Research in Child Development,* 58: 222–242.

Bates, E., D. Thal, and V. Marchman. 1991. "Symbols and syntax: A Darwinian approach to language development." In *Biological and Behavioral Determinants of Language Development,* ed. N. A. Krasnegor, D. M. Rumbaugh, R. L. Schiefelbusch, and M. Studdert-Kennedy, pp. 29–65, Hillsdale, N.J.: Erlbaum.

Bateson, P. 1979. "How do sensitive periods arise and what are they for?" *Animal Behaviour,* 27: 470–486.

Beckmann, H. 1924. "Die Entwicklung der Zahlleistung bei 2–6 jahrigen Kindern." *Zeitschrift für Angewandte Psychologie,* 22: 1–72.

Beecher, M. D., P. K. Stoddard, and P. Loesche. 1985. "Recognition of parents' voices by young cliff swallows." *Auk,* 102: 600–605.

Beecher, M. D., P. K. Stoddard, S. E. Campbell, and C. L. Horning. 1996. "Repertoire matching between neighbouring song sparrows." *Animal Behaviour*, 51: 917–923.

Beer, C. 1976. "Some complexities in the communication behavior of gulls." In *Origins and Evolution of Speech*, Annals of the New York Academy of Sciences, vol. 280, ed. S. R. Harnad, H. D. Steklis, and J. Lancaster, pp. 413–432. New York: New York Academy of Sciences.

Bellugi, U. 1967. "The acquisition of negation." Ph.D. thesis, Harvard University.

Benedict, H. 1979. "Early lexical development: Comprehension and production." *Journal of Child Language*, 6: 183–200.

Bennett, A. T. D., and I. C. Cuthill. 1994. "Ultraviolet vision in birds: What is its function?" *Vision Research*, 34: 1471–1478.

Beran, M. J., D. M. Rumbaugh, and E. S. Savage-Rumbaugh. 1998. "Chimpanzee *(Pan troglodytes)* counting in a computerized testing paradigm." *Psychological Record*, 48: 3–19.

Berko-Gleason, J. 1973. "Code switching in children's language." In *Cognitive Development and the Acquisition of Language*, ed. T. E. Moore, pp. 159–167. New York: Academic Press.

———1977. "Talking to children: Some notes on feedback." In *Talking to Children*, ed. C. E. Snow and C. A. Ferguson, pp. 199–205. Cambridge: Cambridge University Press.

Berman, R. 1979. "The re-emergence of a bilingual: A case study of a Hebrew-English speaking child." *Working Papers in Bilingualism*, 19: 157–180.

Bertram, B. C. R. 1970. "The vocal behavior of the Indian Hill mynah, *Gracula religiosa.*" *Animal Behaviour Monographs*, 3: 79–192.

Bhatt, R. S., E. A. Wasserman, W. F. Reynolds, Jr., and K. S. Knauss. 1988. "Conceptual behavior in pigeons: Categorization of both familiar and novel examples from four classes of natural and artificial stimuli." *Journal of Experimental Psychology: Animal Behavior Processes*, 14: 219–234.

Bickerton, D. 1990. *Language and Species.* Chicago: University of Chicago Press.

Bitterman, M. E. 1975. "The comparative analysis of learning." *Science*, 188: 699–709.

Bjork, E. L., and E. M. Cummings. 1984. "Infant search errors: Stage concept development or stage memory development." *Memory & Cognition*, 12: 1–19.

Black, R. 1979. "Crib talk and mother-child interaction: A comparison of form and function." *Papers and Reports on Child Language Development*, 17: 90–97.

Blank, M. 1974. "Cognitive function of language in the preschool years." *Developmental Psychology*, 10: 229–245.

Bloom, L. 1970. *Language Development: Form and Function in Emerging Grammars.* Cambridge, Mass.: MIT Press.

———1973. *One Word at a Time: The Use of Single Word Utterances before Syntax.* The Hague: Mouton.

———1991. "On the acquisition of negation in Tamil and English." *Journal of Child Language*, 18: 715–716.

Bloom, L., and M. Lahey. 1978. *Language Development and Language Disorder.* New York: John Wiley.

Bloom, L., K. Lifter, and J. Broughton. 1985. "The convergence of early cognition and language in the second year of life: Problems in conceptualization and measurement." In *Children's Single Word Speech,* ed. M. Barrett, pp. 149–180. New York: John Wiley.

Bloom, L., C. Margulis, E. Tinker, and N. Fujita. 1998. "Early conversations and word learning: Contributions from child and adult." *Child Development,* 67: 3154–3175.

Bock, W. J. 1978. "Morphology of the larynx of *Corvus brachyrhynchos* (Passeriformes: Corvidae)." *Wilson Bulletin,* 90: 553–565.

Boesch, C. 1991. "Teaching among wild chimpanzees." *Animal Behaviour,* 41: 530–532.

Bohannon, J. N. III, R. J. Padgett, K. E. Nelson, and M. Melvin. 1996. "Useful evidence on negative evidence." *Developmental Psychology,* 32: 551–555.

Boosey, E. J. 1947. "The African Grey parrot." *Aviculture Magazine,* 53: 39–40.

Borden, G. J., and K. S. Harris. 1984. *Speech Science Primer: Physiology, Acoustics, and Perception of Speech.* Baltimore: Williams and Wilkins.

Bossema, I. 1979. "Jays and oaks: An eco-ethological study of a symbiosis." *Behaviour,* 70: 1–17.

Bower, T. G. R. 1982. *Development in Infancy.* San Francisco: W. H. Freeman.

Bowerman, M. 1978. "The acquisition of word meaning: An investigation of some current conflicts." In *Proceedings of the Third International Child Language Symposium: The Development of Communication,* ed. N. Waterson and C. E. Snow, pp. 263–287. New York: John Wiley.

———1982. "Reorganization processes in lexical and syntactic development." In *Language Acquisition: The State of the Art,* ed. E. Wanner and L. R. Gleitman, pp. 319–346. Cambridge: Cambridge University Press.

Bowmaker, J. K. 1986. "Avian color vision and the environment." Paper presented at the Nineteenth Meeting of the International Ornithological Congress, Ottawa, Ontario, Canada, June.

Bowmaker, J. K., L. A., Heath, D. Das, and D. M. Hunt. 1994. "Spectral sensitivity and opsin structure of avian rod and cone visual pigments." *Investigative Ophthalmology & Visual Science,* 35: 1708.

Bowmaker, J. K., L. A. Heath, S. E. Wilkie, D. Das, and D. M. Hunt. 1996. "Middle-wave cone and rod visual pigments in birds: Spectral sensitivity and opsin structure." *Investigative Ophthalmology & Visual Science,* 37: S804.

Boysen, S. T. 1993. "Counting in chimpanzees: Nonhuman principles and emergent properties of number." In *The Development of Numerical Competence,* ed. S. T. Boysen and E. J. Capaldi, pp. 39–59. Hillsdale, N.J.: Erlbaum.

Boysen, S. T., and G. G. Berntson. 1989. "Numerical competence in a chimpanzee *(Pan troglodytes).*" *Journal of Comparative Psychology,* 103: 23–31.

———1990. "The development of numerical skills in the chimpanzee." In *"Language" and Intelligence in Monkeys and Apes: Comparative Developmental Perspectives,* ed. S. T. Parker and K. R. Gibson, pp. 435–450. Cambridge: Cambridge University Press.

Brackenbury, J. H. 1978. "A comparison of the origin and temporal arrangement of pulsed sounds in the songs of the Grasshopper and Sedge warblers, *Locusta naeiva* and *Acrocephalus schoenobaenus.*" *Journal of Zoology,* 184: 187–206.

———1982. "The structural basis of voice production and its relationship to sound characteristics." In *Acoustic Communication in Birds,* vol. 1: *Production, Perception, and Design Features of Sound,* ed. D. E. Kroodsma and E. H. Miller, pp. 53–73. New York: Academic Press.

———1989. "Functions of the syrinx and the control of sound production." In *Form and Function in Birds,* ed. A. J. King and J. McLelland, pp. 193–220. London: Academic Press.

Bradshaw, G. 1993. "Beyond animal language." In *Language and Communication: Comparative Perspectives,* ed. H. L. Roitblat, L. M. Herman, and P. E. Nachtigall, pp. 25–44. Hillsdale, N.J.: Erlbaum.

Braine, M. D. S. 1976. "Children's first word combinations." *Monographs of the Society for Research in Child Development,* 41: 1–104.

Brannon, E. M., and H. S. Terrace. 1998. "Ordering of the numerosities 1 to 9 by monkeys." *Science,* 282: 746–749.

Braun, H. 1952. "Uber das Unterscheidungsvermögen unbenannter Anzahlen bei Papageien." *Zeitschrift für Tierpsychologie,* 9: 40–91.

Brauth, S. E., J. T. Heaton, S. E. Durand, W. Liang, and W. S. Hall. 1994. "Functional anatomy of forebrain vocal control pathways in the budgerigar *(Melopsittacus undulatus).*" *Brain, Behavior, and Evolution,* 44: 210–233.

Brehm, A. E. 1866. *Tierleben,* vol. 1. Leipzig: Verlag des Bibliographischen Instituts.

Breland, K., and M. Breland. 1961. "The misbehavior of organisms." *American Psychologist,* 16: 661–664.

Bremmer, J. G. 1978a. "Egocentric versus allocentric spatial coding in nine-month-old infants: Factors influencing choice of code." *Developmental Psychology,* 14: 346–355.

———1978b. "Spatial errors made by infants: Inadequate spatial cues or evidence of egocentrism." *British Journal of Psychology,* 69: 77–84.

Brenowitz, E. A. 1997. "Comparative approaches to the avian song system." *Journal of Neurobiology,* 33: 517–531.

Brittan-Powell, E. F., R. J. Dooling, O. N. Larsen, and J. T. Heaton. 1997. "Mechanisms of vocal production in budgerigars *(Melopsittacus undulatus).*" *Journal of the Acoustical Society of America,* 101: 578–589.

Britton, J. 1970. *Language and Learning.* London: Penguin Books.

Brodigan, D. L., and G. B. Peterson. 1976. "Two-choice conditional discrimination performance of pigeons as a function of reward expectancy, prechoice delay, and domesticity." *Animal Learning & Behavior,* 4: 121–124.

Bronowski, J., and U. Bellugi. 1970. "Language, name, and concept." *Science,* 168: 669–673.

Brown, E. D., and S. M. Farabaugh. 1997. "What birds with complex social relationships can tell us about vocal learning: Vocal sharing in avian groups." In *Social Influences on Vocal Development,* ed. C. T. Snowdon and M. Hausberger, pp. 98–127. Cambridge: Cambridge University Press.

Brown, I. 1976. "Role of referent concreteness in the acquisition of passive sentence comprehension through abstract modeling." *Journal of Experimental Child Psychology,* 22: 185–199.

Brown, R. 1958. *Words and Things.* New York: Free Press.

———1973. *A First Language: The Early Stages.* Cambridge, Mass.: Harvard University Press.

———1977. "Early social interaction and language acquisition." In *Studies in Mother-Infant Interaction,* ed. H. R. Schaffer, pp. 271–289. London: Academic Press.

———1978. "Learning how to do things with words." In *Human Growth and Development,* ed. J. S. Bruner and A. Garton, pp. 62–84. Oxford: Oxford University Press.

Bruner, J. S. 1964. "The course of cognitive development." *American Psychologist,* 19: 1–16.

———1977. "Early social interaction and language acquisition." In *Studies in Mother-Infant Interaction,* ed. H. R. Schaffer, pp. 271–289. London: Academic Press.

———1978. "Acquiring the uses of language." *Canadian Journal of Psychology,* 32: 204–218.

———1983. *Child's Talk: Learning to Use Language.* New York: Norton.

Burdyn, L. E., and R. K. Thomas. 1984. "Conditional discrimination with conceptual simultaneous and successive cues in the squirrel monkey *(Saimiri sciureus)." Journal of Comparative Psychology,* 98: 405–413.

Burghardt, G. M. 1984. "Animal awareness: Current perceptions and historical perspective." *American Psychologist,* 40: 905–919.

Burns, R. A. 1988. "Subitizing and rhythm in serial numerical investigations with animals." *Behavioral and Brain Sciences,* 11: 581–582.

Burton, P. J. K. 1974. "Jaw and tongue features in Psittaciformes and other orders with special reference to the anatomy of the Tooth-billed Pigeon *(Didunculus stringirostris)." Journal of Zoology* (London), 174: 255–276.

Busnel, R. G., and H. D. Mebes. 1975. "Hearing and communication in birds: The cocktail party effect in intraspecific communication of *Agapornis roseicollis.*" *Life Science,* 17: 1567–1569.

Butterworth, B. 1995. Editorial. *Mathematical Cognition,* 1: 1–2.

Butterworth, G., N. Jarrett, and L. Hicks. 1982. "Spatiotemporal identity in infancy: Perceptual competence or conceptual deficit?" *Developmental Psychology,* 18: 435–449.

Byers, B. E. 1992. "Development of two song categories by chestnut-sided warblers." *Animal Behaviour,* 44: 799–810.

———1996. "Messages encoded in the songs of chestnut-sided warblers." *Animal Behaviour,* 52: 691–705.

Byrne, R. 1995. *The Thinking Ape: Evolutionary Origins of Intelligence.* Oxford: Oxford University Press.

Caine, N. G., R. L. Addington, and T. L. Windfelder. 1995. "Factors affecting the rates of food calls given by red-bellied tamarins." *Animal Behaviour,* 50: 53–60.

Calculator, S. N. 1985. "Describing and treating discourse problems in mentally retarded children: The myth of mental retardese." In *School Discourse Problems,* ed. D. Ripich, D. Newberry, and F. Spinelli, pp. 125–147. San Diego: College-Hill Press.

Call, J., and M. Tomasello. 1994. "Production and comprehension of referential pointing by orangutans *(Pongo pygmaeus)." Journal of Comparative Psychology,* 108: 307–317.

———1998. "Distinguishing intentional from accidental actions in orangutans *(Pongo pygmaeus),* chimpanzees *(Pan troglodytes)* and human children *(Homo sapiens)." Journal of Comparative Psychology,* 112: 192–206.

Capaldi, E. J. 1993. "Animal numerical abilities: Implications for a hierarchical approach to instrumental learning." In *The Development of Numerical Competence,* ed. S. T. Boysen and E. J. Capaldi, pp. 211–223. Hillsdale, N.J.: Erlbaum.

Capaldi, E. J., and D. J. Miller. 1988. "Counting in rats: Its functional significance and the independent cognitive processes that comprise it." *Journal of Experimental Psychology: Animal Behavior Processes,* 14: 3–17.

Carey, S. 1978. "The child as word learner." In *Linguistic Theory and Psychological Reality,* ed. M. Halle, J. Bresnan, and A. Miller, pp. 264–293. Cambridge, Mass.: MIT Press.

———1982. "Semantic development." In *Language Acquisition: The State of the Art,* ed. E. Wanner and L. R. Gleitman, pp. 347–389. Cambridge: Cambridge University Press.

Caro, T. M., and M. D. Hauser. 1992. "Is there teaching in nonhuman animals?" *Quarterly Review of Biology,* 67: 151–174.

Carr, E., and E. Kologinsky. 1983. "Acquisition of sign language by autistic children. II: Spontaneity and generalization effects." *Journal of Applied Behavior Analysis,* 16: 297–314.

Caselli, M. C., E. Bates, P. Casadio, J. Fenson, L. Fenson, L. Sanderl, and J. Weir. 1995. "A cross-linguistic study of early lexical development." *Cognitive Development,* 10: 159–200.

Casey, R. M., and M. C. Baker. 1993. "Aggression and song development in White-crowned Sparrows." *Condor,* 95: 723–728.

Cazden, C. B. 1972. *Child Language and Education.* New York: Holt, Rinehart & Winston.

Chapman, M. 1987. "A longitudinal study of cognitive representation in symbolic play, self-recognition, and object permanence during the second year." *International Journal of Behavioural Development,* 10: 151–170.

Charlop, M. H., and J. P. Milstein. 1989. "Teaching autistic children conversational speech using video modeling." *Journal of Applied Behavior Analysis,* 22: 275–285.

Cheney, D. L., and R. M. Seyfarth. 1988. "How vervet monkeys perceive their grunts: Field playback experiments." *Animal Behaviour,* 30: 739–751.

———1990. *How Monkeys See the World: Inside the Mind of Another Species.* Chicago: University of Chicago Press.

Chevalier-Skolnikoff, S. 1976. "The ontogeny of primate intelligence and its implications for communicative potential: A preliminary report." In *Origins and Evolution of Speech,* Annals of the New York Academy of Sciences, vol. 280, ed. S. R. Harnad, H. D. Steklis, and J. Lancaster, pp. 173–211. New York: New York Academy of Sciences.

——1981. "The Clever Hans phenomenon, cuing, and ape signing: A Piagetian analysis of methods for instructing animals." In *The Clever Hans Phenomenon,* Annals of the New York Academy of Sciences, vol. 364, ed. T. A. Sebeok and R. Rosenthal, pp. 60–93. New York: New York Academy of Sciences.

——1989. "Spontaneous tool use and sensori-motor intelligence in *Cebus* compared with other monkeys and apes." *Behavioral and Brain Sciences,* 12: 561–627.

Chevalier-Skolnikoff, S., and F. E. Poirier. 1977. *Primate Bio-Social Development: Biological, Social, and Ecological Determinants.* New York: Garland.

Chiba, T., and M. Kajiyama. 1941. *The Vowel: Its Nature and Structure.* Tokyo: Kaiseikan.

Choi, S., and A. Gopnik. 1995. "Early acquisition of verbs in Korean: A cross-linguistic study." *Journal of Child Language,* 22: 497–529.

Chomsky, N. 1966. *Cartesian Linguistics.* New York: Harper & Row.

Church, R. M., and W. H. Meck. 1984. "The numerical attribute of stimuli." In *Animal Cognition,* ed. H. L. Roitblat, T. G. Bever, and H. S. Terrace, pp. 445–464. Hillsdale, N.J.: Erlbaum.

Clark, A. C. 1963. *Dolphin Island.* New York: Holt, Rinehart & Winston.

Clark, E. V. 1973. "What's in a word? On the child's acquisition of semantics in his first language." In *Cognitive Development and the Acquisition of Language,* ed. T. E. Moore, pp. 65–110. New York: Academic Press.

——1982. "The young word-maker: A case study of innovation in the child's lexicon." In *Language Acquisition: The State of the Art,* ed. E. Wanner and L. R. Gleitman, pp. 390–425. Cambridge: Cambridge University Press.

——1987. "The principle of contrast: A constraint on language acquisition." In *Mechanisms of Language Acquisition,* ed. B. MacWhinney, pp. 1–33. Baltimore: University Park Press.

——1998. "Lexical creativity in French-speaking children." *Cahiers de Psychologie Cognitive,* 17: 513–530.

Clark, R. A. 1978. "The transition from action to gesture." In *Action, Gesture and Symbol: The Emergence of Language,* ed. A. Lock, pp. 231–257. London: Academic Press.

Clayton, N. S., and A. Dickinson. 1998. "Episodic-like memory during cache recovery by scrub jays." *Nature,* 395: 272–274.

Clench, M. H. 1978. "Tracheal elongation in birds of paradise." *Condor,* 80: 423–430.

Cobb, S. 1960. "Observations on the comparative anatomy of the avian brain." In *Perspectives in Biology and Medicine,* vol. 3, ed. D. I. Ingle and S. O. Waife, pp. 383–408. Chicago: University of Chicago Press.

Cole, M., and S. Scribner. 1974. *Culture and Thought: A Psychological Introduction.* New York: John Wiley.

Collis, G. M. 1977. "Visual co-orientation and maternal speech." In *Studies in Mother-Infant Interaction*, ed. H. R. Schaffer, pp. 355–375. London: Academic Press.

———1981. "Social interaction with objects: A perspective on human infancy." In *Behavioural Development*, ed. K. Immelmann, G. W. Barlow, L. Petrinovich, and M. Main, pp. 603–320. Cambridge: Cambridge University Press.

Conti-Ramsden, G., and M. Jones. 1997. "Verb use in specific language impairment." *Journal of Speech and Hearing Research*, 40: 1298–1313.

Cook, R. G., K. K. Cavoto, and B. R. Cavoto. 1995. "Same-different texture discrimination and concept learning by pigeons." *Journal of Experimental Psychology: Animal Behavior Processes*, 21: 253–260.

———1996. "Mechanisms of multidimensional grouping, fusion, and search in avian texture discrimination." *Animal Learning & Behavior*, 24: 150–167.

Cornell, E. H. 1978. "Learning to find things: A reinterpretation of object permanence studies." In *Alternatives to Piaget: Critical Essays on the Theory*, ed. L. S. Siegel and C. J. Brainerd, pp. 1–10. New York: Academic Press.

Craig, H. K., and T. M. Gallagher. 1979. "The structural characteristics of monologues in the speech of normal children: Syntactic nonconversational aspects." *Journal of Speech and Hearing Research*, 22: 46–62.

Crain, S. 1991. "Language acquisition in the absence of experience." *Behavioral and Brain Sciences*, 14: 597–650.

Crelin, E. 1987. *The Human Vocal Tract: Anatomy, Function, Development, and Evolution.* New York: Vantage Press.

Cross, T. G. 1977. "Mother's speech adjustments: The contribution of selected child listener variables." In *Talking to Children: Language Input and Acquisition*, ed. C. E. Snow and C. A. Ferguson, pp. 151–188. Cambridge: Cambridge University Press.

Cruickshank, A. J., J.-P. Gautier, and C. Chappuis. 1993. "Vocal mimicry in wild African Grey Parrots *Psittacus erithacus*." *Ibis*, 135: 293–299.

Curio, E. 1988. "Cultural transmission of enemy recognition by birds." In *Social Learning: Psychological and Biological Perspectives*, ed. T. R. Zentall and B. G. Galef, Jr., pp. 75–97. Hillsdale, N.J.: Erlbaum.

Curio, E., U. Ernst, and W. Vieth. 1978. "Cultural transmission of enemy recognition: One function of mobbing." *Science*, 202: 899–901.

Cynx, J. 1990. "Experimental determination of a unit of song production in the zebra finch *(Taeniopygia guttata).*" *Journal of Comparative Psychology*, 104: 3–10.

Cynx, J., S. H. Hulse, and S. Polyzois. 1986. "A psychophysical measure of pitch discrimination loss resulting from a frequency range constraint in European starlings *(Sturnus vulgaris).*" *Journal of Experimental Psychology: Animal Behavior Processes*, 12: 394–402.

Davis, H. 1984. "Discrimination of the number three by a raccoon *(Procyon lotor).*" *Animal Learning & Behavior*, 4: 121–124.

———1993. "Numerical competence in animals: Life beyond Clever Hans." In *The Development of Numerical Competence*, ed. S. T. Boysen and E. J. Capaldi, pp. 109–125. Hillsdale, N.J.: Erlbaum.

Davis, H., and M. Albert. 1986. "Numerical discrimination by rats using sequential auditory stimuli." *Animal Learning & Behavior,* 14: 57–59.

———1987. "Failure to transfer or train a numerical discrimination using sequential visual stimuli in rats." *Bulletin of the Psychonomic Society,* 25: 472–474.

Davis, H., and S. A. Bradford. 1986. "Counting behavior in rats in a simulated natural environment." *Ethology,* 73: 265–280.

Davis, H., and J. Memmot. 1982. "Counting behavior in animals: A critical evaluation." *Psychological Bulletin,* 92: 547–571.

Davis, H., and R. Pérusse. 1988. "Numerical competence in animals: Definitional issues, current evidence, and a new research agenda." *Behavioral and Brain Sciences,* 11: 561–615.

Dawkins, M. S., and A. Woodington. 1997. "Distance and the presentation of visual stimuli to birds." *Animal Behaviour,* 54: 1019–1025.

de Blois, S. T., and M. A. Novak. 1994. "Object permanence in rhesus monkeys *(Macaca mulatta)."* *Journal of Comparative Psychology,* 108: 318–327.

de Blois, S. T., M. A. Novak, and M. Bond. 1998. "Object permanence in orangutans *(Pongo pygmaeus)* and squirrel monkeys *(Saimiri sciureus)."* *Journal of Comparative Psychology,* 112: 137–152.

Dehaene, S. 1992. "Varieties of numerical abilities." *Cognition,* 44: 1–42.

Dehaene, S., and J. P. Changeaux. 1993. "Development of elementary numerical abilities: A neuronal model." *Journal of Cognitive Neuroscience,* 5: 390–407.

Dehaene, S., and L. Cohen. 1994. "Dissociable mechanisms of subitizing and counting: Neuropsychological evidence from simultanoagnosis patients." *Journal of Experimental Psychology: Human Perception and Performance,* 20: 958–975.

———1995. "Toward an anatomical and functional model of number processing." *Mathematical Cognition,* 1: 83–120.

Dehaene, S., G. Dehaene-Lambertz, and L. Cohen. 1998. "Abstract representation of numbers in the animal and human brain." *Trends in Neurosciences,* 21: 355–361.

DeLoache, J. S., and N. M. Burns. 1994. "Early understanding of the representational function of pictures." *Cognition,* 52: 83–110.

Dennett, D. C. 1971. "Intentional systems." *Journal of Philosophy,* 68: 68–87.

———1978a. "Beliefs about beliefs." *Behavioral and Brain Sciences,* 1: 568–570.

———1978b. *Brainstorms.* Cambridge, Mass.: MIT Press.

———1983. "Intentional systems in cognitive ethology: The 'Panglossian paradigm' defended." *Behavioral and Brain Sciences,* 6: 343–355.

———1987. *The Intentional Stance.* Cambridge, Mass.: MIT Press.

———1988. "The intentional stance in theory and practice." In *Machiavellian Intelligence: Social Expertise and the Evolution of Intellect in Monkeys, Apes, and Humans,* ed. R. W. Byrne and A. Whiten, pp. 180–202. Oxford: Oxford University Press.

Densmore, F. 1918. *Teton Sioux Music.* Bureau of American Ethnology Bulletin 93. Washington, D. C.: Government Printing Office.

Dent, M. L., E. F. Brittan-Powell, R. J. Dooling, and A. Pierce. 1997. "Perception of synthetic /ba/-/wa/ speech continuum by budgerigars *(Melopsittacus undulatus)."* *Journal of the Acoustical Society of America,* 102: 1891–1897.

de Villiers, J. G., and P. A. de Villiers. 1978. *Language Acquisition.* Cambridge, Mass.: Harvard University Press.

de Villiers, P. A., and J. G. de Villiers. 1979. *Early Language.* Cambridge, Mass.: Harvard University Press.

deVoogd, T. J., J. R. Krebs, S. D. Healy, and A. Purvis. 1993. "Evolutionary correlation between repertoire size and a brain nucleus amongst passerine birds." *Proceedings of the Royal Society, London,* B254: 75–82.

de Waal, F. B. M. 1989. *Peacemaking among the Primates.* Cambridge, Mass.: Harvard University Press.

Diamond, A. 1990a. "Developmental time course in human infants and infant monkeys, and the neural basis of, inhibitory control of reaching." In *The Development and Neural Bases of Higher Cognitive Functions,* Annals of the New York Academy of Sciences, vol. 608, ed. A. Diamond, pp. 637–669. New York: New York Academy of Sciences.

——1990b. "The development and neural bases of memory functions as indexed by the AB̄ and delayed response tasks in human infants and infant monkeys." In *The Development and Neural Bases of Higher Cognitive Functions,* Annals of the New York Academy of Sciences, vol. 608, ed. A. Diamond, pp. 267–309. New York: New York Academy of Sciences.

Dickson, D. R., and W. Maue-Dickson. 1982. *Anatomical and Physiological Bases of Speech.* Boston: Little, Brown.

Dilger, W. C. 1960. "The comparative ethology of the African parrot genus *Agapornis.*" *Zeitschrift für Tierpsychologie,* 17: 649–685.

Dodd, B. J. 1972. "Effects of social and vocal stimulation on infant babbling." *Developmental Psychology,* 7: 80–83.

Dollaghan, C. A. 1987. "Fast mapping in normal and language-impaired children." *Journal of Speech and Hearing Disorders,* 52: 218–222.

Dollinger, S. J., and M. H. Thelen. 1978. "Leadership and imitation in children." *Personality & Social Psychology Bulletin,* 4: 487–490.

Dooley, G. B., and T. V. Gill. 1977. "Acquisition and use of mathematical skills by a linguistic chimpanzee." In *Language Learning by a Chimpanzee,* ed. D. M. Rumbaugh, pp. 247–260. New York: Academic Press.

Dooling, R. J., and S. D. Brown. 1990. "Speech perception by budgerigars *(Melopsittacus undulatus):* Spoken vowels." *Perception & Psychophysics,* 47: 568–574.

Dooling, R. J., and J. C. Saunders. 1975. "Hearing in the parakeet *(Melopsittacus undulatus):* Absolute thresholds, critical ratios, frequency difference limens, and vocalizations." *Journal of Comparative and Physiological Psychology,* 88: 1–20.

Dooling, R. J., C. T. Best, and S. D. Brown. 1995. "Discrimination of synthetic full-formant and sinewave /ra-la/ continua by budgerigars *(Melopsittacus undulatus)* and zebra finches *(Taeniopygia guttata).*" *Journal of the Acoustical Society of America,* 97: 1839–1846.

Dooling, R. J., K. Okanoya, and S. D. Brown. 1989. "Speech perception by budgerigars *(Melopsittacus undulatus):* The voiced-voiceless distinction." *Perception & Psychophysics,* 46: 65–71.

Dooling, R. J., S. D. Brown, T. J. Park, and S. D. Soli. 1987. "Perceptual organization of acoustic stimuli by budgerigars *(Melopsittacus undulatus):* I. Pure tones." *Journal of Comparative Psychology,* 101: 139-149.

Doré, F. Y. 1986. "Object permanence in adult cats *(Felis catus)." Journal of Comparative Psychology,* 100: 340-347.

———1990. "Search behaviour of cats *(Felis catus)* in an invisible displacement test: Cognition and experience." *Canadian Journal of Psychology,* 44: 359-370.

———1991. *Search Behaviour of Cats in an Invisible Displacement Test: Object Permanence and Retroactive Interference.* Cahiers de Recherche de l'Ecole de Psychologie (Université Laval), vol. 117.

Doré, F. Y., and C. Dumas. 1987. "Psychology of animal cognition: Piagetian studies." *Psychological Bulletin,* 102: 219-233.

Doré, F. Y., S. Fiset, S. Goulet, M-C. Dumas, and S. Gagnon. 1996. "Search behavior in cats and dogs: Interspecific differences in working memory and spatial cognition." *Animal Learning & Behavior,* 24: 142-149.

Doré, J. 1985. "Holophrases revisited: Their 'logical' development from dialog." In *Children's Single Word Speech,* ed. M. Barrett, pp. 23-58. New York: John Wiley.

———1987. "An analysis of crib monologues from 22 months to three years." Paper presented at the Twelfth Annual Boston University Conference on Language Development, Boston, October.

———1989. "Monologue as reenvoicement of dialogue." In *Narratives from the Crib,* ed. K. Nelson, pp. 231-260. Cambridge, Mass.: Harvard University Press.

Dorman, M. F., M. Studdert-Kennedy, and L. J. Raphael. 1977. "Stop consonant recognition: Release bursts and formant transitions as functionally equivalent, context-dependent cues." *Perception & Psychophysics,* 22: 109-122.

Dücker, G. 1976. "Erlernen von drei verschiedenen Positionen durch Vogel." *Zeitschrift für Tierpsychologie,* 42: 301-314.

Dücker, G., and B. Rensch. 1977. "The solution of patterned string problems by birds." *Behaviour,* 62: 164-173.

Dumas, C. 1992. "Object permanence in cats *(Felis catus):* An ecological approach to the study of invisible displacements." *Journal of Comparative Psychology,* 106: 404-410.

Dumas, C., and C. Brunet. 1994. "Permanence de l'objet chez le singe capucin *(Cebus apella):* Etude de déplacements invisibles." *Canadian Journal of Experimental Psychology,* 48: 341-358.

Dumas, C., and F. Y. Doré. 1989. "Cognitive development of kittens: A cross-sectional study of object permanence." *Journal of Comparative Psychology,* 103: 191-200.

———1991. "Cognitive development in kittens *(Felis catus):* An observational study of object permanence and sensorimotor intelligence." *Journal of Comparative Psychology,* 105: 357-365.

Dumas, C., and D. M. Wilkie. 1995. "Object permanence in ring doves *(Streptopelia risoria)." Journal of Comparative Psychology,* 109: 142-150.

Dunham, P. J., F. Dunham, and A. Curwin. 1993. "Joint-attentional states and lexical acquisition at 18 months." *Developmental Psychology,* 29: 827-831.

Durand, S. E., J. T. Heaton, S. K. Amateau, and S. E. Brauth. 1997. "Vocal control pathways through the anterior forebrain of a parrot *(Melopsittacus undulatus)." Journal of Comparative Neurology,* 377: 179–206.

Edwards, C. A., J. A. Jagielo, and T. R. Zentall. 1983. "Same/different symbol use by pigeons." *Animal Learning & Behavior,* 11: 349–355.

Elowson, A. M., P. T. Tannenbaum, and C. T. Snowdon. 1991. "Food associated calls correlate with food preferences in cotton-top tamarins." *Animal Behaviour,* 42: 931–937.

Emmerton, J. 1998. "Numerosity differences and effects of stimulus density on pigeons' discrimination performance." *Animal Learning & Behavior,* 26: 243–256.

Emmerton, J., A. Lohmann, and J. Niemann. 1997. "Pigeons' serial ordering of numerosity with visual arrays." *Animal Learning & Behavior,* 25: 234–244.

Essock, S. M., T. V. Gill, and D. M. Rumbaugh. 1977. "Language relevant object- and color-naming tasks." In *Language Learning by a Chimpanzee,* ed. D. M. Rumbaugh, pp. 193–206. New York: Academic Press.

Etienne, A. S. 1973a. "Developmental stages and cognitive structures as determinants of what is learned." In *Constraints on Learning,* ed. R. A. Hinde and J. Stevenson-Hinde, pp. 371–395. New York: Academic Press.

——1973b. "Searching behavior towards a disappearing prey in the domestic chick as affected by preliminary experience." *Animal Behaviour,* 21: 749–761.

——1976/1977. "L'étude comparative de la permanence d'objet chez l'animal." *Bulletin de Psychologie,* 327: 187–197.

——1984. "The meaning of object permanence at different zoological levels." *Human Development,* 27: 309–320.

Evans, C. S., and P. Marler. 1991. "On the use of video images as social stimuli in birds: Audience effects on alarm calling." *Animal Behaviour,* 41: 17–26.

Falls, J. B. 1985. "Song matching in western meadowlarks." *Canadian Journal of Zoology,* 63: 2520–2524.

Falls, J. B., J. R. Krebs, and P. K. McGregor. 1982. "Song matching in the great tit *(Parus major):* The effect of similarity and familiarity." *Animal Behaviour,* 30: 997–1009.

Fant, G. 1970. *Acoustic Theory of Speech Production,* 2nd ed. The Netherlands: Mouton.

——1991. "Discussion summaries: Composite models." *Journal of Phonetics,* 19: 489–490.

Farabaugh, S. M. 1982. "The ecological and social significance of duetting." In *Acoustic Communication in Birds,* vol. 2: *Song Learning and Its Consequences,* ed. D. E. Kroodsma and E. H. Miller, pp. 85–124. New York: Academic Press.

Fay, W. H., and A. L. Schuler. 1980. *Emerging Language in Autistic Children.* Baltimore: University Park Press.

Fazio, B. B. 1996. "Mathematical abilities of children with specific language impairment: A 2-year follow-up." *Journal of Speech and Hearing Research,* 39: 839–849.

Fazio, R. H., S. J. Sherman, and P. M. Herr. 1982. "The feature-positive effect in the self-perception process. Does not doing matter as much as doing?" *Journal of Personality and Social Psychology,* 42: 404–411.

Fee, M. S., B. Shraiman, B. Pesaran, and P. P. Mitra. 1998. "The role of nonlinear dynamics of the syrinx in the vocalizations of a songbird." *Nature,* 395: 67–71.

Feldman, C. F. 1989. "Monologue as problem-solving narrative." In *Narratives from the Crib,* ed. K. Nelson, pp. 98–119. Cambridge, Mass.: Harvard University Press.

Ferguson, C. A., L. Menn, and C. Stoel-Gammon, eds. 1992. *Phonological Development: Models, Research, and Implications.* Timonium, Md.: York Press.

Ferrier, L. 1978. "Word, context, and imitation." In *Action, Gesture and Symbol: The Emergence of Language,* ed. A. Lock, pp. 471–483. London: Academic Press.

Ferster, C. B. 1964. "Arithmetic behavior in chimpanzees." *Scientific American,* 210: 98–106.

Ferster, C. B., and C. E. Hammer. 1966. "Synthesizing the components of arithmetic behavior." In *Operant Behavior: Areas of Research and Application,* ed. W. K. Honig, pp. 634–676. New York: Appleton, Century, Crofts.

Fey, M. E. 1986. *Language Intervention with Young Children.* San Diego: College-Hill Press.

Filipek, P. A., D. N. Kennedy, V. S. Caviness, S. L. Rossnick, T. A. Spraggins, and P. M. Starerwicz. 1989. "Magnetic resonance imaging-based brain morphometry: Development and application to normal subjects." *Annals of Neurology,* 25: 61–67.

Finn, F. 1927. *Bird Behavior: Psychical and Physiological.* New York: Dodd, Mead.

Fisher-Jorgensen, E. 1954. "Acoustic analysis of stop consonants." *Miscellanea Phonetica,* 2: 42–49.

Flavell, J. H. 1985. *Cognitive Development.* Englewood Cliffs, N.J.: Prentice-Hall.

Flege, J. E., O.-S. Bohn, and S. Jang. 1997. "Effects of experience on non-native speakers' production and perception of English vowels." *Journal of Phonetics,* 25: 437–470.

Fletcher, N. H. 1988. "Bird song—a quantitative acoustic model." *Journal of Theoretical Biology,* 135: 455–481.

Forshaw, J. M. 1989. *Parrots of the World,* 3rd ed. Willoughby, Australia: Weldon Publishing.

Fouts, R. S. 1973a. "Acquisition and testing of gestural signs in four young chimpanzees." *Science,* 180: 978–980.

——1973b. "Capacities for language in the great apes." In *Proceedings of the Ninth International Congress of Anthropological and Ethnological Sciences,* ed. S. Tax and G. C. Neuberger. The Hague: Mouton.

——1974. "Language: Origins, definition, and chimpanzees." *Journal of Human Evolution,* 3: 475–482.

——1978. "Sign language in chimpanzees: Implications of the visual mode and the comparative approach." In *Sign Language and Language Acquisition in Man and Ape,* ed. F. C. C. Peng, pp. 121–136. Boulder, Colo.: Westview Press.

Fouts, R. S., and R. Rigby. 1977. "Man-chimpanzee communication." In *How Animals Communicate,* ed. T. Sebeok, pp. 1034–1054. Bloomington: Indiana University Press.

Fouts, R. S., B. Chown, and L. Goodin. 1976. "Transfer of signed responses in American Sign Language from vocal English stimuli to physical object stimuli by a chimpanzee *(Pan)." Learning and Motivation,* 7: 458–475.

Frith, C. B. 1994. "Adaptive significance of tracheal elongation in manucodes (Paradisaeidae)." *Condor,* 96: 552–555.

Fromkin, V., and R. Rodman. 1974. *An Introduction to Language.* New York: Holt, Rinehart & Winston.

———1983. *An Introduction to Language,* 3rd ed. New York: Holt, Rinehart & Winston.

Frydman, O. 1995. "The concept of number and the acquisition of counting concepts: The 'when', the 'how', and the 'what' of it." *Cahiers de Psychologie Cognitive,* 14: 653–684.

Funk, M. S. 1996. "Development of object permanence in the New Zealand parakeet *(Cyanoramphus auriceps)." Animal Learning & Behavior,* 24: 375–383.

Furness, W. H. 1916. "Observations on the mentality of chimpanzees and orangutans." *Proceedings of the American Philosophical Society,* 55: 281.

Furrow, D. 1984. "Social and private speech at two years." *Child Development,* 55: 355–362.

Furrow, D., and K. Nelson. 1986. "A further look at the motherese hypothesis: A reply to Gleitman, Newport, and Gleitman." *Journal of Child Language,* 13: 163–176.

Fuson, K. C. 1979. "The development of self-regulating aspects of speech." In *The Development of Self-Regulation through Private Speech,* ed. G. Zivin, pp. 135–217. New York: John Wiley.

———1988. *Children's Counting and Concepts of Number.* New York: Springer-Verlag.

———1995. "Aspects and uses of counting: An AUC framework for considering research on counting to update the Gelman/Gallistel counting principles." *Cahiers de Psychologie Cognitive,* 14: 724–731.

Fuson, K. C., and J. W. Hall. 1983. "The acquisition of early number word meanings: A conceptual analysis and review." In *The Development of Mathematical Thinking,* ed. H. P. Ginsburg, pp. 49–107. New York: Academic Press.

Gagnon, S., and F. Y. Doré. 1992. "Search behavior in various breeds of adult dogs *(Canis familiaris):* Object permanence and olfactory cues." *Journal of Comparative Psychology,* 106: 58–68.

———1993. "Search behavior of dogs *(Canis familiaris)* in invisible displacement problems." *Animal Learning & Behavior,* 21: 246–254.

———1994. "Cross-sectional study of object permanence in domestic puppies *(Canis familiaris)." Journal of Comparative Psychology,* 108: 220–232.

Gallagher, T. M., and H. K. Craig. 1978. "Structural characteristics of monologues in the speech of normal children: Semantics and conversational aspects." *Journal of Speech and Hearing Research,* 21: 103–117.

Gallistel, C. R. 1988. "Counting versus subitizing versus the sense of number." *Behavioral and Brain Sciences,* 11: 585–586.

———1993. "A conceptual framework for the study of numerical estimation and arithmetic reasoning in animals." In *The Development of Numerical Competence,* ed. S. T. Boysen and E. J. Capaldi, pp. 211–223. Hillsdale, N.J.: Erlbaum.

Gallistel, C. R., and R. Gelman. 1991. "Subitizing: The pre-verbal counting process." In *Thoughts, Memories, and Emotions: Essays in Honor of George Mandler,* ed. W. E. Kessen, A. Ortony, and F. I. M. Craik, pp. 65–81. Hillsdale, N.J.: Erlbaum.

———1992. "Preverbal and verbal counting and computation." *Cognition,* 44: 43–74.

Gallup, G. G., Jr. 1989. Editorial. *Journal of Comparative Psychology,* 103: 3.

Gannon, P. J., R. L. Holloway, D. C. Broadfield, and A. R. Braun. 1997. "Asymmetry of chimpanzee planum temporale: Humanlike pattern of Wernicke's brain language area homolog." *Science,* 279: 220–222.

Garcia, J., and R. A. Koelling. 1966. "Relation of cue to consequence in avoidance learning." *Psychonomic Science,* 4: 123–124.

Gardner, B. T., and R. A. Gardner. 1975. "Evidence for sentence constituents in early utterances of child and chimpanzee." *Journal of Experimental Psychology: General,* 104: 244–267.

Gardner, R. A., and B. T. Gardner. 1969. "Teaching sign language to a chimpanzee." *Science,* 187: 644–672.

———1978. "Comparative psychology and language acquisition." In *Psychology: The State of the Art,* ed. K. Salzinger and F. L. Denmark, pp. 37–78. New York: New York Academy of Sciences.

———1989. "Early signs of language in cross-fostered chimpanzees." *Human Evolution,* 4: 337–365.

Gardner, R. A., T. E. Van Cantfort, and B. T. Gardner. 1992. "Categorical replies to categorical questions by cross-fostered chimpanzees." *American Journal of Psychology,* 105: 27–57.

Garvey, C. 1977a. *Play.* Cambridge, Mass.: Harvard University Press.

———1977b. "Play with language and speech." In *Child Discourse,* ed. S. Ervin-Tripp and R. Mitchel-Keenan, pp. 27–47. New York: Academic Press.

Gast, H. 1957. "Der Umgang mit Zahlen und Zahlgebilden in der frühen Kindheit." *Zeitschrift für Psychologie,* 161: 1–90.

Gaunt, A. S. 1983. "An hypothesis concerning the relationship of syringeal structure to vocal abilities." *Auk,* 100: 853–862.

Gaunt, A. S., and S. L. L. Gaunt. 1985. "Electromyographic studies of the syrinx in parrots (Aves: Psittacidae)." *Zoomorphology,* 105: 1–11.

Gay, T., L-J. Boé, P. Perrier, G. Feng, and E. Swayne. 1991. "The acoustic sensitivity of vocal tract constrictions: A preliminary report." *Journal of Phonetics,* 19: 445–452.

Gelman, R. 1980. "What children know about numbers." *Educational Psychologist,* 15: 54–68.

Gelman, R., and C. R. Gallistel. 1986. *The Child's Understanding of Number,* 2nd ed. Cambridge, Mass.: Harvard University Press.

Gentner, D. 1982. "Why nouns are learned before verbs: Linguistic relativity versus natural partitioning." In *Language Development: Language, Cognition, and Culture,* ed. S. A. Kuczaj, pp. 38–62. Frankfurt: Springer-Verlag.

Geschwind, N. 1979. "Specializations of the human brain." *Scientific American,* 241: 180–199.

Ginsburg, N. 1960. "Conditioned vocalization in the budgerigar." *Journal of Comparative and Physiological Psychology,* 53: 183–186.

————1963. "Conditioned talking in the mynah bird." *Journal of Comparative and Physiological Psychology,* 56: 1061–1063.

Girolametto, L. E. 1988. "Improving the social-conversational skills of developmentally-delayed children: An intervention study." *Journal of Speech and Hearing Disorders,* 53: 156–167.

Glantz, K., N. I. Durlach, R. C. Barnett, and W. A. Aviles. 1996. "Virtual reality (VR) for psychotherapy: From the physical to the social environment." *Psychotherapy,* 33: 464–473.

Glanville, A. D., and K. M. Dallenbach. 1929. "The range of attention." *American Journal of Psychology,* 41: 207–236.

Gleitman, L. R. 1990. "The structural sources of verb meaning." *Language Acquisition,* 1: 3–55.

Gleitman, L. R., E. L. Newport, and H. Gleitman. 1984. "The current status of the motherese hypothesis." *Journal of Child Language,* 11: 43–79.

Gnam, R. 1988. "Preliminary results on the breeding biology of Bahama amazon." *Parrot Letter,* 1: 23–26.

Godard, R. 1991. "Long-term memory of individual neighbours in a migratory songbird." *Nature,* 350: 228–229.

Godard, R., and R. H. Wiley. 1995. "Individual recognition of song repertoires in two wood warblers." *Behavioral Ecology and Sociobiology,* 37: 119–123.

Goetestam, K. G., and D. Berntzen. 1997. "Use of the modelling effect in one-session exposure." *Scandinavian Journal of Behaviour Therapy,* 26: 97–101.

Goldfield, E. C., and D. J. Dickerson. 1981. "Keeping track of locations during movement in 8- to 10-month-old infants." *Journal of Experimental Child Psychology,* 32: 48–64.

Goldin-Meadow, S. 1997. "The resilience of language in humans." In *Social Influences on Vocal Development,* ed. C. T. Snowdon and M. Hausberger, pp. 293–311. Cambridge: Cambridge University Press.

Goldin-Meadow, S., and C. Mylander. 1990. "Beyond the input given: The child's role in the acquisition of language." *Language,* 66: 323–356.

Goldin-Meadow, S., M. E. P. Seligman, and R. Gelman. 1976. "Language in the two-year-old." *Cognition,* 4: 189–202.

Goldstein, H. 1984. "The effects of modeling and corrected practice on generative language and learning of preschool children." *Journal of Speech and Hearing Disorders,* 49: 389–398.

Goller, F. 1998. "Vocal gymnastics and the bird brain." *Nature,* 395: 11–12.

Gollin, E. S. 1985. "Ontogeny, phylogeny, and causality." In *The Comparative Development of Adaptive Skills: Evolutionary Implications,* ed. E. S. Gollin, pp. 1–17. Hillsdale, N.J.: Erlbaum.

Gomez, J. C. 1990. "The emergence of intentional communication as a problem-solving strategy in the gorilla." In *"Language" and Intelligence in Monkeys and Apes: Comparative Developmental Perspectives,* ed. S. T. Parker and K. R. Gibson, pp. 333–355. Cambridge: Cambridge University Press.

Goodall, J. 1986. *The Chimpanzees of Gombe: Patterns of Behavior.* Cambridge, Mass.: Harvard University Press.

Goodwin, R. 1985. "A word in edgeways? The development of conversation in the single word period." In *Children's Single Word Speech,* ed. M. Barrett, pp. 113–147. New York: John Wiley.

Gopnik, A., and A. N. Meltzoff. 1986. "Relations between semantic and cognitive development in the one-word stage: The specificity hypothesis." *Child Development,* 57: 1040–1053.

Gopnik, A., S. Choi, and T. Baumberger. 1996. "Cross-linguistic difference in early semantic and cognitive development." *Cognitive Development,* 11: 197–227.

Gorsuch, R. L., and A. J. Figueredo. 1991. "Sequential canonical analysis as an exploratory form of path analysis." Paper presented at the American Evaluation Association Conference, Chicago, October.

Gossette, R. L. 1969. Personal communication to O. H. Mowrer, 1980, in Mowrer, *Psychology of Language and Learning,* pp. 105–106. New York: Plenum Press.

Gossette, R. L., and M. F. Gossette. 1967. "Examination of the reversal index (RI) across fifteen different mammalian and avian species." *Perceptual and Motor Skills,* 24: 987–990.

Gossette, R. L., M. F. Gossette, and W. Riddell. 1966. "Comparisons of successive discrimination reversal performances among closely and remotely related avian species." *Animal Behaviour,* 14: 560–564.

Goulet, S., F. Y. Doré, and R. Lehotkay. 1996. "Activations of locations in working memory in cats." *The Quarterly Journal of Experimental Psychology,* 49B: 81–92.

Goulet, S., F. Y. Doré, and R. Rousseau. 1994. "Object permanence and working memory in cats *(Felis catus)*." *Journal of Experimental Psychology: Animal Behavior Processes,* 20: 347–365.

Gramza, A. F. 1970. "Vocal mimicry in captive budgerigars *(Melopsittacus undulatus)*." *Zeitschrift für Tierpsychologie,* 27: 971–983.

Granier-Deferre, C., and Y. Kodratoff. 1986. "Iterative and recursive behaviors in chimpanzees during problem solving: A new descriptive model inspired from the artificial intelligence approach." *Cahiers de Psychologie Cognitive,* 6: 483–500.

Grant, D. S., and S. E. MacDonald. 1986. "Matching to element and compound samples in pigeons: The role of sample coding." *Journal of Experimental Psychology: Animal Behavior Processes*, 12: 160–171.

Greenewalt, C. H. 1968. *Bird Song: Acoustics and Physiology.* Washington, D. C.: Smithsonian Institution Press.

Greenfield, P. M. 1971. "Goal as environmental variable in the development of intelligence." In *Intelligence: Genetic and Environmental Influences*, ed. R. Cancro, pp. 252–261. New York: Grune and Stratton.

———1978. "Developmental processes in the language learning of child and chimp." *Behavioral and Brain Sciences*, 4: 573–574.

———1980. "Toward an operational and logical analysis of intentionality: The use of discourse in early child language." In *The Social Foundations of Language and Thought*, ed. D. R. Olson, pp. 254–279. New York and London: Norton.

———1991. "Language, tools and brain: The ontogeny and phylogeny of hierarchically organized sequential behavior." *Behavioral and Brain Sciences*, 14: 531–595.

Greenfield, P. M., and E. S. Savage-Rumbaugh. 1984. "Perceived variability and symbol use: A common language-cognition interface in children and chimpanzees *(Pan troglodytes).*" *Journal of Comparative Psychology*, 98: 201–218.

———1990. "Grammatical combination in *Pan paniscus:* Processes of learning and invention in the evolution and development of language." In *"Language" and Intelligence in Monkeys and Apes: Comparative Developmental Perspectives*, ed. S. T. Parker and K. R. Gibson, pp. 540–578. Cambridge: Cambridge University Press.

———1991. "Imitation, grammatical development, and the invention of protogrammar by an ape." In *Biological and Behavioral Determinants of Language Development*, ed. N. A. Krasnegor, D. M. Rumbaugh, R. L. Schiefelbusch, and M. Studdert-Kennedy, pp. 235–258. Hillsdale, N.J.: Erlbaum.

———1993. "Comparing communicative competence in child and chimp: The pragmatics of repetition." *Journal of Child Language*, 20: 1–26.

Greeno, J. G., M. S. Riley, and R. Gelman. 1984. "Conceptual competence and children's counting." *Cognitive Psychology*, 16: 94–143.

Griffin, D. R. 1976. *The Question of Animal Awareness: Evolutionary Continuity of Mental Experience.* New York: Rockefeller University Press.

———1984. *Animal Thinking.* Cambridge, Mass.: Harvard University Press.

———1985. "The cognitive dimensions of animal communication." In *Experimental Behavioral Ecology and Sociobiology*, ed. B. Hölldobler and M. Lindauer, pp. 471–482. New York: Fischer-Verlag.

———1992. *Animal Minds.* Chicago: University of Chicago Press.

Griffiths, P. 1985. "The communicative functions of children's single word speech." In *Children's Single Word Speech*, ed. M. D. Barrett, pp. 87–112. New York: John Wiley.

Grosslight, J. H., and W. C. Zaynor. 1967. "Verbal behavior and the mynah bird." In *Research in Verbal Behavior and Some Neurophysiological Implications*, ed. K. Salzinger and S. Salzinger, pp. 5–9. New York: Academic Press.

Grosslight, J. H., W. C. Zaynor, and B. L. Lively. 1964. "Speech as a stimulus for differential vocal behavior in the mynah bird *(Gracula religiosa)." Psychonomic Science,* 1: 7–8.

Gruber, H. E., J. S. Girgus, and A. Banuazizi. 1971. "The development of object permanence in the cat." *Developmental Psychology,* 4: 9–15.

Güttinger, H. R. 1979. "The integration of learnt and genetically programmed behaviour: A study of hierarchical organization in songs of canaries, greenfinches, and their hybrids." *Zeitschrift für Tierpsychologie,* 49: 285–303.

Gwinner, E., and J. Kneutgen. 1962. "Uber die biologische Bedeutung der 'zweckdienlichen' Anwendung erlernter Laute bei Vogeln." *Zeitschrift für Tierpsychologie,* 19: 692–696.

Hallock, M. B., and J. Worobey. 1984. "Cognitive development in chimpanzee infants *(Pan troglodytes)." Journal of Human Evolution,* 13: 441–447.

Hampson, J., and K. Nelson. 1993. "The relation of maternal language to variation in rate and style of language acquisition." *Journal of Child Language,* 20: 313–342.

Harré, R. 1984. "Vocabularies and theories." In *The Meaning of Primate Signals,* ed. R. Harré and V. Reynolds, pp. 90–110. Cambridge: Cambridge University Press.

Harriman, A., and R. Berger. 1986. "Olfactory acuity in the common raven *(Corvus corax)." Physiology and Behavior,* 36: 257–262.

Harris, C. L., W. B. Gross, and A. Robeson. 1968. "Vocal acoustics of the chicken." *Poultry Science,* 47: 107–112.

Harris, M., D. Jones, and J. Grant. 1983. "The nonverbal context of mothers' speech to infants." *First Language,* 4: 21–30.

Harris, M., D. Jones, S. Brookes, and J. Grant. 1986. "Relations between non-verbal context of maternal speech and rate of language development." *British Journal of Developmental Psychology,* 4: 261–268.

Harris, P. L. 1973. "Perseverative errors in search by young infants." *Child Development,* 44: 28–33.

——— 1975. "Development of search and object permanence during infancy." *Psychological Bulletin,* 82: 332–344.

——— 1983. "Infant cognition." In *Handbook of Child Psychology: Infancy and Developmental Psychobiology,* vol. 2, ed. M. M. Haith and J. J. Campos, pp. 689–782. New York: John Wiley.

Harris, R. 1984. "Must monkeys mean?" In *The Meaning of Primate Signals,* ed. R. Harré and V. Reynolds, pp. 116–137. Cambridge: Cambridge University Press.

Harris, S., C. Kasari, and M. D. Sigman. 1996. "Joint attention and language gains in children with Down syndrome." *American Journal on Mental Retardation,* 100: 608–619.

Hausberger, M., J. M. Black, and J. P. Richard. 1991. "Bill opening and sound spectrum in Barnacle Goose loud calls: Individuals with 'wide mouths' have higher pitched voices." *Animal Behaviour,* 42: 319–322.

Hayes, K. J., and C. H. Nissen. 1956/1971. "Higher mental functions of a home-raised chimpanzee." In *Behavior of Nonhuman Primates,* vol. 4, ed. A. M. Schrier and F. Stollnitz, pp. 57–115. New York: Academic Press.

Healy, S. D. 1996. "Ecological specialization in the avian brain." In *Neuroethological Studies of Cognitive and Perceptual Processes,* ed. C. F. Moss and S. J. Shettleworth, pp. 84–110. Boulder, Colo.: Westview Press.

Hearst, E. 1984. "Absence as information: Some implications for learning, performance, and representational processes." In *Animal Cognition,* ed. H. L. Roitblat, T. G. Bever, and H. S. Terrace, pp. 311–332. Hillsdale, N.J.: Erlbaum.

Hecker, M. H. L. 1971. "Speaker recognition by visual comparison of spectrograms." In *Speaker Recognition,* American Speech and Hearing Association Monographs, no. 16, ed. M. H. L. Hecker, pp. 50–73. Rockville, Md.

Hediger, H. K. P. 1981. "The Clever Hans phenomenon from an animal psychologist's point of view." In *The Clever Hans Phenomenon,* Annals of the New York Academy of Sciences, vol. 364, ed. T. A. Sebeok and R. Rosenthal, pp. 1–17. New York: New York Academy of Sciences.

Held, R. 1965. "Plasticity in sensory-motor systems." *Scientific American,* 213: 84–94.

Hensley, G. 1980. "Encounters with a hookbill—II." *American Cage Bird Magazine,* 52: 11–12, 59.

Herman, L. M. 1980. "Cognitive characteristics of dolphins." In *Cetacean Behavior,* ed. L. M. Herman, pp. 408–409. New York: John Wiley.

——1986. "Cognition and language competencies of bottlenosed dolphins." In *Dolphin Cognition and Behavior: A Comparative Approach,* ed. R. J. Schusterman, J. A. Thomas, and F. G. Wood, pp. 221–252. Hillsdale, N.J.: Erlbaum.

——1987. "Receptive competencies of language-trained animals." In *Advances in the Study of Behavior,* vol. 17, ed. J. S. Rosenblatt, C. Beer, M.-C. Busnel, and P. J. B. Slater, pp. 1–60. New York: Academic Press.

——1988. "The language of animal language research: A reply to Schusterman and Gisiner." *The Psychological Record,* 38: 349–362.

——1989. "In which Procrustean bed does the sea lion sleep tonight?" *Psychological Record,* 39: 19–50.

Herman, L. M., and P. H. Forestell. 1985. "Reporting presence or absence of named objects by a language-trained dolphin." *Neuroscience and Biobehavioral Reviews,* 9: 667–681.

Herman, L. M., P. Morrel-Samuels, and A. A. Pack. 1990. "Bottlenosed dolphin and human recognition of veridical and degraded video displays of an artificial gestural language." *Journal of Experimental Psychology: General,* 119: 215–230.

Herman, L. M., D. G. Richards, and J. P. Wolz. 1984. "Comprehension of sentences by bottlenosed dolphins." *Cognition,* 16: 129–219.

Herman, L. M., J. R. Hovancik, J. D. Gory, and G. L. Bradshaw. 1989. "Generalization of visual matching by a bottlenosed dolphin *(Tursiops truncatus):* Evidence for invariance of cognitive performance with visual and auditory materials." *Journal of Experimental Psychology: Animal Behavior Processes,* 15: 124–136.

Herrick, C. J. 1924. *Neurological Foundations of Behavior.* New York: Holt.

Herrnstein, R. J. 1984. "Objects, categories, and discriminative stimuli." In *Animal Cognition*, ed. H. L. Roitblat, T. G. Bever, and H. S. Terrace, pp. 233–261. Hillsdale, N.J.: Erlbaum.

Herrnstein, R. J., D. H. Loveland, and C. Cable. 1976. "Natural concepts in pigeons." *Journal of Experimental Psychology: Animal Behavior Processes*, 2: 285–301.

Herron, C. 1994. "An investigation of the effectiveness of using an advance organizer to introduce video in the foreign language classroom." *The Modern Language Journal*, 78: 190–198.

Hersch, G. L. 1966. "Bird voices and resonant tuning in helium-air mixtures." Ph.D. thesis, University of California, Berkeley.

Hicks, L. H. 1956. "An analysis of number-concept formation in the rhesus monkey." *Journal of Comparative and Physiological Psychology*, 49: 212–218.

Hienz, R. D., M. B. Sachs, and J. M. Sinnott. 1981. "Discrimination of steady-state vowels by blackbirds and pigeons." *Journal of the Acoustical Society of America*, 70: 699–706.

Highsmith, R. T. 1989. "The singing behavior of golden-winged warblers." *Wilson Bulletin*, 101: 36–50.

Hill, J. H. 1995. "Do apes have language?" In *Research Frontiers in Anthropology*, ed. C. R. Ember and P. N. Peregrine, pp. 1–19. Englewood Cliffs, N.J.: Prentice-Hall.

Hinde, R. A., and J. Stevenson-Hinde, eds. 1973. *Constraints on Learning: Limitations and Predispositions*. New York: Academic Press.

Hodos, W. 1982. "Some perspectives on the evolution of intelligence and the brain." In *Animal Mind—Human Mind*, ed. D. R. Griffin, pp. 33–56. Berlin: Springer-Verlag.

Hoffman, E. A., D. K. Gnanaprakasam, B. Gupta, J. D. Hoford, S. D. Kugelmass, and R. S. Kulawiec. 1992. "VIDA: An environment for multidimensional image display and analysis." In *Biomedical Image Processing and 3-D Microscopy*, no. 1660. San Jose: SPIE.

Holmes, P. W. 1979. "Transfer of matching performance in pigeons." *Journal of the Experimental Analysis of Behavior*, 31: 103–114.

Homberger, D. G. 1979. "Functional morphology of the larynx in the parrot *Psittacus erithacus.*" *American Zoologist*, 19: 988.

———1980. *Funktionell-morphologische Untersuchungen zur Radiation der Ernährungs und Trinkmethoden der Papageien.* Bonner Zoologische Monographien, no. 13. Bonn: Zoologisches Forschungsinstitut und Museum Alexander Koenig.

———1986. *The Lingual Apparatus of the African Grey Parrot, "Psittacus erithacus Linne" (Aves: Psittacidae) Description and Theoretical Mechanical Analysis.* Ornithological Monographs, no. 39. Washington, D.C.: The American Ornithologists' Union.

———1999. "The avian linguo-buccal system: Multiple functions in nutrition and vocalization." In *Proceedings of the Twenty-second International Ornithological Congress*, ed. N. Adams and R. Slotow, pp. xx–yy. Durban: University of Natal.

———In press. "Psittacidae." In *Fauna of Australia,* vol. 2, ed. G. B. Ross, pp. xx-yy. Canberra: Australian Government Publishing Service.

Honig, W. K., and W. R. Matheson. 1995. "Discrimination of relative numerosity and stimulus mixture by pigeons with comparable tasks." *Journal of Experimental Psychology: Animal Behavior Processes,* 21: 348-363.

Honig, W. K., and K. E. Stewart. 1989. "Discrimination of relative numerosity by pigeons." *Animal Learning & Behavior,* 17: 134-146.

———1993. "Relative numerosity as a dimension of stimulus control: The peak shift." *Animal Learning & Behavior,* 21: 346-354.

Hopkins, W. J., and E. S. Savage-Rumbaugh. 1991. "Vocal communication as a function of differential rearing experiences in *Pan paniscus:* A preliminary report." *International Journal of Primatology,* 12: 559-583.

Hulse, S. H. 1978. "Cognitive structures and serial pattern learning by animals." In *Cognitive Processes in Animal Behavior,* ed. S. H. Hulse, H. Fowler, and W. K. Honig, pp. 311-340. Hillsdale, N.J.: Erlbaum.

Hulse, S. H., and J. Cynx. 1985. "Relative pitch perception is constrained by absolute pitch in songbirds *(Mimus, Molothrus,* and *Sturnus)." Journal of Comparative Psychology,* 99: 176-196.

Hulse, S. H., J. Cynx, and J. Humpal. 1984. "Absolute and relative pitch discrimination in serial pitch perception by birds." *Journal of Experimental Psychology: General,* 113: 28-54.

Hulse, S. H., H. Fowler, and W. K. Honig, eds. 1968. *Cognitive Processes in Animal Behavior.* Hillsdale, N.J.: Erlbaum.

Hulse, S. H., S. C. Page, and R. F. Braaten. 1990. "Frequency range size and the frequency range constraint in auditory perception by European starlings *(Sturnus vulgaris)." Animal Learning & Behavior,* 18: 238-245.

Hultsch, H. 1990. "Recombination of acquired songs as a correlate of package formation." In *Brain-Perception-Cognition,* ed. N. Elsner and G. Roth, p. 433. Stuttgart: Thieme Verlag.

Humphrey, N. K. 1976. "The social function of intellect." In *Growing Points in Ethology,* ed. P. P. G. Bateson and R. A. Hinde, pp. 303-317. Cambridge: Cambridge University Press.

Hurford, J. R. 1987. *Language and Number.* Oxford: Basil Blackwell.

Hurlock, E. 1934. "Experimental investigations of childhood play." *Psychological Bulletin,* 31: 47-66.

Hurly, T. A., L. Ratcliffe, and R. Weisman 1990. "Relative pitch recognition in white-throated sparrows, *Zonotrichia albicollis." Animal Behaviour,* 40: 176-181.

Hurly, T. A., R. Weisman, L. Ratcliffe, and I. Johnsrude. 1991. "Absolute and relative pitch production in the songs of the white-throated sparrow *(Zonotrichia albicollis)." Bioacoustics,* 3: 81-91.

Huttenlocker, J. 1974. "The origins of language comprehension." In *Theories in Cognitive Psychology: The Loyola Symposium,* ed. R. L. Solso, pp. 331-368. Hillsdale, N.J.: Erlbaum.

———1998. "Language input and language growth." *Preventive Medicine,* 27: 195-199.

Ilyichev, V., and O. Silayeva. 1992. *Talking Birds.* Trans. A. Taruts. Moscow: Nauka Publishers.

Itakura, S., and J. R. Anderson. 1996. "Learning to use experimenter-given cues during an object-choice task by a capuchin monkey." *Current Psychology of Cognition,* 15: 103–112.

Jarvis, E. 1998. "Windows into the molecular mechanisms of song learning and vocal communication." Talk presented at the Twenty-second International Ornithological Congress, Durban, South Africa.

Jenkins, H. M., and R. S. Sainsbury. 1969. "The development of stimulus control through differential reinforcement." In *Fundamental Issues in Associative Learning,* ed. N. J. Mackintosh and W. K. Honig, pp. 121–161. Halifax, N.S.: Dalhousie University Press.

———1970. "Discrimination learning with the distinctive feature on positive or negative trials." In *Attention: Contemporary Theory and Analysis,* ed. D. Mostofsky, pp. 239–273. New York: Appleton-Century-Crofts.

Jevons, W. S. 1871. "The power of numerical discrimination." *Nature,* 3: 281–282.

Johnson, K. 1997. *Acoustic and Auditory Phonetics.* Oxford: Blackwell.

Jones, A. E., and P. J. B. Slater. 1996. "The role of aggression in song tutor choice in the zebra finch: Cause or effect?" *Behaviour,* 133: 103–115.

Jones, A. E., C. ten Cate, and P. J. B. Slater. 1996. "Early experience and plasticity of song in adult male zebra finches *(Taeniopygia guttata).*" *Journal of Comparative Psychology,* 110: 354–369.

Jouanjean-L'Antoëne, A. 1997. "Reciprocal interactions and the development of communication between parents and children." In *Social Influences on Vocal Development,* ed. C. T. Snowdon and M. Hausberger, pp. 312–327. Cambridge: Cambridge University Press.

Kagawa, Y., R. Shimoyama, T. Yamabuchi, T. Murai, and K. Takarada. 1992. "Boundary element models of the vocal tract and radiation field and their response characteristics." *Journal of Sound Vibration,* 157: 385–403.

Kako, E. 1999. "Elements of syntax in the systems of three language-trained animals." *Animal Learning & Behavior,* 27: 1–14.

Kalischer, O. 1901. "Weitere Mittheilung zur Grosshirnlocalization bei den Vögeln." *Sitzungsberichte der Koniglich Preussischen Akademie der Wissenschaften zu Berlin,* 1: 428–439.

Kamhi, A. G. 1982. "Overextensions and underextensions: How different are they?" *Journal of Child Language,* 9: 243–247.

Kamil, A. C. 1984. "Adaptation and cognition: Knowing what comes naturally." In *Animal Cognition,* ed. H. L. Roitblat, T. G. Bever, and H. S. Terrace, pp. 533–544. Hillsdale, N.J.: Erlbaum.

———1988. "A synthetic approach to the study of animal intelligence." In *Nebraska Symposium on Motivation: Comparative Perspectives in Modern Psychology,* vol. 7, ed. D. Leger, pp. 257–308. Lincoln: University of Nebraska Press.

———1998. "On the proper definition of cognitive ethology." In *Animal Cognition in Nature,* ed. R. P. Balda, I. M. Pepperberg, and A. C. Kamil, pp. 1–28. London: Academic Press.

Kamil, A. C., and R. P. Balda. 1985. "Cache recovery and spatial memory in Clark's nutcrackers *(Nucifraga columbiana)." Journal of Experimental Psychology: Animal Behavior Processes,* 11: 95-111.

Kamil, A. C., and M. W. Hunter III. 1970. "Performance on object discrimination learning set by the Greater Hill mynah, *Gracula religiosa." Journal of Comparative and Physiological Psychology,* 13: 68-73.

Kamil, A. C., and H. L. Roitblat. 1985. "The ecology of foraging behavior: Implications for animal learning and memory." *Annual Review of Psychology,* 36: 141-169.

Kamil, A. C., and T. D. Sargent. 1981. *Foraging Behavior: Ecological, Ethological, and Psychological Approaches.* New York: Garland Press.

Kane, J. W., and J. M. Sternheim. 1988. *Physics,* 3rd ed. New York: John Wiley.

Karlan, G. R., A. Lentz, B. Brenn-White, P. Hodur, D. Egger, and D. Frankoff. 1982. "Establishing generalized, productive verb-noun phrase usage in a manual language system with moderately handicapped children." *Journal of Speech and Hearing Disorders,* 47: 31-42.

Kasari, C., M. Sigman, P. Mundy, and N. Yirmiya. 1990. "Affective sharing in the context of joint attention interactions of normal, autistic, and mentally retarded children." *Journal of Autism and Developmental Disorders,* 20: 87-100.

Kaufman, E. L., M. W. Lord, T. W. Reese, and J. Volkmann. 1949. "The discrimination of visual number." *American Journal of Psychology,* 62: 498-525.

Kellogg, W. N. 1968. "Communication and language in the home-raised chimpanzee." *Science,* 162: 423-438.

Kewley-Port, D., X. Li, Y. Zheng, and A. T. Neel. 1996. "Fundamental frequency effects on thresholds for vowel formant discrimination." *Journal of the Acoustical Society of America,* 100: 2462-2470.

King, A. P., and M. J. West. 1983. "Epigenesis of cowbird song—a joint endeavor of males and females." *Nature,* 305: 704-706.

———1989. "Presence of female cowbirds *(Molothrus ater ater)* affects vocal imitation and improvisation in males." *Journal of Comparative Psychology,* 103: 39-44.

King, A. P., T. M. Freeberg, and M. J. West. 1996. "Social experience affects the process and outcome of vocal ontogeny in two populations of cowbirds *(Molothrus ater)." Journal of Comparative Psychology,* 110: 276-285.

King, A. S. 1989. "Functional anatomy of the syrinx." In *Form and Function in Birds,* ed. A. S. King and J. McClelland, pp. 105-192. London: Academic Press.

King-Smith, D. 1984. *Harry's Mad.* London: Victor Gollantz.

Klahr, D. 1973. "Quantification processes." In *Visual Information Processing,* ed. W. G. Chase, pp. 3-34. New York: Academic Press.

Klahr, D., and J. G. Wallace. 1973. "The role of quantification operators in the development of conservation of quantity." *Cognitive Psychology,* 4: 301-327.

———1976. *Cognitive Development: An Information Processing View.* Hillsdale, N.J.: Erlbaum.

Klatt, D. H., and R. A. Stefanski. 1974. "How does a mynah bird imitate human speech?" *Journal of the Acoustical Society of America,* 55: 822-832.

Kluender, K. R., R. Diehl, and P. R. Killeen. 1987. "Japanese quail can learn phonetic categories." *Science*, 237: 1195-1197.

Koegel, R. L., K. Dyer, and L. K. Bell. 1987. "The influence of child-preferred activities on autistic children's social behavior." *Journal of Applied Behavior Analysis*, 20: 243-252.

Koehler, O. 1943. " 'Zähl'-Versuche an einem Kolkraben und Vergleichsversuche an Menschen." *Zeitschrift für Tierpsychologie*, 5: 575-712.

———1950. "The ability of birds to 'count.' " *Bulletin of the Animal Behaviour Society*, 9: 41-45.

———1953. "Thinking without words." *Proceedings of the Fourteenth International Congress of Zoology*, 75-88.

———1972. "Der Sprachebegabung der Papagein." In *Grzimek's Tierleben VIII*, Munich: Kindler-Verlag.

Köhler, W. 1929. *Gestalt Psychology*. New York: H. Liveright.

Krashen, S. D. 1976. "Formal and informal linguistic environments in language learning and language acquisition." *TESOL Quarterly*, 10: 157-168.

———1980. "The input hypothesis." In *Current Issues in Bilingual Education*, ed. J. E. Alatis, pp. 168-180. Washington, D.C.: Georgetown University Press.

———1982. *Principles and Practices in Second Language Acquisition*. Oxford: Pergamon Press.

———1983. "Second language acquisition theory and the preparation of teachers." *Georgetown University Round Table on Languages and Linguistics*, 11: 255-263.

Krebs, J. 1977. "Song and territory in the Great Tit." In *Evolutionary Ecology*, ed. B. Stonehouse and C. Perrins, pp. 47-62. New York: Macmillan.

Kroodsma, D. E. 1974. "Song learning, dialects, and dispersal in the Bewick's wren." *Zeitschrift für Tierpsychologie*, 35: 352-380.

———1978. "Aspects of learning in the ontogeny of bird song: Where, from whom, when, how many, which, and how accurately?" In *The Development of Behavior: Comparative and Evolutionary Aspects*, ed. G. M. Burghardt and M. Bekoff, pp. 215-230. New York: Garland Press.

———1979. "Vocal dueling among male marsh wrens: Evidence for ritualized expressions of dominance/subordinance." *Auk*, 96: 506-515.

———1981. "Winter wren singing behavior: A pinnacle of song complexity." *Condor*, 82: 357-365.

———1988. "Song types and their use: Developmental flexibility of the male blue-winged warbler." *Ethology*, 79: 235-247.

———1990. "How the mismatch between the experimental design and the intended hypothesis limits confidence in knowledge, as illustrated by an example from birdsong dialects." In *Interpretation and Explanation in the Study of Animal Behavior: Comparative Perspectives*, ed. M. Bekoff and D. Jamieson, pp. 226-245. Boulder, Colo.: Westview Press.

Kroodsma, D. E., and E. H. Miller, eds. 1982. *Acoustic Communication in Birds*. New York: Academic Press.

———1996. *Ecology and Evolution of Acoustic Commuication in Birds*. Ithaca: Cornell University Press.

Kroodsma, D. E., and R. Pickert. 1984a. "Repertoire size, auditory templates, and selective vocal learning in songbirds." *Animal Behaviour,* 32: 395-399.

———1984b. "Sensitive phases for song learning: Effects of social interaction and individual variation." *Animal Behaviour,* 32: 389-394.

Kroodsma, D. E., W. R. Meservey, and R. Pickert. 1983. "Vocal learning in Parulinae." *Wilson Bulletin,* 95: 138-140.

Kroodsma, D. E., R. C. Bereson, B. E. Byers, and E. Minear. 1989. "Use of song types by the Chestnut-sided Warbler: Evidence for both intrasexual and intersexual functions." *Canadian Journal of Zoology,* 67: 447-456.

Krushinskii, L. V. 1960. *Animal Behavior: Its Normal and Abnormal Development.* Trans. Basil Haigh. New York: Consultants Bureau.

Kuczaj, S. A. 1977. "The acquisition of regular and irregular past tense forms." *Journal of Verbal Learning and Verbal Behavior,* 16: 589-600.

———1982a. "On the nature of syntactic development." In *Language Development: Syntax and Semantics,* ed. S. A. Kuczaj, pp. 37-71. Hillsdale, N.J.: Erlbaum.

———1982b. "Language play and language acquisition." In *Advances in Child Development and Behavior,* ed. H. Reese, pp. 197-233. New York: Academic Press.

———1983. *Crib Speech and Language Play.* New York: Springer-Verlag.

———1998. "Is an evolutionary theory of language play possible?" *Cahiers de Psychologie Cognitive,* 17: 135-154.

Kuczaj, S. A., and A. Bean. 1982. "The development of non-communicative speech systems." In *Language Development: Language, Thought, and Culture,* ed. S. A. Kuczaj, pp. 279-300. Hillsdale, N.J.: Erlbaum.

Kuczaj, S. A., and V. M. Kirkpatrick. 1993. "Similarities and differences in human and animal language research: Toward a comparative psychology of language." In *Language and Communication: Comparative Perspectives,* ed. H. L. Roitblat, L. M. Herman, and P. E. Nachtigall, pp. 45-63. Hillsdale, N.J.: Erlbaum.

Kuczaj, S. A., R. H. Borys, and M. Jones. 1989. "On the interaction of language and thought: Some thoughts and developmental data." In *Cognition and Social Worlds,* ed. A. Gellatly, D. Rogers, and J. Slaboda, pp. 168-189. Oxford: Oxford University Press.

Kuhl, P. K., and J. D. Miller. 1978. "Speech perception by the chinchilla: Identification functions for synthetic VOT stimuli." *Journal of the Acoustical Society of America,* 63: 905-917.

Kuhl, P. K., and D. M. Padden. 1982. "Enhanced discriminability at the phonetic boundaries for the voicing feature in macaques." *Perception & Psychophysics,* 32: 542-550.

Kuhl, P. K., K. A. Williams, and A. N. Meltzoff. 1991. "Cross modal speech perception in adults and infants using nonspeech auditory stimuli." *Journal of Experimental Psychology: Human Perception and Performance,* 17: 829-840.

Kuhn, G. M. 1975. "On the front cavity resonance and its possible role in speech perception." *Journal of the Acoustical Society of America,* 58: 428-433.

Ladefoged, P. 1982. *A Course in Phonetics.* San Diego: Harcourt Brace Jovanovich.

Laidler, K. 1978. "Language in the Orang-utan." In *Action, Gesture and Symbol: The Emergence of Language,* ed. A. Lock, pp. 133–155. London: Academic Press.

Lambert, W. E. 1981. "Bilingualism and language acquisition." In *Native Language and Foreign Language Acquisition,* Annals of the New York Academy of Sciences, vol. 379, ed. H. Winitz, pp. 9–22. New York: New York Academy of Sciences.

Landers, W. F. 1968. "The effects of different amounts and types of experience on infant object concepts." Ph.D. thesis, University of Houston.

Landry, S. H., K. E. Smith, C. L. Miller-Loncar, and P. R. Swank. 1997. "Predicting cognitive-language and social growth curves from early maternal behaviors in children at varying degrees of biological risk." *Developmental Psychology,* 33: 1040–1053.

Langley, C., and D. A. Riley. 1988. "Feature versus conjunctive search in pigeons." Paper presented at the Annual Meeting of the Psychonomic Society, Chicago, November.

Lashley, K. S. 1913. "Reproduction of inarticulate sounds in the parrot." *Journal of Animal Behaviour,* 3: 361–366.

Lauter, J. L., N. Pearl, and C. M. Baldwin. 1986. "VOT variability: Within-subject and between-subject measurements of stop-consonant productions by female speakers of English, Japanese, Navajo, and Spanish." Paper presented to the Acoustical Society of America, Anaheim, Calif. Abstract in *Journal of the Acoustical Society of America,* 80: S62.

Lawrence, D. H., and J. DeRivera. 1954. "Evidence for relational transposition." *Journal of Comparative and Physiological Psychology,* 47: 465–471.

Laws, J. M. 1994. "The effect of social interaction on song development in yearling Lazuli Buntings *(Passerina amoena).*" Paper presented at the 1994 North American Ornithological Conference, Missoula, Mont., June.

Lawson, R. W., and D. V. Lanning. 1980. "Nesting and status of the Maroon-fronted parrot *(Rhynchopsitta terrisi)*." In *Conservation of New World Parrots,* International Council for Bird Preservation, Technical Publication no. 1, ed. R. F. Pasquier, pp. 385–392. Washington, D.C.: Smithsonian Institution Press.

Lea, S. E. G. 1984. "Complex general process learning in nonmammalian vertebrates." In *The Biology of Learning,* ed. P. Marler and H. S. Terrace, pp. 373–397. Berlin: Springer-Verlag.

LeCompte, G. K., and G. Gratch. 1972. "Violation of a rule as a method of diagnosing infants' levels of object concept." *Child Development,* 43: 385–396.

Lefebvre, L., and L. A. Giraldeau. 1994. "Cultural transmission in pigeons is affected by the number of tutors and bystanders present." *Animal Behaviour,* 47: 331–337.

Lemish, D., and M. L. Rice. 1986. "Television as a talking picture book: A prop for language acquisition." *Journal of Child Language,* 13: 251–274.

Lemon, R. E. 1975. "How birds develop song dialects." *Condor,* 77: 385–406.

Lenneberg, E. H. 1967. *Biological Foundations of Language.* New York: John Wiley.

———1971. "Of language, knowledge, apes, and brains." *Journal of Psycholinguistic Research,* 1: 1–29.

———1973. "Biological aspects of language." In *Communication, Language, and Meaning,* ed. G. A. Miller, pp. 49–60. New York: Basic Books.

Leonard, L. 1989. "Language learnability and specific language impairment in children." *Applied Psycholinguistics,* 10: 179–202.

Leonard, L., K. Chapman, L. Rowan, and A. Weiss. 1983. "Three hypotheses concerning young children's imitations of lexical items." *Developmental Psychology,* 19: 591–601.

Leonard, L., R. Schwartz, K. Chapman, L. Rowan, P. Prelock, B. Terrell, A. Weiss, and C. Messick. 1982. "Early lexical acquisition in children with specific language impairment." *Journal of Speech and Hearing Research,* 25: 554–564.

Levinson, S. T. 1980. "The social behavior of the White-fronted Amazon *(Amazona albifrons)."* In *Conservation of New World Parrots,* International Council for Bird Preservation, Technical Publication no. 1, ed. R. F. Pasquier, pp. 403–417. Washington, D.C.: Smithsonian Institution Press.

Liberman, A. M., P. Delattre, F. S. Cooper, and L. J. Gerstman. 1954. "The role of consonant-vowel in the perception of the unvoiced stop consonants." *American Journal of Psychology,* 65: 497–516.

Lieberman, P. 1984. *The Biology and Evolution of Language.* Cambridge, Mass.: Harvard University Press.

———1991. *Uniquely Human: The Evolution of Speech, Thought, and Selfless Behavior.* Cambridge, Mass.: Harvard University Press.

———1996. "Some biological constraints on the analysis of prosody." In *Signal to Syntax: Bootstrapping from Speech to Grammar in Early Acquisition,* ed. J. L. Morgan and K. Demuth, pp. 55–65. Hillsdale, N.J.: Erlbaum.

Lieberman, P., and S. E. Blumstein. 1988. *Speech Physiology, Speech Perception, and Acoustic Phonetics.* Cambridge: Cambridge University Press.

Lilly, J. C. 1967. "Dolphin's vocal mimicry as a unique ability and a step toward understanding." In *Research in Verbal Behavior and Some Neurophysiological Implications,* ed. K. Salzinger and S. Salzinger, pp. 21–27. New York: Academic Press.

Lin, G., and W. Gong. 1989. "Conditional abstract judgement with numerousness concept by rhesus monkeys." *Acta Psychologia Sinica,* 21: 299–305.

Lippit, R., N. Polansky, and S. Rosen. 1952. "The dynamics of power." *Human Relations,* 5: 37–64.

Litowitz, B. E. 1993. "Deconstruction in the zone of proximal development." In *Contexts for Learning: Sociocultural Dynamics in Children's Development,* ed. E. A. Forman, N. Minick, and C. A. Stone, pp. 184–196. Oxford: Oxford University Press.

———1998. "An expanded developmental line for negation: Rejection, refusal, denial." *Journal of the American Psychoanalytic Association,* 46: 121–148.

Lock, A. 1980. *The Guided Reinvention of Language.* London: Academic Press.

———1991. "The role of social interaction in early language development." In *Biological and Behavioral Determinants of Language Development,* ed.

N. S. Krasnegor, D. M. Rumbaugh, R. L. Schielfelbusch, and M. Studdert-Kennedy, pp. 287-300. Hillsdale, N.J.: Erlbaum.

Locke, J., and C. Snow. 1997. "Social influences on vocal learning in human and nonhuman primates." In *Social Influences on Vocal Development,* ed. C. T. Snowdon and M. Hausberger, pp. 274-292. Cambridge: Cambridge University Press.

Loesche, P., M. D. Beecher, and P. K. Stoddard. 1993. "Perception of cliff swallow calls by birds *(Hirundo pyrrhonota* and *Sturnus vulgaris)* and humans *(Homo sapiens)." Journal of Comparative Psychology,* 106: 239-247.

Lofting, H. 1948. *The Voyages of Dr. Doolittle.* Philadelphia: Lippincott.

Lögler, P. 1959. "Versuche zur Frage des 'Zähl'-Vermögens an einem Graupapagei und Vergleichsversuche an Menschen." *Zeitschrift für Tierpsychologie,* 16: 179-217.

Lombardi, C. M., C. C. Fachinelli, and J. D. Delius. 1984. "Oddity of visual patterns conceptualized by pigeons." *Animal Learning & Behavior,* 12: 1-6.

Lorenz, K. Z. 1952. *King Solomon's Ring.* Trans. M. K. Wilson. New York: Harper & Row.

———1970. *Studies in Animal and Human Behavior,* vol. 1. Cambridge, Mass.: Harvard University Press.

Lovaas, I. 1977. *The Autistic Child: Language Development through Behavior Modification.* New York: Irvington.

Lucanus, F. von. 1923. "Das Sprechen der Papageien und ihre geistign Fahigkeiten." *Ornithologische Monatsberichte,* 31: 97-102, 121-127.

———1925. *Das Leben der Vögel.* Berlin: A. Scherl.

MacDougal-Shackleton, S. A., and S. H. Hulse. 1995. "Concurrent absolute and relative pitch processing by European starlings *(Sturnus vulgaris)." Journal of Comparative Psychology,* 110: 139-146.

Mackintosh, N. J., B. Wilson, and R. A. Boakes. 1985. "Differences in mechanism of intelligence among vertebrates." *Philosophical Transactions of the Royal Society, London,* B308: 53-65.

Macphail, E. M. 1982. *Brain and Intelligence in Vertebrates.* Oxford: Clarendon Press.

———1985. "Vertebrate intelligence: The null hypothesis." *Philosophical Transactions of the Royal Society, London,* B308: 37-51.

———1987. "The comparative psychology of intelligence." *Behavioral and Brain Sciences,* 10: 645-695.

Maeda, S. 1991. "Discussion summaries: Toward better models of speech production." *Journal of Phonetics,* 19: 493-495.

Maertens, N. W., R. C. Jones, and A. Waite. 1977. "Elemental groupings help children perceive cardinality: A two-phase research study." *Journal of Research in Mathematical Education,* 8: 181-193.

Manabe, K., T. Kawashima, and J. E. R. Staddon. 1995. "Differential vocalization in budgerigars: Towards an experimental analysis of naming." *Journal of the Experimental Analysis of Behavior,* 63: 111-126.

Mandler, G., and B. J. Shebo. 1982. "Subitizing: An analysis of its component processes." *Journal of Experimental Psychology: General,* 111: 1–22.

Mandler, J. M. 1990. "A new perspective on cognitive development in infancy." *American Scientist,* 78: 236–243.

Mann, N. I., and P. J. B. Slater. 1994. "What causes young male zebra finches, *Taeniopygia guttata,* to choose their father as song tutor?" *Animal Behaviour,* 47: 671–677.

———1995. "Song tutor choice by zebra finches in aviaries." *Animal Behaviour,* 49: 811–820.

Maratsos, M. P. 1991. "How the acquisition of nouns may be different from that of verbs." In *Biological and Behavioral Determinants of Language Development,* ed. N. A. Krasnegor, D. M. Rumbaugh, R. L. Schiefelbusch, and M. Studdert-Kennedy, pp. 67–88. Hillsdale, N.J.: Erlbaum.

Margoliash, D., C. A. Staicer, and S. A. Inoue. 1994. "Stereotyped and plastic song in adult indigo buntings, *Passerina cyanea.*" *Animal Behaviour,* 42: 367–388.

Marler, P. 1967. "Comparative study of song development in sparrows." In *Proceedings of the Fourteenth International Ornithological Congress,* ed. D. W. Snow, pp. 231–244. Oxford: Blackwell Scientific Publications.

———1970. "A comparative approach to vocal learning: Song development in white-crowned sparrows." *Journal of Comparative and Physiological Psychology,* 71: 1–25.

———1973. "Speech development and bird song: Are there any parallels?" In *Communication, Language, and Meaning,* ed. G. A. Miller, pp. 73–83. New York: Basic Books.

———1974. "Animal communication." In *Nonverbal Communication,* ed. L. Krames, P. Pliner, and T. Alloway, pp. 25–50. New York: Plenum Press.

———1996. "Social cognition: Are primates smarter than birds?" In *Current Ornithology,* vol. 13, ed. V. Nolan, Jr., and E. D. Ketterson, pp. 1–32. New York: Plenum Press.

Marler, P., and S. Peters. 1982a. "Developmental overproduction and selective attrition: New processes in the epigenesis of birdsong." *Developmental Psychobiology,* 15: 369–378.

———1982b. "Subsong and plastic song: Their role in the vocal learning process." In *Acoustic Communication in Birds,* vol. 2: *Song Learning and Its Consequences,* ed. D. E. Kroodsma and E. H. Miller, pp. 25–50. New York: Academic Press.

Marschark, M. 1983. "A code by any other name . . ." *Behavioral and Brain Sciences,* 6: 151–152.

Marschark, M., V. S. Everhart, J. Martin, and S. A. West. 1987. "Identifying linguistic creativity in deaf and hearing children." *Metaphor and Symbolic Activity,* 2: 281–306.

Masur, E. F. 1988. "Infants' imitation of novel and familiar behaviors." In *Social Learning: Psychological and Biological Perspectives,* ed. T. R. Zentall and B. G. Galef, Jr., pp. 301–318. Hillsdale, N.J.: Erlbaum.

Mathieu, M., and G. Bergeron. 1981. "Piagetian assessment of cognitive development in chimpanzees *(Pan troglodytes)*." In *Primate Behavior and Sociobiology*, ed. A. B. Chiarell and R. S. Corrucini, pp. 142–147. Berlin: Springer-
Verlag.

Mathieu, M., M.-A. Bouchard, L. Granger, and J. Herscovitch. 1976. "Piagetian ob-
ject-permanence in *Cebus capucinus, Lagothrica flavicauda*, and *Pan troglodytes*." *Animal Behaviour*, 24: 585–588.

Matsuzawa, T. 1985. "Use of numbers by a chimpanzee." *Nature*, 315: 57–59.

Matsuzawa, T., T. Asano, K. Kubota, and K. Murofushi. 1996. "Acquisition and
generalization of numerical labeling by a chimpanzee." In *Current Perspectives in Primate Social Dynamics: Selected Proceedings of the Ninth Con-
gresss of the International Primatological Society*, ed. D. M. Taub and F. A.
King, pp. 416–430. New York: Van Nostrand Reinhold.

Mattick, I. 1972. "The teacher's role in helping young children develop language."
In *Language in Early Childhood Education*, ed. C. B. Cazden, pp. 107–116.
Washington, D.C.: National Association for the Education of Young Children.

Matyniak, K. A., G. L. Wheller, and L. J. Stettner. 1971. "Reversal learning in the
crow, *Corvus americanus*." *Communications in Behavioral Biology*, 6:
177–185.

McClelland, J. 1989. "Larynx and trachea." In *Form and Function in Birds*, vol.
4, ed. A. S. King and J. McClelland, pp. 69–103. London: Academic Press.

McClure, M. K., and J. Helland. 1979. "A chimpanzee's use of dimensions in responding same and different." *The Psychological Record*, 29: 371–378.

McCowan, B., and D. Reiss. 1995. "Quantitative comparison of whistle repertoires
from captive adult bottlenose dolphins (Delphinidae, *Tursiops truncatus*): A
re-evaluation of the signature whistle hypothesis." *Ethology*, 100: 194–209.

———1997. "Vocal learning in captive bottlenose dolphins: A comparison with
human and nonhuman animals." In *Social Influences on Vocal Development*,
ed. C. T. Snowdon and M. Hausberger, pp. 178–207. Cambridge: Cambridge
University Press.

McCune, L., and M. Vihman. 1987. "Vocal motor schemes." *Papers and Reports
on Child Language Development*, 26: 72–79.

McGonigle, B. O., and B. T. Jones. 1978. "Levels of stimulus processing by the
squirrel monkey: Relative and absolute judgements compared." *Perception*,
7: 635–659.

McGowan, K. J., and G. E. Woolfenden. 1989. "A sentinel system in the Florida
scrub jay." *Animal Behaviour*, 37: 1000–1006.

McGowan, R. S. 1991. "Nonlinearities for one-dimensional propagation in the vocal tract." *Journal of Phonetics*, 19: 425–432.

McGregor, P. K., and J. R. Krebs. 1984. "Song learning and deceptive mimicry."
Animal Behaviour, 32: 280–287.

Mebes, H. D. 1978. "Pair-specific duetting in the Peach-faced Lovebird." *Natur-
wissenschaften*, 65: 66–67.

Meck, W. H., and R. M. Church. 1983. "A mode control model of counting and
timing processes." *Journal of Experimental Psychology: Animal Behavior
Processes*, 9: 320–334.

Meltzoff, A. N. 1988. "Imitation, objects, tools, and the rudiments of language in human ontogeny." *Human Evolution,* 3: 45–64.

Menault, E. 1875. *The Intelligence of Animals with Illustrative Anecdotes.* New York: Scribner, Armstrong.

Menn, L. 1983. "Development of articulatory, phonetic, and phonological capabilities." In *Language Production,* vol. 2, ed. B. Butterworth, pp. 3–50. London: Academic Press.

Menyuk, P., J. W. Liebergott, and M. C. Schultz. 1995. *Early Language Development in Full-term and Premature Infants.* Hillsdale, N.J.: Erlbaum.

Menzel, E. W., Jr. 1969. "Responsiveness to food and signs of food in chimpanzee discrimination learning." *Journal of Comparative and Physiological Psychology,* 68: 484–489.

———1973. "Leadership and communication in young chimpanzees." In *Precultural Primate Behavior,* ed. E. W. Menzel, pp. 192–225. Basel: Karger.

Menzel, E. W., Jr., and W. A. Draper. 1965. "Primate selection of food by size: Visible versus invisible rewards." *Journal of Comparative and Physiological Psychology,* 59: 231–239.

Menzel, E. W., Jr., and C. Juno. 1982. "Marmosets *(Saguinus fuscicollis):* Are learning sets learned?" *Science,* 217: 750–752.

———1985. "Social foraging in marmoset monkeys and the question of intelligence." *Philosophical Transactions of the Royal Society, London,* B308: 145–158.

Merrill, D. C., B. J. Reiser, S. K. Merrill, and S. Landes. 1996. "Guided learning by doing." *Cognition and Instruction,* 13: 315–372.

Mervis, C. B., and J. Bertrand. 1993. "Acquisition of early object labels: The roles of operating principles and input." In *Enhancing Children's Communication: Research Foundations for Intervention,* ed. A. P. Aiser and D. B. Gray, pp. 287–316. Baltimore: Brookes.

Mervis, C. B., C. A. Mervis, K. E. Johnson, and J. Bertrand. 1992. "Studying early lexical development: The value of the systematic diary method." *Advances in Infancy Research,* 7: 291–378.

Messick, C. K. 1984. "Phonetic and contextual aspects of the transition to early words." Ph.D. thesis, Purdue University.

Michael, J., P. Whitely, and B. Hesse. 1983. "The pigeon parlance project." *Verbal Behavior News,* 2: 6–9.

Miles, H. L. 1978. "Language acquisition in apes and children." In *Sign Language and Language Acquisition in Man and Ape,* ed. F. C. C. Peng, pp. 108–120. Boulder, Colo.: Westview Press.

———1983. "Apes and language." In *Language in Primates,* ed. J. de Luce and H. T. Wilder, pp. 43–61. New York: Springer-Verlag.

———1990. "The cognitive foundations for reference in a signing orangutan." In *"Language" and Intelligence in Monkeys and Apes: Comparative Developmental Perspectives,* ed. S. T. Parker and K. R. Gibson, pp. 511–539. Cambridge: Cambridge University Press.

Miller, D. B. 1977. "Two-voiced phenomenon in birds: Further evidence." *Auk,* 94: 567–572.

Miller, G. A. 1967. *The Psychology of Communication.* New York: Basic Books.

———1983. "Cognition and comparative psychology." *Behavioral and Brain Sciences,* 6: 152-153.

Miller, G. A., G. Galanter, and K. H. Pribram. 1960. *Plans and the Structure of Behavior.* New York: Holt.

Miller, J. D. 1989. "Auditory perceptual interpretation of the vowel." *Journal of the Acoustical Society of America,* 85: 2114-2134.

Miller, J. L. 1981. "Effects of speaking rate on segmental distinctions." In *Perspectives on the Study of Speech,* ed. P. D. Eimas and J. L. Miller, pp. 39-74. Hillsdale, N.J.: Erlbaum.

Miller, R. E., and J. V. Murphy. 1964. "Influence of the spatial relationships between the cue, reward, and response in discrimination learning." *Journal of Experimental Psychology,* 67: 120-123.

Mischel, W., and R. M. Liebert. 1967. "The role of power in the adoption of self-reward patterns." *Child Development,* 38: 673-683.

Mitchell, R. W., P. Yao, and P. T. Sherman. 1985. "Discriminant responding of a dolphin to differentially rewarded stimuli." *Journal of Comparative Psychology,* 99: 218-225.

Moerk, E. L. 1977. *Pragmatic and Semantic Aspects of Early Language Development.* Baltimore: University Park Press.

———1994. "Corrections in first language acquisition: Theoretical controversies and factual evidence." *International Journal of Psycholinguistics,* 10: 33-58.

———1996. "Input and learning processes in first language acquisition." *Advances in Child Development and Behavior,* 26: 181-228.

Moore, C. A. 1992. "The correspondence of vocal tract resonance with volumes obtained from magnetic resonance images." *Journal of Speech and Hearing Research,* 35: 1009-1024.

Morales, M., P. Mundy, and J. Rojas. 1998. "Following the direction of gaze and language development in 6-month-olds." *Infant Behavior and Development,* 21: 373-377.

Moran, G., E. Joch, and L. Sorenson. 1983. "The response of meerkats *(Suricata suricatta)* to changes in olfactory cues on established scent posts." Paper presented at the Annual Meeting of the Animal Behavior Society, Lewisburg, Penn., June.

Morgan, C. L. 1894. *An Introduction to Psychology.* London: W. Scott.

Morgan, J. L. 1996. "Finding relations between input and outcome in language acquisition." *Developmental Psychology,* 32: 556-559.

Morgan, J. L., K. M. Bonamo, and L. L. Travis. 1995. "Negative evidence on negative evidence." *Developmental Psychology,* 31: 180-197.

Morgane, P. J., M. S. Jacobs, and A. Galaburda. 1986. "Evolutionary morphology of the dolphin brain." In *Dolphin Cognition and Behavior: A Comparative Approach,* ed. R. J. Schusterman, J. A. Thomas, and F. G. Wood, pp. 5-29. Hillsdale, N.J.: Erlbaum.

Morse, D. H. 1970. "Territorial and courtship songs of birds." *Nature,* 226: 659-661.

Morse, P. A., and C. T. Snowdon. 1975. "An investigation of categorical speech discrimination by rhesus monkeys." *Perception & Psychophysics,* 17: 9-16.

Morton, E. S. 1977. "On the occurrence and significance of motivational-structural rules in some bird and mammal sounds." *American Naturalist,* 111: 855-869.

———1982. "Grading, discreteness, redundancy, and motivational-structural rules." In *Acoustic Communication in Birds,* vol. 1: *Production, Perception, and Design Features of Sound,* ed. D. E. Kroodsma and E. H. Miller, pp. 183-212. New York: Academic Press.

Mowrer, O. H. 1950. *Learning Theory and Personality Dynamics.* New York: Ronald Press.

———1952. "The autism theory of speech development and some clinical applications." *Journal of Speech and Hearing Disorders,* 17: 263-268.

———1954. "A psychologist looks at language." *American Psychologist,* 9: 660-694.

———1958. "Hearing and speaking: An analysis of language learning." *Journal of Speech and Hearing Disorders,* 23: 143-152.

———1969. "Theory and research—a review." Urbana: University of Illinois. Mimeographed. Cited in O. H. Mowrer, 1980, *Psychology of Language and Learning,* pp. 105-106. New York: Plenum Press.

Mundinger, P. 1970. "Vocal imitation and individual recognition of finch calls." *Science,* 168: 480-482.

———1979. "Call learning in the Carduelinae: Ethological and systematic considerations." *Systematic Zoology,* 28: 270-283.

Murphey, C. M., and D. J. Messer. 1977. "Mothers, infants and pointing: A study of gesture." In *Studies in Mother-Infant Interaction,* ed. H. R. Schaffer, pp. 325-354. London: Academic Press.

Myers, J. A. 1917. "Studies on the syrinx of *Gallus domesticus.*" *Journal of Morphology,* 29: 165-214.

Naguib, M., and D. Todt. 1998. "Recognition of neighbors' song in a species with large and complex song repertoires: The Thrush Nightingale." *Journal of Avian Biology,* 29: 155-160.

Natale, F., and F. Antinucci. 1989. "Stage 6 object-concept and representation." In *Cognitive Structure and Development in Nonhuman Primates,* ed. F. Antinucci, pp. 97-112. Hillsdale, N.J.: Erlbaum.

Natale, F., F. Antinucci, G. Spinozzi, and P. Potì. 1986. "Stage 6 object-concept and representation in nonhuman primate cognition: A comparison between gorilla *(Gorilla gorilla gorilla)* and Japanese macaque *(Macaca fuscata).*" *Journal of Comparative Psychology,* 100: 335-339.

Nauta, W. J. H., and H. J. Karten. 1971. "A general profile of the vertebrate brain, with sidelights on the ancestry of cerebral cortex." In *The Neurosciences Second Study Program: Evolution of Brain and Behavior,* ed. F. O. Schmitt, pp. 7-26. New York: Rockefeller University Press.

Neal, K. B. 1996. "The development of a vocalization in an African Grey parrot *(Psittacus erithacus).*" Senior thesis, University of Arizona.

Neapolitan, D. M., I. M. Pepperberg, and L. Schinke-Llano. 1988. "Second language acquisition: Possible insights from how birds acquire song." *Studies in Second Language Acquisition,* 10: 1–11.

Nearey, T. 1978. *Phonetic Features for Vowels.* Bloomington: Indiana University Linguistics Club.

———1989. "Static, dynamic, and relational properties in vowel perception." *Journal of the Acoustical Society of America,* 85: 2088–2113.

Nelson, D. A. 1985. "The syntactic and semantic organization of pigeon guillemot *(Cepphus columba)* vocal behavior." *Zeitschrift für Tierpsychologie,* 67: 97–130.

———1987. "Song syllable discrimination by song sparrows *(Melospiza melodia)." Journal of Comparative Psychology,* 101: 25–32.

———1988. "Feature weighting in species song recognition by the field sparrow *(Spizella pusilla)." Behaviour,* 106: 158–182.

———1992. "Song overproduction and selective attrition lead to song sharing in the field sparrow *(Spizella pusilla)." Behavioral Ecology and Sociobiology,* 30: 415–424.

———1998. "External validity and experimental design: The sensitive phase for song learning." *Animal Behaviour,* 56: 487–491.

Nelson, D. A., and P. Marler. 1989. "Categorical perception of a natural stimulus continuum: Birdsong." *Science,* 244: 976–978.

Nelson, D. A., P. Marler, and M. L. Morton. 1996. "Overproduction in song development: An evolutionary correlate with migration." *Animal Behaviour,* 51: 1127–1140.

Nelson, K. 1974. "Concept, word, and sentence: Interrelations in acquisition and development." *Psychological Review,* 81: 267–285.

———1985. *Making Sense: The Acquisition of Shared Meaning.* Orlando: Academic Press.

———1987. "What's in a name? Reply to Seidenberg and Petitto." *Journal of Experimental Psychology: General,* 116: 293–296.

———1989. "Monologues in the crib." In *Narratives from the Crib,* ed. K. Nelson, pp. 1–23. Cambridge, Mass.: Harvard University Press.

———1995. "The dual category problem in lexical acquisition." In *Beyond Names for Things,* ed. M. Tomasello and W. E. Merriman, pp. 223–250. Hillsdale, N.J.: Erlbaum.

———1996. *Language in Cognitive Development: The Emergence of the Mediated Mind.* Cambridge: Cambridge University Press.

Nelson, K. E. 1978. "Toward a rare-event cognitive comparison theory of syntax acquisition." Paper presented at the First International Congress for the Study of Child Language, Tokyo, August.

Nelson, K. E., J. Welsh, S. M. Camarata, L. Butkovsky, and M. Camarata. 1995. "Available input for language-impaired children and younger children of matched language levels." *First Language,* 15: 1–17.

Nottebohm, F. 1970. "Ontogeny of bird song." *Science,* 167: 950–956.

———1976. "Phonation in the Orange-winged Amazon parrot." *Journal of Comparative Physiology A,* 108: 157–170.

————1980. "Brain pathways for vocal learning in birds: A review of the first ten years." *Progress in Psychobiology and Physiological Psychology,* 9: 85–124.

Nottebohm, F., and M. Nottebohm. 1969. "The parrots of Bush Bush." *Animal Kingdom,* 72: 19–23.

Nottebohm, F., S. Kasparian, and C. Pandazis. 1981. "Brain space for a learned task." *Brain Research,* 213: 99–109.

Nowicki, S. 1987. "Vocal tract resonances in oscine bird sound production: Evidence from bird song in a helium atmosphere." *Nature,* 325: 53–55.

Nowicki, S., and P. Marler. 1988. "How do birds sing?" *Music Perception,* 5: 391–426.

Nowicki, S., and D. A. Nelson. 1990. "Defining natural categories in acoustic signals: Comparison of three methods applied to 'chick-a-dee' call notes." *Ethology,* 86: 89–101.

Olive, J. P., A. Greenwood, and J. Coleman. 1993. *Acoustics of American English Speech: A Dynamic Approach.* New York: Springer-Verlag.

Olthof, A., C. M. Iden, and W. A. Roberts. 1997. "Judgments of ordinality and summation of number symbols by squirrel monkeys *(Saimiri sciureus)." Journal of Experimental Psychology: Animal Behavior Processes,* 23: 325–333.

O'Sullivan, C., and C. P. Yeager. 1989. "Communicative context and linguistic competence: The effects of social setting on a chimpanzee's conversational skill." In *Teaching Sign Language to Chimpanzees,* ed. R. A. Gardner, B. T. Gardner, and T. E. Van Cantfort, pp. 269–279. Albany, N.Y.: SUNY Press.

Oyama, T., T. Kikuchi, and S. Ichihara. 1981. "Span of attention, backward masking, and reaction time." *Perception & Psychophysics,* 29: 106–112.

Page, S. C., S. H. Hulse, and J. Cynx. 1989. "Relative pitch perception in the European starling *(Sturnus vulgaris):* Further evidence for an elusive phenomenon." *Journal of Experimental Psychology: Animal Behavior Processes,* 15: 137–146.

Park, T. J., and R. J. Dooling. 1985. "Perception of species-specific contact calls by budgerigars *(Melopsittacus undulatus)." Journal of Comparative Psychology,* 99: 391–402.

Parkel, D. A., and S. T. Smith, Jr. 1979. "Application of computer-assisted language designs." In *Language Intervention from Ape to Child,* ed. R. L. Schiefelbusch and J. H. Hollis, pp. 441–464. Baltimore: University Park Press.

Parker, S. T. 1977. "Piaget's sensorimotor series in an infant macaque: A model for comparing unstereotyped behavior and intelligence in human and nonhuman primates." In *Primate Bio-social Development: Biological, Social, and Ecological Determinants,* ed. S. Chevalier-Skolnikoff and F. E. Poirier, pp. 43–112. New York: Garland.

Parker, S. T., and K. R. Gibson. 1977. "Object manipulation, tool use and sensorimotor intelligence as feeding adaptations in Cebus monkeys and great apes." *Journal of Human Evolution,* 6: 623–641.

Pasnak, R., and S. L. Kurtz. 1987. "Brightness and size transposition by rhesus monkeys." *Bulletin of the Psychonomic Society,* 25: 109–112.

Pastore, N. 1955. "Discrimination and delayed response learning in the canary." *Psychological Reports,* 1: 307-315.

——1961. "Number sense and 'counting' ability in the canary." *Zeitschrift für Tierpsychologie,* 18: 561-573.

Paton, J. A., K. R. Manogue, and F. Nottebohm. 1981. "Bilateral organization of the vocal control pathway in the budgerigar, *Melopsittacus undulatus." Journal of Neuroscience,* 1: 1276-1288.

Patterson, D. K., and I. M. Pepperberg. 1994. "A comparative study of human and parrot phonation: Acoustic and articulatory correlates of vowels." *Journal of the Acoustical Society of America,* 96: 634-648.

Patterson, D. K., and I. M. Pepperberg. 1998. "Acoustic and articulatory correlates of stop consonants in a parrot and a human subject." *Journal of the Acoustical Society of America,* 103: 2197-2215.

Patterson, D. K., I. M. Pepperberg, B. H. Story, and E. Hoffman. 1997. "How parrots talk: Insights based on CT scans, image processing, and mathematical models." In *SPIE Proceedings: Physiology and Function from Multidimensional Images,* vol. 3033, ed. E. Hoffman, pp. 14-24. Bellingham, Wash.: SPIE.

Patterson, F. G. 1978. "Linguistic capabilities of a lowland gorilla." In *Sign Language and Language Acquisition in Man and Ape,* ed. F. C. C. Peng, pp. 161-201. Boulder, Colo.: Westview Press.

Payne, R. B. 1978. "Microgeographic variation in songs of Splendid Sunbirds *Nectarinia coccinigaster:* Population phenetics, habitats, and song dialects." *Behaviour,* 65: 282-308.

——1981. "Song learning and social interaction in indigo buntings." *Animal Behaviour,* 29: 688-697.

——1982. "Ecological consequences of song sharing: Breeding success and intraspecific song mimicking in indigo buntings." *Ecology,* 63: 401-411.

——1983. "The social context of song mimicry: Song-matching dialects in indigo buntings *(Passerina cyanea)." Animal Behaviour,* 31: 788-805.

Payne, R. B., and K. D. Groschupf. 1984. "Sexual selection and interspecific competition: A field experiment on territorial behavior of nonparental finches *(Vidua spp.)." Auk,* 101: 140-145.

Payne, R. B., L. L. Payne, and S. M. Doehlert. 1984. "Interspecific song learning in a wild chestnut-sided warbler." *Wilson Bulletin,* 96: 292-294.

Payne, R. B., L. L. Payne, and S. Whitesell. 1988. "Interspecific learning and cultural transmission of song in House Finches." *Wilson Bulletin,* 100: 667-670.

Pea, R. D. 1980. "The development of negation in early child language." In *The Social Foundations of Language and Thought: Essays in Honor of Jerome S. Bruner,* ed. D. R. Olson, pp. 156-186. New York: John Wiley.

Peek, F. W. 1972. "An experimental study of the territorial function of vocal and visual displays in the male Red-winged Blackbird *(Agelaius phoeniceus)." Animal Behaviour,* 20: 112-118.

Pepperberg, I. M. 1978. "Object identification by an African Grey parrot *(Psittacus erithacus)."* Paper presented at the meeting of the Midwest Animal Behavior Society, W. Lafayette, Ind., March.

———1981. "Functional vocalizations by an African Grey parrot *(Psittacus eritha-cus)." Zeitschrift für Tierpsychologie,* 55: 139–160.

———1983a. "Cognition in the African Grey parrot: Preliminary evidence for au-ditory/vocal comprehension of the class concept." *Animal Learning & Be-havior,* 11: 179–185.

———1983b. "Interspecies communication: Innovative vocalizations of an African Grey parrot." Paper presented at the Annual Meeting of the Animal Behavior Society, Lewisburg, Penn., June.

———1985. "Social modeling theory: A possible framework for understanding avian vocal learning." *Auk,* 102: 854–864.

———1986a. "Acquisition of anomalous communicatory systems: Implications for studies on interspecies communication." In *Dolphin Cognition and Behav-ior: A Comparative Approach,* ed. R. J. Schusterman, J. A. Thomas, and F. G. Wood, pp. 289–302. Hillsdale, N.J.: Erlbaum.

———1986b. "Sensitive periods, social interaction, and song acquisition: The di-alectics of dialects?" *Behavioral and Brain Sciences,* 9: 756–757.

———1987a. "Acquisition of the same/different concept by an African Grey parrot *(Psittacus erithacus):* Learning with respect to categories of color, shape, and material." *Animal Learning & Behavior,* 15: 423–432.

———1987b. "Evidence for conceptual quantitative abilities in the African Grey parrot: Labeling of cardinal sets." *Ethology,* 75: 37–61.

———1987c. "Interspecies communication: A tool for assessing conceptual abili-ties in the African Grey parrot *(Psittacus erithacus)."* In *Cognition, Lan-guage, and Consciousness: Integrative Levels,* ed. G. Greenberg and E. To-bach, pp. 31–56. Hillsdale, N.J.: Erlbaum.

———1988a. "An interactive modeling technique for acquisition of communica-tion skills: Separation of 'labeling' and 'requesting' in a psittacine subject." *Applied Psycholinguistics,* 9: 59–76.

———1988b. "Comprehension of 'absence' by an African Grey parrot: Learning with respect to questions of same/different." *Journal of the Experimental Analysis of Behavior,* 50: 553–564.

———1988c. "The importance of social interaction and observation in the acqui-sition of communicative competence: Possible parallels between avian and human learning." In *Social Learning: Psychological and Biological Perspec-tives,* ed. T. R. Zentall and B. G. Galef, Jr., pp. 279–299. Hillsdale, N.J.: Erlbaum.

———1990a. "Cognition in an African Grey parrot *(Psittacus erithacus):* Further evidence for comprehension of categories and labels." *Journal of Compar-ative Psychology,* 104: 41–52.

———1990b. "Conceptual abilities of some nonprimate species, with an emphasis on an African Grey parrot." In *"Language" and Intelligence in Apes and Monkeys: Comparative Developmental Perspectives,* ed. S. T. Parker and K. R. Gibson, pp. 469–507. Cambridge: Cambridge University Press.

———1990c. "Referential mapping: A technique for attaching functional signifi-cance to the innovative utterances of an African Grey parrot." *Applied Psy-cholinguistics,* 11: 23–44.

————1990d. "Some cognitive capacities of an African Grey Parrot *(Psittacus er-ithacus)."* In *Advances in the Study of Behavior,* vol. 19, ed. P. J. B. Slater, J. S. Rosenblatt, and C. Beer, pp. 357-409. New York: Academic Press.

————1991. "A communicative approach to animal cognition: A study of concep-tual abilities of an African Grey parrot." In *Cognitive Ethology: The Minds of Other Animals,* ed. C. A. Ristau, pp. 153-186. Hillsdale, N.J.: Erlbaum.

————1992a. "Proficient performance of a conjunctive, recursive task by an Afri-can Grey parrot *(Psittacus erithacus)." Journal of Comparative Psychology,* 106: 295-305.

————1992b. "Social interaction as a condition for learning in avian species." In *The Inevitable Bond,* ed. H. Davis and D. Balfour, pp. 178-204. Cambridge: Cambridge University Press.

————1993a. "Cognition and communication in an African Grey parrot *(Psittacus erithacus):* Studies on a nonhuman, nonprimate, nonmammalian subject." In *Language and Communication: Comparative Perspectives,* ed. H. L. Roitblat, L. M. Herman, and P. E. Nachtigall, pp. 221-248. Hillsdale, N.J.: Erl-baum.

————1993b. "A review of the effects of social interaction on vocal learning in African Grey parrots *(Psittacus erithacus)." Netherlands Journal of Zoology,* 43: 104-124.

————1994a. "Evidence for numerical competence in an African Grey parrot *(Psit-tacus erithacus)." Journal of Comparative Psychology,* 108: 36-44.

————1994b. "Vocal learning in Grey Parrots *(Psittacus erithacus):* Effects of so-cial interaction, reference, and context." *Auk,* 111: 300-313.

————1996. "Categorical class formation by an African Grey parrot *(Psittacus er-ithacus)."* In *Stimulus Class Formation in Humans and Animals,* ed. T. R. Zentall and P. R. Smeets, pp. 71-90. Amsterdam: Elsevier.

————1997. "Social influences on the acquisition of human-based codes in parrots and nonhuman primates." In *Social Influences on Vocal Development,* ed. C. T. Snowdon and M. Hausberger, pp. 157-177. Cambridge: Cambridge Uni-versity Press.

————1998. "The African Grey Parrot: How cognitive processing might affect al-lospecific vocal learning." In *Animal Cognition in Nature,* ed. R. P. Balda, I. M. Pepperberg, and A. C. Kamil, pp. 381-409. London: Academic Press.

Pepperberg, I. M., and M. V. Brezinsky. 1991. "Acquisition of a relative class con-cept by an African Grey parrot *(Psittacus erithacus):* Discriminations based on relative size." *Journal of Comparative Psychology,* 105: 286-294.

Pepperberg, I. M., and M. S. Funk. 1990. "Object permanence in four species of psittacine birds: An African Grey parrot *(Psittacus erithacus),* an Illiger ma-caw *(Ara maracana),* a parakeet *(Melopsittacus undulatus),* and a cockatiel *(Nymphus hollandicus)." Animal Learning & Behavior,* 18: 97-108.

Pepperberg, I. M., and F. A. Kozak. 1986. "Object permanence in the African Grey parrot *(Psittacus erithacus)." Animal Learning & Behavior,* 14: 322-330.

Pepperberg, I. M., and M. A. McLaughlin. 1996. "Effect of avian-human joint atten-tion on allospecific vocal learning by Grey parrots *(Psittacus erithacus)." Journal of Comparative Psychology,* 110: 286-297.

Pepperberg, I. M., and D. M. Neapolitan. 1988. "Second language acquisition: A framework for studying the importance of input and interaction in exceptional song acquisition." *Ethology,* 77: 150-168.

Pepperberg, I. M., and L. Schinke-Llano. 1991. "Language acquisition and use in a bilingual environment: A framework for studying birdsong in zones of sympatry." *Ethology,* 89: 1-28.

Pepperberg, I. M., K. J. Brese, and B. J. Harris. 1991. "Solitary sound play during acquisition of English vocalizations by an African Grey parrot *(Psittacus erithacus):* Possible parallels with children's monologue speech." *Applied Psycholinguistics,* 12: 151-178.

Pepperberg, I. M., L. I. Gardiner, and L. J. Luttrell. 1999. "Limited contextual vocal learning in the Grey parrot *(Psittacus erithacus):* The effect of interactive coviewers on videotaped instruction." *Journal of Comparative Psychology,* 113: 158-172.

Pepperberg, I. M., J. R. Naughton, and P. A. Banta. 1998. "Allospecific vocal learning by Grey parrots *(Psittacus erithacus):* A failure of videotaped instruction under certain conditions." *Behavioural Processes,* 42: 139-158.

Pepperberg, I. M., M. R. Willner, and L. B. Gravitz. 1997. "Development of Piagetian object permanence in a Grey parrot *(Psittacus erithacus)." Journal of Comparative Psychology,* 111: 63-75.

Pepperberg, I. M., S. E. Garcia, E. C. Jackson, and S. Marconi. 1995. "Mirror use by African Grey parrots *(Psittacus erithacus)." Journal of Comparative Psychology,* 109: 182-195.

Pepperberg, I. M., K. S. Howell, P. A. Banta, D. K. Patterson, and M. Meister. 1998. "Measurement of the trachea of the Grey parrot *(Psittacus erithacus)* via Magnetic Resonance Imaging, dissection, and Electron Beam Computed Tomography." *Journal of Morphology,* 238: 81-91.

Peters, A. M. 1996. "The development of collaborative story retelling by a two-year-old blind child and his father." In *Social Interaction, Social Context, and Language: Essays in Honor of Susan Ervin-Tripp,* ed. D. I. Slobin, J. Gehrhardt, A. Kyratzis, and J. Guo, pp. 391-416. Mahwah, N.J.: Erlbaum.

Peterson, G. B. 1984. "How expectancies guide behavior." In *Animal Cognition,* ed. H. L. Roitblat, T. G. Bever, and H. S. Terrace, pp. 135-148. Hillsdale, N.J.: Erlbaum.

Peterson, G. E., and H. L. Barney. 1952. "Control methods used in a study of the identification of vowels." *Journal of the Acoustical Society of America,* 24: 175-184.

Petitto, L. A., and P. F. Marentette. 1991. "Babbling in the manual mode: Evidence for the ontogeny of language." *Science,* 251: 1943-1946.

Petrinovich, L. 1972. "Psychological mechanisms in language development." In *Advances in Psychobiology,* vol. 1, ed. G. Newton and A. H. Riesen, pp. 259-285. New York: Wiley-Interscience.

———1985. "Factors influencing song development in white-crowned sparrows *(Zonotrichia leucophrys)." Journal of Comparative Psychology,* 99: 15-29.

————1988. "The role of social factors in white-crowned sparrow song development." In *Social Learning: Psychological and Biological Perspectives,* ed. T. R. Zentall and B. G. Galef, Jr., pp. 255–278. Hillsdale, N.J.: Erlbaum.

Petrinovich, L., and L. F. Baptista. 1987. "Song development in the white-crowned sparrow: Modification of learned song." *Animal Behaviour,* 35: 961–974.

Pezdek, K., and E. Stevens. 1984. "Children's memory for auditory and visual information on television." *Developmental Psychology,* 20: 212–218.

Piaget, J. 1952. *The Origins of Intelligence in Children.* Trans. Margaret Cook. New York: International Universities Press. (Original written in French in 1936.)

————1954a. "Language and thought from the genetic point of view." *Acta Psychologica,* 10: 51–60.

————1954b. *The Construction of Reality in the Child.* Trans. Margaret Cook. New York: Basic Books. (Original written in French in 1937.)

————1962. *Play, Dreams, and Imitation.* New York: Norton.

————1966. *Psychology of Intelligence.* Totawa, N.J.: Littlefield, Adams.

————1971. *Biology and Knowledge.* Chicago: University of Chicago Press.

————1978. *Behavior and Evolution.* New York: Pantheon.

————1980. *Adaptation and Intelligence: Organic Selection and Phenocopy.* Chicago: University of Chicago Press.

Pickert, S. 1981. "Imaginative dialogues in children's private speech." *First Language,* 2: 5–20.

Pierce, K., and L. Schreibman. 1995. "Increasing complex social behaviors in children with autism: Effects of peer implemented pivotal response training." *Journal of Applied Behavior Analysis,* 28: 285–295.

Pine, J. M. 1994. "Referential style and maternal directiveness: Different measures yield different results." *Applied Psycholinguistics,* 15: 135–148.

Pinker, S. 1989. "Resolving a learnability paradox in the acquisition of the verb lexicon." In *The Teachability of Language,* ed. M. L. Rice and R. L. Schiefelbusch, pp. 13–62. Baltimore: Brookes.

Pisacreta, R., E. Redwood, and K. Witt. 1984. "Transfer of matching-to-figure samples in the pigeon." *Journal of the Experimental Analysis of Behavior,* 42: 223–237.

Plante, E., and L. Turkstra. 1991. "Sources of error in the quantitative analysis of MRI scans." *MRI,* 9: 589–595.

Plooij, F. 1978. "Some basic traits of language in wild chimpanzees." In *Action, Gesture and Symbol: The Emergence of Language,* ed. A. Lock, pp. 111–132. London: Academic Press.

Portmann, A. 1950. "Système nerveux." In *Traité de Zoologie,* vol. 15, ed. P. P. Grassé, pp. 185–203. Paris: Masson.

Portmann, A., and W. Stingelin. 1961. "The central nervous system." In *Biology and Comparative Physiology of Birds,* vol. 2, ed. A. J. Marshall, pp. 1–36. New York: Academic Press.

Potì, P. 1989. "Early sensorimotor development in macaques *(Macaca fuscata, Macaca fascicularis).*" In *Cognitive Structure and Development in Nonhuman Primates,* ed. F. Antinucci, pp. 39–53. Hillsdale, N.J.: Erlbaum.

Power, D. M. 1966a. "Agnostic behavior and vocalizations of Orange-chinned Parakeets in captivity." *Condor,* 68: 562–581.

———1966b. "Antiphonal duetting and evidence for auditory reaction time in the Orange-chinned Parakeet." *Auk,* 83: 314–319.

Premack, D. 1971. "On the assessment of language-competence in the chimpanzee." In *Behavior of Nonhuman Primates,* vol. 4, ed. A. M. Schrier and F. Stollnitz, pp. 186–228. New York: Academic Press.

———1976. *Intelligence in Ape and Man.* Hillsdale, N.J.: Erlbaum.

———1978. "On the abstractness of human concepts: Why it would be difficult to talk to a pigeon." In *Cognitive Processes in Animal Behavior,* ed. S. H. Hulse, H. Fowler, and W. K. Honig, pp. 421–451. Hillsdale, N.J.: Erlbaum.

———1983. "The codes of man and beasts." *Behavioral and Brain Sciences,* 6: 125–167.

———1986. *Gavagai.* Cambridge, Mass.: MIT Press.

Price, P. H. 1979. "Developmental determinants of structure in Zebra Finch song." *Journal of Comparative and Physiological Psychology,* 93: 260–277.

Pullum, G. K., and W. A. Ladusaw. 1986. *Phonetic Symbol Guide.* Chicago: University of Chicago Press.

Putney, R. T. 1985. "Do willful apes know what they are aiming at?" *Psychological Record,* 35: 49–62.

Pyke, G. H., H. R. Pulliam, and E. L. Charnov. 1977. "Optimal foraging: A selective review of theory and tests." *Quarterly Review of Biology,* 52: 137–154.

Quine, W. V. O. 1960. *Word and Object.* Cambridge, Mass.: MIT Press.

Radeau, M. 1994. "Auditory-visual spatial interaction and modularity." *Current Psychology of Cognition,* 13: 3–51.

Rakic, P., J.-P. Bourgeois, N. Zecevic, M. F. Eckenhoff, and P. S. Goldman-Rakic. 1986. "Concurrent overproduction in synapses in diverse regions of the primate cerebral cortex." *Science,* 232: 232–235.

Rasmussen, K. 1972. "The Netsilik Eskimos." In *Shaking the Pumpkin,* ed. J. Rothenberg, p. 45. Garden City, N.Y.: Doubleday.

Redhead, E. S., and J. M. Pierce. 1995. "Similarity and discrimination learning." *Quarterly Journal of Experimental Psychology,* 48: 46–66.

Redshaw, M. 1978. "Cognitive development in human and gorilla infants." *Journal of Human Evolution,* 7: 133–141.

Reese, H. W. 1968. *The Perception of Stimulus Relations: Discrimination Learning and Transposition.* New York: Academic Press.

———1972. "Acquired distinctiveness and equivalence of cues in young children." *Journal of Experimental Child Psycholology,* 13: 171–182.

Reichle, J., R. Rogers, and C. Barrett. 1984. "Establishing pragmatic discriminations among the communicative functions of requesting, rejecting, and commenting in an adolescent." *Journal of the Association of the Severely Handicapped,* 9: 31–36.

Reiser, R. A., M. S. Tessmer, and P. C. Phelps. 1984. "Adult-child interaction in children's learning from 'Sesame Street.' " *Educational Communication and Technical Journal,* 32: 217–223.

Reiss, D., and B. McCowan. 1993. "Spontaneous vocal mimicry and production by bottlenose dolphins *(Tursiops truncatus):* Evidence for vocal learning." *Journal of Comparative Psychology,* 107: 301–312.

Remez, R., P. Rubin, L. Nygaard, and W. Howell. 1987. "Perceptual normalization of vowels produced by sinusoidal voices." *Journal of Experimental Psychology: Human Perception and Performance,* 13: 40–61.

Rensch, B., and G. Dücker. 1973. "Discrimination of patterns indicating four and five degrees of reward by birds." *Behavioural Biology,* 9: 279–288.

Repp, B. H. 1983. "Trading relations among acoustic cues in speech perception are largely a result of phonetic categorization." *Speech Communication,* 2: 341–361.

Rescorla, R. A. 1981. "Within-signal learning in autoshaping." *Animal Learning & Behavior,* 9: 245–252.

Reynolds, P. 1976. "Play, language, and human evolution." In *Play: Its Role in Development and Evolution,* ed. J. Bruner, A. Jolly, and K. Sylva, pp. 621–635. New York: Basic Books.

Rice, M. L. 1980. *Cognition to Language: Categories, Word Meanings, and Training.* Baltimore: University Park Press.

———1991. "Children with specific language impairment: Toward a model of teachability." In *Biological and Behavioral Determinants of Language Development,* ed. N. A. Krasnegor, D. M. Rumbaugh, R. L. Schiefelbusch, and M. Studdert-Kennedy, pp. 447–480. Hillsdale, N.J.: Erlbaum.

Rice, M. L., and L. Woodsmall. 1988. "Lessons from television: Children's word learning when viewing." *Child Development,* 59: 420–429.

Richards, D. G., J. P. Woltz, and L. M. Herman. 1984. "Vocal mimicry of computer-generated sounds and labeling of objects by a bottlenose dolphin *(Tursiops truncatus)." Journal of Comparative Psychology,* 98: 10–28.

Riley, D. A. 1968. *Discrimination Learning.* Boston: Allyn and Bacon.

Rilling, M. 1993. "Invisible counting animals: A history of contributions from comparative psychology, ethology, and learning theory." In *The Development of Numerical Competence,* ed. S. T. Boysen and E. J. Capaldi, pp. 3–37. Hillsdale, N.J.: Erlbaum.

Ristau, C. A. 1991. "Aspects of cognitive ethology of an injury-feigning bird, the piping plover." In *Cognitive Ethology: The Minds of Other Animals,* ed. C. A. Ristau, pp. 91–126. Hillsdale, N.J.: Erlbaum.

Ristau, C. A., and D. Robbins. 1982. "Language in the great apes: A critical review." In *Advances in the Study of Behavior,* vol. 12, ed. J. S. Rosenblatt, R. A. Hinde, C. Beer, and M.-C. Busnel, pp. 142–255. New York: Academic Press.

Robisson, P. 1992. "Vocalizations in *Aptenodytes* penguins: Application of the two-voice theory." *Auk,* 109: 654–658.

Rocissano, L., and Y. Yatchmink. 1984. "Joint attention in mother-toddler interaction: A study of individual variation." *Merrill-Palmer Quarterly,* 30: 11–31.

Rogoff, B. 1984. "Introduction: Thinking and learning in a social context." In *Everyday Cognition: Its Development in Social Contexts,* ed. B. Rogoff and J. Lave, pp. 1–8. Cambridge, Mass.: Harvard University Press.

———1990. *Apprenticeship in Thinking: Cognitive Development in Social Context.* Oxford: Oxford University Press.

Rogoff, B., and W. Gardner. 1984. "Adult guidance of cognitive development." In *Everyday Cognition: Its Development in Social Contexts,* ed. B. Rogoff and J. Lave, pp. 95–116. Cambridge, Mass.: Harvard University Press.

Roitblat, H. L. 1987. *Introduction to Comparative Cognition.* New York: W. H. Freeman.

Roitblat, H. L., and L. von Fersen. 1992. "Comparative cognition: Representations and processes in learning and memory." *Annual Review of Psychology,* 43: 671–710.

Roitblat, H. L., L. M. Herman, and P. E. Nachtigall, eds. 1993. *Language and Communication: Comparative Perspectives.* Hillsdale, N.J.: Erlbaum.

Romanes, G. J. 1883/1977. *Animal Intelligence.* Washington, D.C.: University Publications of America.

Rossing, T. D. 1982. *The Science of Sound.* Reading, Mass.: Addison-Wesley.

Roush, R. S., and C. T. Snowdon. 1994. "Ontogeny of food-associated calls in cotton-top tamarins." *Animal Behaviour,* 47: 263–273.

Rowley, I. 1980. "Parent-offspring recognition in a cockatoo, the galah, *Cacatua roseicapilla.*" *Australian Journal of Zoology,* 28: 445–456.

Rozin, P. 1976. "The evolution of intelligence and access to the cognitive unconscious." In *Progress in Psychobiology and Physiological Psychology,* vol. 6, ed. J. M. Sprague and A. N. Epstein, pp. 245–280. New York: Academic Press.

Rubinstein, R. A. 1979. "The cognitive consequences of bilingual education in northern Belize." *American Ethnologist,* 6: 583–601.

Rumbaugh, D. M., ed. 1977. *Language Learning by a Chimpanzee.* New York: Academic Press.

Rumbaugh, D. M., and T. V. Gill. 1977. "Lana's acquisition of language skills." In *Language Learning by a Chimpanzee,* ed. D. M. Rumbaugh, pp. 165–192. New York: Academic Press.

Rumbaugh, D. M., and J. L. Pate. 1984. "Primates' learning by levels." In *Behavioral Evolution and Integrative Levels,* ed. G. Greenberg and E. Tobach, pp. 221–240. Hillsdale, N.J.: Erlbaum.

Rumbaugh, D. M., E. C. von Glasersfeld, H. Warner, P. Pisani, T. V. Gill, J. V. Brown, and C. L. Bell. 1973. "A computer-controlled language training system for investigating the language skills of young apes." *Behavior Research Methodology and Instrumentation,* 5: 385–392.

Rummelhart, D. E., and D. A. Norman. 1978. "Accretion, tuning, and restructuring: Three modes of learning." In *Semantic Factors in Cognition,* ed. J. Cooton and R. Klatzky, pp. 37–53. Hillsdale, N.J.: Erlbaum.

Rüppell, W. 1933. "Physiologie und Akustik der Vögelstimme." *Journal für Ornithologie,* 81: 433–542.

Russ, J. C. 1992. *The Imaging Processing Handbook.* Boca Raton, Fla.: CRC Press.

Russell, I. S. 1979. "Brain size and intelligence: A comparative perspective." In *Brain, Behavior, and Evolution,* ed. D. A. Oakley and H. C. Plotkin, pp. 126–153. London: Methuen.

Ryalls, J. H., and P. Lieberman. 1984. "Fundamental frequency and vowel perception." *Journal of the Acoustical Society of America,* 72: 1631–1634.

Ryan, C. M., and S. E. G. Lea. 1994. "Images of conspecifics as categories to be discriminated by pigeons and chickens: Slides, video tapes, stuffed birds and live birds." *Behavioural Processes,* 33: 155–176.

Ryan, J. 1973. "Interpretation and imitation in early language development." In *Constraints on Learning,* ed. R. A. Hinde and J. Stephenson-Hinde, pp. 427–443. New York: Academic Press.

Salmon, C. M., L. E. Rowan, and P. R. Mitchell. 1998. "Facilitating prelinguistic communication: Impact of adult prompting." *Infant-Toddler Intervention,* 8: 11–27.

Salomon, G. 1977. "Effects of encouraging Israeli mothers to co-observe 'Sesame Street' with their five-year olds." *Child Development,* 48: 1146–1151.

Santiago, H. C., and A. A. Wright. 1984. "Pigeon memory: *Same/different* concept learning, serial probe recognition acquisition, and probe delay effects on the serial-position function." *Journal of Experimental Psychology: Animal Behavior Processes,* 10: 498–512.

Sarich, V. M., and J. E. Cronin. 1977. "Generation length and rates of homind evolution." *Nature,* 269: 354–355.

Saunders, D. A. 1983. "Vocal repertoire and individual vocal recognition in the short-billed white-tailed Black cockatoo, *Calyptorhynchus funereus latirostris* Carnaby." *Australian Wildlife Research,* 10: 527–536.

Saunders, R., and W. Sailor. 1979. "A comparison of the strategies of reinforcement in two-choice learning problems with severely retarded children." *American Association for the Education of the Severely (Profoundly) Handicapped Review,* 4: 323–334.

Savage-Rumbaugh, E. S. 1984a. "Acquisition of functional symbol use in apes and children." In *Animal Cognition,* ed. H. L. Roitblat, T. G. Bever, and H. S. Terrace, pp. 291–310. Hillsdale, N.J.: Erlbaum.

———1984b. "Verbal behavior at a procedural level in the chimpanzee." *Journal of the Experimental Analysis of Behavior,* 41: 223–250.

———1986. *Ape Language: From Conditioned Response to Symbol.* New York: Columbia University Press.

———1987. "Communication, symbolic communication, and language: Reply to Seidenberg and Petitto." *Journal of Experimental Psychology: General,* 116: 288–292.

———1991. "Language learning in the bonobo: How and why they learn." In *Biological and Behavioral Determinants of Language Development,* ed. N. A. Krasnegor, D. M. Rumbaugh, R. L. Schiefelbusch, and M. Studdert-Kennedy, pp. 209–233. Hillsdale, N.J.: Erlbaum.

Savage-Rumbaugh, E. S., and D. M. Rumbaugh. 1978. "Symbolization, language, and chimpanzees: A theoretical reevaluation based on initial language acquisition processes in four young *Pan troglodytes.*" *Brain and Language,* 6: 265–300.

Savage-Rumbaugh, E. S., K. Brakke, and S. Hutchins. 1992. "Linguistic development: Contrasts between co-reared *Pan troglodytes* and *Pan paniscus.*" In *Topics in Primatology: Human Origins,* ed. T. Nishida, W. C. McGrew, P. Marler, M. Pickford, and F. B. M. de Waal, pp. 51–66. Tokyo: University of Tokyo Press.

Savage-Rumbaugh, E. S., D. M. Rumbaugh, and S. Boysen. 1980a. "Do apes use language?" *American Scientist,* 68: 49–61.

Savage-Rumbaugh, E. S., D. M. Rumbaugh, and K. McDonald. 1985. "Language-learning in two species of apes." *Neuroscience Biobehavioral Review,* 9: 653–665.

Savage-Rumbaugh, E. S., D. M. Rumbaugh, S. T. Smith, and J. Lawson. 1980b. "Reference: The linguistic essential." *Science,* 210: 922–925.

Savage-Rumbaugh, E. S., R. A. Sevcik, D. M. Rumbaugh, and E. Rubert. 1985. "The capacity of animals to acquire language: Do species differences have anything to say to us?" *Philosophical Transactions of the Royal Society, London,* B308: 177–185.

Savage-Rumbaugh, E. S., K. McDonald, R. A. Sevcik, W. D. Hopkins, and E. Rubert. 1986. "Spontaneous symbol acquisition and communicative use by pygmy chimpanzees *(Pan paniscus).*" *Journal of Experimental Psychology: General,* 112: 457–492.

Savage-Rumbaugh, E. S., J. L. Pate, J. Lawson, S. T. Smith, and S. Rosenbaum. 1983. "Can a chimpanzee make a statement?" *Journal of Experimental Psychology: General,* 112: 457–492.

Savage-Rumbaugh, E. S., J. Murphy, R. A. Sevcik, K. E. Brakke, S. L. Williams, and D. M. Rumbaugh. 1993. "Language comprehension in ape and child." *Monographs of the Society for Research in Child Development,* 233: 1–258.

Sayigh, L. S., P. L. Tyack, R. S. Wells, and M. D. Scott. 1990. "Signature whistles of free-ranging bottlenose dolphins *(Tursiops truncatus):* Stability and mother-offspring comparisons." *Behavioral Ecology and Sociobiology,* 26: 247–260.

Scanlan, J. 1988. "Analysis of avian 'speech': Patterns and production." Ph.D. thesis, University College, London.

Schaffer, H. R., and C. K. Crook. 1979. "Maternal control techniques in a directed play situation." *Child Development,* 50: 989–996.

Schindlinger, M. D. 1995. "The evolution of vocal repertoires in parrots: Evidence from the wild." Paper presented at the Twenty-fourth International Ethological Congress, Honolulu, August.

Schino, G., G. Spinozzi, and L. Berlinguer. 1990. "Object concept and mental representation in *Cebus apella* and *Macaca fascicularis.*" *Primates,* 31: 537–544.

Schoener, T. W. 1971. "Theory of feeding strategies." *Annual Review of Ecology and Systematics,* 2: 369–404.

Schuler, A. L. 1979. "Echolalia: Issues and clinical application." *Journal of Speech and Hearing Disorders,* 44: 413–434.

Schusterman, R. J., and R. C. Gisiner. 1988. "Artificial language comprehension in dolphins and sea lions: The essential cognitive skills." *The Psychological Record,* 38: 311–348.

———1989. "Please parse the sentence: Animal cognition in the Procrustean bed of linguistics." *Psychological Record,* 39: 1–18.

Schusterman, R. J., and K. Krieger. 1984. "California sea lions are capable of semantic comprehension." *Psychological Record,* 34: 3–24.

————1986. "Artificial language comprehension and size transposition by a California sea lion *(Zalophus californianus)." Journal of Comparative Psychology*, 100: 348–355.

Schusterman, R. J., R. C. Gisiner, and E. B. Hanggi. 1988. "Priming short-term memory on a language task in sea lions." Paper presented at the annual meeting of the Psychonomic Society, Chicago, November.

Schwartz, R., and L. Leonard. 1985. "Lexical imitation and acquisition in language-impaired children." *Journal of Speech and Hearing Disorders*, 50: 141–149.

Schwartz, R. G., and B. Y. Terrell. 1983. "The role of input frequency in lexical acquisition." *Journal of Child Language*, 10: 571–588.

Scollon, R. 1976. *Conversations with a One Year Old.* Honolulu: University Press of Hawaii.

Searcy, W. A., and E. A. Brenowitz. 1988. "Sexual differences in species recognition of avian song." *Nature*, 332: 152–154.

Searcy, W. A., P. Marler, and S. S. Peters. 1985. "Songs of isolation-reared sparrows function in communication, but are significantly less effective than learned songs." *Behavioral Ecology and Sociobiology*, 17: 223–229.

Searle, J. R. 1969. *Speech Acts.* Cambridge: Cambridge University Press.

Sebeok, T. A., and D. J. Umiker-Sebeok. 1980. "Questioning apes." In *Speaking of Apes: A Critical Anthology of Two-Way Communication with Man*, ed. T. A. Sebeok and D. J. Umiker-Sebeok, pp. 1–59. New York: Plenum Press.

Secules, T., C. Herron, and M. Tomasello. 1992. "The effect of video context on foreign language learning." *The Modern Language Journal*, 76: 480–490.

Seibt, U. 1982. "Zahlbegriff und Zählverhalten bei Tieren: Neue Versuche und Deutungen." *Zeitschrift für Tierpsychologie*, 60: 325–341.

Seidenberg, M. S., and L. A. Petitto. 1987. "Communication, symbolic communication, and language: Comment on Savage-Rumbaugh, McDonald, Sevcik, Hopkins, and Rubert (1986)." *Journal of Experimental Psychology: General*, 116: 279–287.

Seligman, M. E. P., and J. L. Hager. 1972. *Biological Boundaries of Learning.* New York: Appleton-Century-Crofts.

Seyfarth, R. M., D. L. Cheney, and P. Marler. 1980. "Vervet monkey alarm calls: Semantic communication in a free-ranging primate." *Animal Behaviour*, 28: 1070–1094.

Shackleton, S. A., L. Ratcliffe, and D. Weary. 1992. "Relative frequency parameters and song recognition in Black-capped Chickadees." *Condor*, 94: 782–785.

Shatz, M. 1982. "On mechanisms of language acquisition: Can features of the communicative environment account for development?" In *Language Acquisition: The State of the Art*, ed. E. Wanner and L. R. Gleitman, pp. 102–127. Cambridge: Cambridge University Press.

Sherry, D. F. 1982. "Food storage, memory, and marsh tits." *Animal Behaviour*, 30: 631–633.

————1984. "Food storage by black-capped chickadees: Memory for the location and contents of caches." *Animal Behaviour*, 32: 451–464.

Shettleworth, S., and J. Krebs. 1982. "How marsh tits find their hoards: The roles of site preference and spatial memory." *Journal of Experimental Psychology: Animal Behavior Processes,* 8: 354–375.

——1986. "Stored and encountered seeds: A comparison of two spatial memory tasks in marsh tits and chickadees." *Journal of Experimental Psychology: Animal Behavior Processes,* 12: 248–257.

Shiovitz, K. A. 1975. "The process of species-specific song recognition by the indigo bunting, *Passerina cyanea,* and its relationship to the organization of avian acoustical behavior." *Behaviour,* 55: 128–179.

Shiovitz, K. A., and R. E. Lemon. 1980. "Species identification of songs by indigo buntings as determined by responses to computer-generated sounds." *Behaviour,* 74: 167–199.

Shy, E., P. K. McGregor, and J. Krebs. 1986. "Discrimination of song types by male great tits." *Behavioural Processes,* 13: 1–12.

Shyan, M. R., A. A. Wright, R. G. Cook, and M. Jitsumori. 1987. "Acquisition of the auditory *same/different* task in a rhesus monkey." *Bulletin of the Psychonomic Society,* 25: 1–4.

Siegel, L. S. 1982. "The development of quantity concepts: Perceptual and linguistic factors." In *Children's Logical and Mathematical Cognition,* ed. C. J. Brainerd, pp. 123–155. New York: Springer-Verlag.

Siegler, R. S. 1991. "In young children's counting, procedures precede principles." *Educational Psychology Review,* 3: 127–135.

Sigurdson, J. 1989. "Frequency-modulated whistles as a medium for communication with the bottle-nose dolphin *(Tursiops truncatus).*" Paper presented at the Animal Language Workshop, Honolulu, April.

Silaeva, O. L. 1995. "Bioacoustic types of avian vocal imitations." *Biological Bulletin,* 22: 608–615.

Silverstone, J. L. 1989. "Numerical abilities in the African Grey Parrot: Sequential numerical tags." Senior honors thesis, Northwestern University.

Skinner, B. F. 1938. *The Behavior of Organisms.* New York: Appelton-Century-Crofts.

——1957. *Verbal Behavior.* New York: Appleton-Century-Crofts.

——1974. *About Behaviorism.* New York: Knopf.

Skuse, D. H. 1988. "Extreme deprivation in early childhood." In *Language Development in Exceptional Circumstances,* ed. D. Bishop and K. Mogford, pp. 29–46. New York: Churchill-Livingston.

Sleigh, B. 1955. *Carbonel, the King of the Cats.* Indianapolis: Bobbs-Merrill.

Smith, D. G. 1979. "Male singing ability and territorial integrity in Red-winged Blackbirds *(Agelaius phoeniceus).*" *Behaviour,* 68: 191–206.

Smith, E. E., and J. Jonides. 1995. "Working memory in humans: Neurophysiological evidence." In *The Cognitive Neurosciences,* ed. M. Gazzaniga, pp. 1009–1020. Cambridge, Mass.: MIT Press.

Smith, J. C., and D. L. Roll. 1967. "Trace conditioning with X-rays as the aversive stimulus." *Psychonomic Science,* 9: 11–12.

Smith, W. J. 1977. *The Behavior of Communicating.* Cambridge, Mass.: Harvard University Press.

————1986. "Signaling behavior: Contributions of different repertoires." In *Dolphin Cognition and Behavior: A Comparative Approach,* ed. R. J. Schusterman, J. A. Thomas, and F. G. Wood, pp. 315–330. Hillsdale, N.J.: Erlbaum.

————1988. "Patterned daytime singing of the eastern wood-pewee, *Contopus virens.*" *Animal Behaviour,* 36: 1111–1123.

————1991. "Animal communication and the study of cognition." In *Cognitive Ethology: The Minds of Other Animals,* ed. C. A. Ristau, pp. 209–230. Hillsdale, N.J.: Erlbaum.

————1997. "The behavior of communicating, after twenty years." In *Perspectives in Ethology,* vol. 12, ed. D. H. Owings, M. D. Beecher, and N. S. Thompson, pp. 7–53. New York: Plenum Press.

Smith, W. J., and A. M. Smith. 1992. "Behavioral information provided by two song forms of the eastern kingbird, *T. tyrannus.*" *Behaviour,* 120: 90–102.

————1996. "Information about behavior provided by Louisiana waterthrush, *Seurus motacilla* (Parulinae), songs." *Animal Behaviour,* 51: 785–799.

Snow, C. E. 1979. "The role of social interaction in language acquisition." In *Children's Language and Communication,* ed. W. A. Collins, pp. 157–182. Hillsdale, N.J.: Erlbaum.

————1984. "Parent-child interaction and the development of communicative ability." In *The Acquisition of Communicative Competence,* ed. R. L. Schiefelbusch and J. Pickar, pp. 69–107. Baltimore: University Park Press.

Snow, C. E., and M. Hoefnagel-Höhle. 1978. "The critical period for language acquisition: Evidence from second language learning." *Child Development,* 49: 1114–1128.

Snow, D. W., and B. K. Snow. 1983. "Territorial song of the Dunnock *Prunella modularis.*" *Bird Study,* 30: 51–56.

Snowdon, C. T., and M. Hausberger, eds. 1997. *Social Influences on Vocal Development.* Cambridge: Cambridge University Press.

Snyder, L. S., E. Bates, and I. Bretherton. 1981. "Content and context in early lexical development." *Journal of Child Language,* 8: 565–682.

Snyder, N. F., J. W. Wiley, and C. B. Kepler. 1987. *The Parrots of Luquillo: Natural History and Conservation of the Puerto Rican Parrot.* Los Angeles: Western Foundation for Vertebrate Zoology.

Sokolov, Y. N. 1963. *Perception and the Conditioned Reflex.* New York: Pergamon.

Sophian, C. 1985. "Understanding the movement of objects: Early developments in spatial cognition." *British Journal of Developmental Psychology,* 3: 321–333.

————1995. "The trouble with competence models." *Cahiers de Psychologie Cognitive,* 14: 753–759.

Sophian, C., and S. Sage. 1983. "Development in infants' search for displaced objects." *Journal of Experimental Child Psychology,* 35: 143–160.

Sorokin, V. N. 1992. "Determination of vocal tract shapes for vowels." *Speech Communication,* 11: 71–85.

Spector, D. A. 1991. "The singing behavior of yellow warblers." *Behaviour,* 117: 29–52.

————1992. "Wood-warbler song systems: A review of paruline singing behavior." *Current Ornithology,* 9: 199–238.

Spector, D. A., L. K. McKim, and D. E. Kroodsma. 1989. "Yellow warblers are able to learn songs and the situations in which to use them." *Animal Behaviour,* 38: 723–725.

Spence, K. W. 1937. "The differential response in animals to stimuli varying within a single dimension." *Psychological Review,* 44: 430–444.

Spinozzi, G. 1989. "Early sensorimotor development in *Cebus (Cebus apella)*." In *Cognitive Structure and Development in Nonhuman Primates,* ed. F. Antinucci, pp. 55–66. Hillsdale, N.J.: Erlbaum.

Spinozzi, G., and F. Natale. 1989. "Early sensorimotor development in *Gorilla.*" In *Cognitive Structure and Development in Nonhuman Primates,* ed. F. Antinucci, pp. 21–38. Hillsdale, N.J.: Erlbaum.

Staicer, C. A. 1996. "Honest advertisement of pairing status: Evidence from a tropical resident wood-warbler." *Animal Behaviour,* 51: 375–390.

Stanford, K. N. 1995. "Studies on two-dimensional object identification by an African Grey parrot." Senior thesis, University of Arizona.

Starkey, P., and R. G. Cooper. 1995. "The development of subitizing in young children." *British Journal of Developmental Psychology,* 13: 399–420.

Stein, R. 1968. "Modulation in bird sounds." *Auk,* 85: 229–243.

Stern, C., and W. Stern. 1928. *Die Kindersprache.* Leipzig: J. A. Barth.

Sternberg, R. J. 1985. "Cognitive approaches to intelligence." In *Handbook of Intelligence,* ed. B. Wolman, pp. 59–118. New York: John Wiley.

Stettner, L. J. 1967. "Brain lesions in birds: Effects on discrimination acquisition and reversal." *Science,* 155: 1689–1692.

————1974. "Avian discrimination and reversal learning." In *Birds: Brain and Behavior,* ed. J. Goodman and M. W. Schein, pp. 165–220. New York: Academic Press.

Stettner, L. J., and K. Matyniak. 1968. "The brain of birds." *Scientific American,* 218: 64–76.

Stevens, K. N., and A. S. House. 1956. "Studies of formant transitions using a vocal tract analog." *Journal of the Acoustical Society of America,* 28: 578–585.

————1961. "An acoustical theory of vowel production and some of its implications." *Journal of Speech and Hearing Research,* 4: 303–320.

Stevens, S. S. 1951. "Mathematics, measurement, and psychophysics." In *Handbook of Experimental Psychology,* ed. S. S. Stevens, pp. 1–49. New York: John Wiley.

Stevens, T. S. 1888. "Notes on an intelligent parrot." *Trenton Natural History Society,* 3: 347–356.

Stoddard, P. K. 1996. "Vocal recognition of neighbors by territorial passerines." In *Ecology and Evolution of Acoustic Communication in Birds,* ed. D. E. Kroodsma and E. H. Miller, pp. 356–374. Ithaca: Cornell University Press.

Stoddard, P. K., M. D. Beecher, and M. S. Willis. 1988. "Responses of territorial male song sparrows to song types and variations." *Behavioral Ecology and Sociobiology,* 22: 125–130.

Stoddard, P. K., M. D. Beecher, P. Loesche, and S. E. Campbell. 1992. "Memory does not constrain individual recognition in a bird with song repertoires." *Behaviour,* 122: 274–287.

Stone, M. 1991. "Imaging the tongue and vocal tract." *British Journal of Disorders of Communication,* 26: 11–23.

Story, B. H., E. A. Hoffman, and I. R. Titze. 1996. "Vocal tract imaging: A comparison of MRI and EBCT." In *SPIE Proceedings in Medical Imaging, Physiology, and Function from Multidimensional Images,* no. 2709. Newport Beach, Calif.: SPIE.

Story, B. H., I. R. Titze, and E. A. Hoffman. 1996. "Vocal tract area functions from magnetic resonance imaging." *Journal of the Acoustical Society of America,* 100: 537–554.

St. Peters, M., A. C. Huston, and J. C. Wright. 1989. "Television and families: Parental coviewing and young children's language development, social behavior, and television processing." Paper presented at the Society for Research in Child Development, Kansas City, Kan., April.

Strange, W. 1989. "Evolving theories of vowel perception." *Journal of the Acoustical Society of America,* 85: 2081–2087.

Striedter, G. 1994. "The vocal control pathways in budgerigars differ from those in songbirds." *Journal of Comparative Neurology,* 343: 35–56.

Studdert-Kennedy, M. 1991. "Language development from an evolutionary perspective." In *Biological and Behavioral Determinants of Language Development,* ed. N. A. Krasnegor, D. M. Rumbaugh, R. L. Schiefelbusch, and M. Studdert-Kennedy, pp. 5–28. Hillsdale, N.J.: Erlbaum.

Subtelny, J., W. L. R. Whitehead, and J. D. Subtelny. 1989. "Cephalometric and cineradiographic study of deviant resonance in hearing impaired speakers." *Journal of Speech and Hearing Disorders,* 54: 249–263.

Sulter, A. M., D. G. Miller, F. W. Rienhart, H. K. Schutte, H. P. Wit, and E. L. Mooyart. 1992. "On the relation between the dimensions and resonance characteristics of the vocal tract: A study with MRI." *MRI,* 10: 365–373.

Sundberg, M. L. 1983. "The beginning of human verbal behavior." Paper presented at the Association of Behavior Analysis Convention, Milwaukee, May.

———1985. "Teaching language to the developmentally disabled." Paper presented at the Association of Behavior Analysis Convention, Columbus, Ohio, May.

Sutherland, C. A., and D. S. McChesney. 1965. "Sound production in two species of geese." *Living Bird,* 4: 99–106.

Sutherland, N. S., and N. J. Mackintosh. 1971. *Mechanisms of Animal Discrimination Learning.* New York: Academic Press.

Swenson, L. C. 1970. "One versus two discrimination by whitenecked ravens *(Corvus cryptoleucus)* with nonnumber dimensions varied." *Animal Behaviour,* 18: 454–460.

Tardif, T. 1996. "Nouns are not always learned before verbs: Evidence from Mandarin speakers' early vocabularies." *Developmental Psychology,* 32: 492–504.

Taves, E. H. 1941. "Two mechanisms for the perception of visual numerousness." *Archives of Psychology,* 37: 1–47.

Temple, Sir Wm. 1692. *Memoirs of What Past in Christendom, from the War Begun 1672 to the Peace Concluded 1679,* pp. 57-60. St. Paul's, London: R. R. for Ric. Chiswell.

Tenaza, R. R. 1976. "Wild mynahs mimic wild primates." *Nature,* 259: 561.

ten Cate, C. 1991. "Behavior-contingent exposure to taped song and zebra finch song learning." *Animal Behaviour,* 42: 857-859.

ten Cate, C., and B. B. Houx. 1998. "Social interactions and song learning: Are behavioural contingencies important?" Paper presented at the Twenty-second International Ornithological Congress, Durban, South Africa.

Terrace, H. S. 1979a. "Is problem-solving language?" *Journal of the Experimental Analysis of Behavior,* 31: 161-175.

———1979b. *Nim.* New York: Knopf.

———1986. Foreword. In *Ape Language: From Conditioned Response to Symbol,* by E. S. Savage-Rumbaugh. New York: Columbia University Press.

Terrace, H. S., L. A. Petitto, R. J. Sanders, and T. G. Bever. 1979. "Can an ape create a sentence?" *Science,* 206: 891-902.

Terrell, S. L., and R. Daniloff. 1996. "Children's word learning using three modes of instruction." *Perceptual and Motor Skills,* 83: 779-787.

Thielcke, G. 1973. "Uniformierung des Gesangs der Tannenmeise *(Parus ater)* durch Lernen." *Journal für Ornithologie,* 114: 443-454.

Thinus-Blanc, C., and P. Scardigli. 1981. "Object permanence in the golden hamster." *Perceptual and Motor Skills,* 53: 1010.

Thinus-Blanc, C., B. Poucet, and N. Chapuis. 1982. "Object permanence in cats: Analysis in locomotor space." *Behavioural Processes,* 7: 81-86.

Thomas, R. K. 1980. "Evolution of intelligence: An approach to its assessment." *Brain, Behavior, and Evolution,* 17: 454-472.

———1991. "Misuse of conditional reasoning in animal research with special reference to the evolution of language." Paper presented at the annual meeting of the Southern Society for Philosophy and Psychology, Atlanta.

———1992. "Interactive models of cognitive abilities of monkeys and humans *(Saimiri sciureus sciureus; S. boliviensus boliviensus; Homo sapiens)."* *International Journal of Comparative Psychology,* 5: 179-190.

———1996. "Investigating cognitive abilities in animals: Unrealized potential." *Cognitive Brain Research,* 3: 157-166.

Thomas, R. K., and L. Chase. 1980. "Relative numerousness judgements by squirrel monkeys." *Bulletin of the Psychonomic Society,* 16: 79-82.

Thomas, R. K., and T. N. Crosby. 1977. "Absolute versus relative class conceptual behavior in squirrel monkeys *(Saimiri sciureus)."* *Animal Learning & Behavior,* 5: 265-271.

Thomas, R. K., and R. B. Lorden. 1993. "Numerical competence in animals: A conservative view." In *The Development of Numerical Competence,* ed. S. T. Boysen and E. J. Capaldi, pp. 127-147. Hillsdale, N.J.: Erlbaum.

Thomas, R. K., and L. M. Noble. 1988. "Visual and olfactory oddity learning in rats: What evidence is necessary to show conceptual behavior?" *Animal Learning & Behavior,* 16: 157-163.

Thomas, R. K., and E. L. Walden. 1985. "The assessment of cognitive development in human and nonhuman primates." In *Nonhuman Primate Models for Hu-*

man Growth and Development, ed. E. Watts, pp. 187–215. Atlanta: Alan R. Liss.

Thomas, R. K., D. Fowlkes, and J. D. Vickery. 1980. "Conceptual numerousness judgments by squirrel monkeys." *American Journal of Psychology*, 93: 247–257.

Thompson, N. S. 1968. "Counting and communication in crows." *Communications in Behavioral Biology*, 2: 223–225.

——1969. "Individual identification and temporal patterning in the cawing of common crows." *Communications in Behavioral Biology*, 4: 29–33.

Thorndike, E. L. 1943. *Man and His Works*. Cambridge, Mass.: Harvard University Press.

Thorpe, W. H. 1959. "Talking birds and the mode of action of the vocal apparatus of birds." *Proceedings of the Zoological Society, London*. 132: 441–455.

——1961. *Bird-song*. Cambridge: Cambridge University Press.

——1964. *Learning and Instinct in Animals*, 2nd ed. Cambridge, Mass.: Harvard University Press.

——1972. *Duetting and Antiphonal Song in Birds. Behaviour*, Supplement 18.

——1974. *Animal and Human Nature*. New York: Doubleday.

Thorpe, W. N., and M. E. W. North. 1965. "Origin and significance of the power of vocal imitation: With special reference to the antiphonal singing of birds." *Nature*, 208: 219–222.

Tinbergen, N., and E. H. Tinbergen. 1972. "Early childhood autism: An ethological approach." *Zeitschrift für Tierpsychologie*, Supplement 10.

Tinklepaugh, O. T. 1928. "An experimental study of representative factors in monkeys." *Journal of Comparative Psychology*, 8: 197–236.

——1932. "Multiple delayed reaction with chimpanzees and monkeys." *Journal of Comparative Psychology*, 13: 207–243.

Todt, D. 1975a. "Social learning of vocal patterns and modes of their application in Grey parrots." *Zeitschrift für Tierpsychologie*, 39: 178–188.

——1975b. "Spontaneous recombinations of vocal patterns in parrots." *Naturwissenshaften*, 62: 399–400.

Todt, D., and H. Hultsch. 1998. "Hierarchical learning, development and representation of song." In *Animal Cognition in Nature*, ed. R. P. Balda, I. M. Pepperberg, and A. C. Kamil, pp. 275–303. London: Academic Press.

Todt, D., H. Hultsch, and D. Heike. 1979. "Conditions affecting song acquisition in nightingales." *Zeitschrift für Tierpsychologie*, 51: 23–35.

Tomasello, M. 1992. "The social bases of language acquisition." *Social Development*, 1: 68–87.

——1996. "The cultural roots of language." In *Communicating Meaning*, ed. B. M. Velichkovsky and D. M. Rumbaugh, pp. 275–307. Mahwah, N.J.: Erlbaum.

Tomasello, M., and J. Call. 1997. *Primate Cognition*. Oxford: Oxford University Press.

Tomasello, M., and M. J. Farrar. 1984. "Cognitive bases of lexical development: Object permanence and relational words." *Journal of Child Language*, 11: 477–493.

————1986. "Object permanence and relational words: A lexical training study." *Journal of Child Language,* 13: 495–505.

Tomasello, M., and J. Todd. 1983. "Joint attention and early lexical acquisition style." *First Language,* 4: 197–212.

Tomasello, M., E. S. Savage-Rumbaugh, and A. C. Kruger. 1993. "Imitative learning of actions on objects by children, chimpanzees, and enculturated chimpanzees." *Child Development,* 64: 1688–1705.

Tomasello, M., R. Strosberg, and N. Akhtar. 1996. "Eighteen-month-old children learn words in non-ostensive contexts." *Journal of Child Language,* 23: 157–176.

Tomasello, M., N. Akhtar, K. Dodson, and L. Rekau. 1997. "Differential productivity in young children's use of nouns and verbs." *Journal of Child Language,* 24: 373–387.

Traunmüller, H. 1988. "Paralinguistic variation and invariance in the characteristic frequencies of vowels." *Phonetica,* 45: 1–29.

Treisman, A., and S. Sato. 1990. "Conjunction search revisited." *Journal of Experimental Psychology: Human Perception and Performance,* 16: 459–478.

Triana, E., and R. Pasnak. 1981. "Object permanence in cats and dogs." *Animal Learning & Behavior,* 9: 135–139.

Trick, L., and Z. Pylyshyn. 1989. "Subitizing and the FNST spatial index model." University of Ontario, COGMEM #44. (Based on a paper presented at the Thirtieth Psychonomic Society Meeting, Atlanta.)

————1991. "A theory of enumeration that grows out of a general theory of vision: Subitizing, counting, and FINSTs." University of Ontario, COGMEM #57.

Trick, L. M., J. T. Enns, and D. A. Brodeur. 1996. "Life span changes in visual enumeration: The number discrimination task." *Developmental Psychology,* 32: 925–932.

Twyman, J. S. 1995. "The functional independence of impure mands and tacts of abstract stimulus properties." *Analysis of Verbal Behavior,* 13: 1–19.

Tyack, P. L. 1986. "Whistle repertoires of two bottlenose dolphins *(Tursiops truncatus):* Mimicry of signature whistles?" *Behavioral Ecology and Sociobiology,* 18: 251–257.

Urcuoli, P. J., and J. A. Nevin. 1975. "Transfer of hue matching in pigeons." *Journal of the Experimental Analysis of Behavior,* 24: 149–155.

Uzgiris, I. C., and J. McV. Hunt. 1966. *"Ordinal Scales of Infant Development, No. 1—Object Permanence"* (film). Champaign, Ill.: University of Illinois.

————1975. *Assessment in Infancy: Ordinal Scales of Psychological Development.* Champaign-Urbana: University of Illinois Press.

Vaidyanathan, R. 1991. "Development of forms and functions of negation in the early stages of language acquisition: A study in Tamil." *Journal of Child Language,* 18: 51–66.

Vanayan, M. H., A. Robertson, and G. B. Biederman. 1985. "Observational learning in pigeons: The effects of model proficiency on observer performance." *Journal of General Psychology,* 112: 349–357.

Vander Wall, S. B. 1982. "An experimental analysis of cache recovery in Clark's nutcrackers." *Animal Behaviour,* 30: 84–94.

Vauclair, J. 1996. *Animal Cognition: An Introduction to Modern Comparative Psychology.* Cambridge, Mass.: Harvard University Press.

Vaughter, R. M., W. Smotherman, and J. M. Ordy. 1972. "Development of object permanence in the infant squirrel monkey." *Developmental Psychology,* 7: 34-38.

Velleman, S. 1987. "Mother-child interactions: A longitudinal micro analysis." Paper presented at the Twelfth Annual Boston University Conference on Language Development, Boston.

Veneziano, E. 1988. "Vocal-verbal interaction and the construction of early lexical knowledge." In *The Emergent Lexicon: The Child's Development of a Linguistic Vocabulary,* ed. M. D. Smith and J. L. Locke, pp. 109-147. San Deigo: Academic Press.

von Glasersfeld, E. 1977. "Linguistic communication: Theory and definition." In *Language Learning by a Chimpanzee,* ed. D. M. Rumbaugh, pp. 55-71. New York: Academic Press.

———1981. "An attention model for the conceptual construction of units and number." *Journal of Research in Mathematical Education,* 12: 83-94.

———1982. "Subitizing: The role of figural patterns in the development of numerical concepts." *Archives de Psychologie,* 50: 191-218.

———1993. "Reflections on number and counting." In *The Development of Numerical Competence,* ed. S. T. Boysen and E. J. Capaldi, pp. 225-243. Hillsdale, N.J.: Erlbaum.

Vygotsky, L. S. 1962. *Thought and Language.* Cambridge, Mass.: MIT Press.

———1966. "Play and its role in the mental development of the child." *Soviet Psychology,* 12: 62-76.

———1978. *Mind in Society: The Development of Higher Mental Processes.* Cambridge, Mass.: Harvard University Press.

Wadsworth, B. J. 1978. *Piaget for the Classroom Teacher.* New York: Longman.

———1984. *Piaget's Theory of Cognitive and Affective Development.* New York: Longman.

Walker, L. C. 1981. "The ontogeny of the neural substrate for language." *Journal of Human Evolution,* 10: 429-441.

Wanker, R., J. Apcin, B. Jennerjahn, and B. Waibel. 1998. "Discrimination of different social companions in spectacled parrotlets *(Forpus conspicillatus):* Evidence for individual vocal recognition." *Behavioral Ecology and Sociobiology,* 43: 197-202.

Warner, R. W. 1971. "The structural basis of the organ of voice in the genera *Anas* and *Aythya* (Aves)." *Journal of Zoology* (London), 164: 197-207.

Warren, D. K., D. K. Patterson, and I. M. Pepperberg. 1996. "Mechanisms of American English vowel production in a Grey Parrot *(Psittacus erithacus)." Auk,* 113: 41-58.

Waser, M. S., and P. Marler. 1977. "Song learning in canaries." *Journal of Comparative and Physiological Psychology,* 91: 1-7.

Wasserman, E. A. 1993. "Comparative cognition: Beginning the second century of the study of animal intelligence." *Psychological Bulletin,* 113: 211-228.

Wasserman, E. A., C. L. DeVolder, and D. J. Coppage. 1992. "Nonsimilarity-based conceptualization in pigeons via secondary or mediated generalization." *Psychological Science,* 6: 374–379.

Wasserman, E. A., J. A. Hugart, and K. Kirkpatrick-Steger. 1995. "Pigeons show same-different conceptualization after training with complex visual stimuli." *Journal of Experimental Psychology: Animal Behavior Processes,* 21: 248–252.

Waters, R. S., and W. A. Wilson, Jr. 1976. "Speech perception by rhesus monkeys: The voicing distinction in synthesized labial and velar stop consonants." *Perception & Psychophysics,* 19: 285–289.

Watkins, B., S. Calvert, A. Huston-Stein, and J. C. Wright. 1980. "Children's recall of television material: Effects of presentation mode and adult labeling." *Developmental Psychology,* 16: 672–674.

Watkins, L. T., J. N. Sprafkin, and D. M. Krolikowski. 1990. "Effects of video based training on spoken and signed language acquisition by students with mental retardation." *Research in Developmental Disabilities,* 11: 273–288.

Watson, J. B. 1929. *Psychology from the Standpoint of a Behaviorist.* Philadelphia: Lippincott.

Waxman, S. R., and D. B. Markow. 1995. "Words as invitations to form categories: Evidence from 12–13-month-old infants." *Cognitive Psychology,* 29: 257–302.

Weary, D. M. 1989. "Categorical perception of bird song: How do great tits *(Parus major)* perceive temporal variation in their song?" *Journal of Comparative Psychology,* 103: 320–325.

———1991. "Use of the relative frequency of notes by veeries in song recognition and production." *Auk,* 108: 977–981.

Weary, D. M., and R. G. Weisman. 1991. "Operant discrimination of frequency and frequency ratio in the Black-capped Chickadee *(Parus atricapillus)." Journal of Comparative Psychology,* 105: 253–259.

Weary, D. M., R. G. Weisman, R. E. Lemon, T. Chin, and J. Mongrain. 1991. "Use of the relative frequency of notes by veeries in song recognition and production." *Auk,* 108: 977–981.

Weeks, T. E. 1979. *Born to Talk.* Rowley, Mass.: Newbury.

Weir, R. 1962. *Language in the Crib.* The Hague: Mouton.

Weisman, R., and L. Ratcliffe. 1989. "Absolute and relative pitch processing in black-capped chickadees, *Parus atricapillus." Animal Behaviour,* 38: 685–692.

Weisman, R., L. Ratcliffe, I. Johnsrude, and T. A. Hurly. 1990. "Absolute and relative pitch production in the song of the Black-capped Chickadee." *Condor,* 92: 118–124.

Wellman, H. M., D. Cross, and K. Bartsch. 1986. *Infant Search and Object Permanence: Meta-Analysis of the A-Not-B Error.* Monographs of the Society for Research in Child Development. Chicago: University of Chicago Press.

Wells, H., and K. Deffenbacher. 1967. "Conjunctive and disjunctive concept learning in humans and squirrel monkeys." *Canadian Journal of Psychology,* 21: 301–308.

Welty, J. C. 1962. *The Life of Birds.* New York: W. B. Saunders.

Wenzel, B. M. 1967. "Olfactory perception in birds." In *Olfaction and Taste,* vol. 2, ed. T. Hayashi, pp. 203–217. Oxford: Pergamon Press.

Wertsch, J. V. 1985. "Introduction." In *Culture, Communication, and Cognition: Vygotskian Perspectives,* ed. J. V. Wertsch, pp. 1–18. Cambridge: Cambridge University Press.

West, M. J., and A. P. King. 1985. "Social guidance of vocal learning by female cowbirds: Validating its functional significance." *Zeitschrift für Tierpsychologie,* 70: 225–235.

———1988. "Female visual displays affect the development of male song in the cowbird." *Nature,* 334: 244–246.

West, M. J., A. P. King, and T. M. Freeberg. 1996. "Social malleability in cowbirds: New measures reveal new evidence of plasticity in the Eastern subspecies *(Molothrus ater ater)." Journal of Comparative Psychology,* 110: 15–26.

West, M. J., A. N. Straud, and A. P. King. 1983. "Mimicry of the human voice by European starlings: The role of social interaction." *Wilson Bulletin,* 95: 635–640.

Westneat, M., J. H. L. Long, W. Hoese, and S. Nowicki. 1993. "Kinematics of birdsong: Functional correlation of cranial movements and acoustic features in sparrows." *Journal of Experimental Biology,* 182: 147–171.

White, S. S. 1968. "Movements of the larynx during crowing in the domestic cock." *Journal of Anatomy,* 103: 390–392.

———1975. "The larynx." In *Sissons and Grossman's the Anatomy of Domestic Animals,* 5th ed., vol. 2, ed. R. Getty, pp. 1891–1897. Philadelphia: Saunders.

Whiteside, S. P. 1998. "Identification of a speaker's sex: A study of vowels." *Perceptual and Motor Skills,* 86: 579–584.

Wickler, W. 1972. "Aufbau und Paarspezifität des Gesangduettes von *Laniarius funebris* (Aves, Passeriformes, Laniidae)." *Zeitschrift für Tierpsychologie,* 30: 464–476.

———1976. "The ethological analysis of attachment." *Zeitschrift für Tierpsychologie,* 42: 12–28.

———1980. "Vocal duetting and the pairbond: I. Coyness and the partner commitment." *Zeitschrift für Tierpsychologie,* 52: 201–209.

———1985. "Coordination of vigilance in bird groups: The 'watchman's song' hypothesis." *Zeitschrift für Tierpsychologie,* 69: 250–253.

Wilkinson, K. M., W. V. Dube, and W. J. McIlvane. 1998. "Fast mapping and exclusion (emergent matching) in developmental language, behavior analysis, and animal cognition research." *Psychological Record,* 48: 407–422.

Williams, D. R., and H. Williams. 1969. "Auto-maintenance in the pigeon: Sustained pecking despite contingent non-reinforcement." *Journal of the Experimental Analysis of Behavior,* 12: 511–520.

Williams, H. 1990. "Models for song learning in the zebra finch: Fathers or others?" *Animal Behaviour,* 39: 745–757.

Williams, H., K. Kilander, and M. L. Sotanski. 1993. "Untutored song, reproductive success and song learning." *Animal Behaviour,* 45: 695–705.

Williams, S. L., K. E. Brakke, and E. S. Savage-Rumbaugh. 1997. "Comprehension skills of language-competent and nonlanguage-competent apes." *Language & Communication,* 17: 301–317.

Wilson, B., N. J. Mackintosh, and R. A. Boakes. 1985. "Matching and oddity learning in the pigeon: Transfer effects and the absence of relational learning." *Quarterly Journal of Experimental Psychology,* 37B: 295–311.

Wise, K. L., L. A. Wise, and R. R. Zimmermann. 1974. "Piagetian object permanence in the infant rhesus monkey." *Developmental Psychology,* 10: 429–437.

Wishart, J. G., and T. G. Bower. 1985. "A longitudinal study of the development of the object concept." *British Journal of Developmental Psychology,* 3: 243–258.

Wolfe, J. M., K. R. Cave, and S. L. Franzel. 1989. "Guided search: An alternative to the feature integration model for visual search." *Journal of Experimental Psychology: Human Perception and Performance,* 15: 419–433.

Wolfgramm, J., and D. Todt. 1982. "Pattern and time specificity in vocal responses of blackbirds *Turdus merula L.*" *Behaviour,* 81: 264–286.

Wolfram, W., and R. Johnson. 1982. *Phonological Analysis: Focus on American English.* Washington, D.C.: Center for Applied Linguistics.

Wolters, G., H. van Kempen, and G. Wijlhuizen. 1987. "Quantification of small numbers of dots: Subitizing or pattern recognition?" *American Journal of Psychology,* 100: 225–237.

Wood, S. A. J. 1996. "Assimilation or coarticulation? Evidence from the temporal co-ordination of tongue gestures for the palatalization of Bulgarian alveolar stops." *Journal of Phonetics,* 24: 139–164.

Wood, S., K. M. Moriarty, B. T. Gardner, and R. A. Gardner. 1980. "Object permanence in child and chimpanzee." *Animal Learning & Behavior,* 8: 3–9.

Woodruff, G., and D. Premack. 1981. "Primitive mathematical concepts in the chimpanzee: Proportionality and numerosity." *Nature,* 293: 568–570.

Woodruff, G., D. Premack, and K. Kennel. 1978. "Conservation of liquid and solid quantity by the chimpanzee." *Science,* 202: 991–994.

Woods, D. L., C. Alain, and K. H. Ogawa. 1998. "Conjoining auditory and visual features during high-rate serial presentation: Processing and conjoining two features can be faster than processing one." *Perception & Psychophysics,* 60: 239–249.

Wright, A. A., and J. D. Delius. 1994. "Scratch and match: Pigeons learn matching and oddity with gravel stimuli." *Journal of Experimental Psychology: Animal Behavior Processes,* 20: 108–112.

Wright, A. A., H. C. Santiago, and S. F. Sands. 1984. "Monkey memory: *Same/ different* concept learning, serial probe acquisition, and probe delay effects." *Journal of Experimental Psychology: Animal Behavior Processes,* 10: 513–529.

Wright, A. A., M. R. Shyan, and M. Jitusmori. 1990. "Auditory *same/different* concept learning by monkeys." *Animal Learning & Behavior,* 18: 287–294.

Wright, A. A., H. C. Santiago, P. J. Urcuioli, and S. F. Sands. 1984. "Monkey and pigeon acquisition of same/different concept using pictorial stimuli." In *Quantitative Analysis of Behavior,* vol. 4, ed. M. L. Commons, R. J. Herrnstein, and A. R. Wagner, pp. 295–317. Cambridge, Mass.: Ballinger.

Wright, T. F. 1996. "Regional dialects in the contact calls of a parrot." *Proceedings of the Royal Society, London,* B263: 867–872.

Wright, T. F., and M. Dorin. (Under review). "Mechanisms for dialect maintenance in the Yellow-naped Amazon: Evidence from playback experiments."

Wynne, C. D., L. von Fersen, and J. E. Staddon. 1992. "Pigeons' inferences are transitive and the outcome of elementary conditioning principles: A response." *Journal of Experimental Psychology: Animal Behavior Processes,* 18: 313-315.

Yamashita, C. 1987. "Field observations and comments on the Indigo macaw *(Anodorhynchus leari),* a highly endangered species from northeastern Brazil." *Wilson Bulletin,* 99: 280-282.

Yerkes, R. M., and A. W. Yerkes. 1929. *The Great Apes: A Study of Anthropoid Life.* New Haven: Yale University Press.

Yoder, P. J., and A. P. Kaiser. 1987. "Exploring the indirect routes by which maternal speech predicts later child language development." Paper presented at the Twelfth Annual Boston University Conference on Language Development, Boston, October.

Yoerg, S. I., and A. C. Kamil. 1991. "Integrating cognitive ethology with cognitive psychology." In *Cognitive Ethology: The Minds of Other Animals,* ed. C. A. Ristau, pp. 273-289. Hillsdale, N.J.: Erlbaum.

Young, M. E., E. A. Wasserman, R. M. Dalrymple. 1997. "Memory-based *same-different* conceptualization by pigeons." *Psychonomic Bulletin & Review,* 4: 552-558.

Zahorian, S. A., and A. J. Jagharghi. 1993. "Spectral-shape features versus formants as acoustic correlates for vowels." *Journal of the Acoustical Society of America,* 94: 1966-1982.

Zajonc, R. B. 1965. "Social facilitation." *Science,* 149: 269-274.

Zajonc, R. B., A. Heingartner, and E. M. Herman. 1969. "Social enhancement and impairment of performance in the cockroach." *Journal of Personality and Social Psychology,* 13: 83-92.

Zann, R. 1997. "Vocal learning in wild and domesticated zebra finches: Signature cues for kin recognition or epiphenomena?" In *Social Influences on Vocal Development,* ed. C. T. Snowdon and M. Hausberger, pp. 85-97. Cambridge: Cambridge University Press.

Zentall, T. R. 1993. "Animal cognition: An approach to the study of animal behavior." In *Animal Cognition: A Tribute to Donald A. Riley,* ed. T. R. Zentall, pp. 3-15. Hillsdale, N.J.: Erlbaum.

——1996. "An analysis of stimulus class formation in animals." In *Stimulus Class Formation in Humans and Animals,* ed. T. R. Zentall and P. R. Smeets, pp. 15-34. Amsterdam: Elsevier.

Zentall, T. R., and D. E. Hogan. 1974. "Abstract concept learning in the pigeon." *Journal of Experimental Psychology,* 102: 393-398.

——1978. "Same/different concept learning in the pigeon: The effect of negative instances and prior adaptation to transfer stimuli." *Journal of the Experimental Analysis of Behavior,* 30: 177-186.

Zentall, T. R., C. A. Edwards, B. S. Moore, and D. E. Hogan. 1981. "Identity: The basis for both matching and oddity learning in pigeons." *Journal of Experimental Psychology: Animal Behavior Processes,* 7: 70-86.

Zentall, T. R., P. Jackson-Smith, J. A. Jagielo, and G. B. Nallan. 1986. "Categorical shape and color naming by pigeons." *Journal of Experimental Psychology: Animal Behavior Processes,* 12: 153–159.

Zorina, Z. A. 1982. "Reasoning ability and adaptivity of behavior in birds." In *Evolution and Environment,* ed. V. J. A. Novak and J. Mlikovsky, pp. 907–912. Praha: CSAV. (Trans. of chapter provided by Z. A. Zorina.)

Zorina, Z. A., T. S. Kalinina, M. E. Mayorova, Y. B. Mikitich, and G. V. Khurtina. 1991. "Relative numerousness judgement in pigeons and crows at urgent comparison of stimuli, earlier connected with different quantities of reinforcement." *Zhurnal Vysshei Nervnoi Deyatelnosti Imeni I P Pavolova,* 41: 306–313.

GLOSSARY

Allospecific   Referring to a behavior or attribute of a species other than the one under study, or a species different from the one under study. Synonym for heterospecific. Contrast with conspecific.

Conspecific   Referring to a behavior or attribute of the given species, or the given species itself. Contrast with allospecific.

Insight detour problems   Tasks in which an animal must figure out, without trial-and-error learning, how to get around a barrier to obtain an object.

Match-to-sample problems   Tasks in which an animal is given a sample, such as a green light, which is then followed immediately by two other samples, e.g., a red and a green light. To be correct (and gain a reward), the animal must choose the green light, that is, match the original sample to one of the subsequent choices.

Nonmatch-to-sample problems   Tasks in which an animal is given a sample, such as a green light, which is then followed immediately by two other samples, e.g., a red and a green light. To be correct (and gain a reward), the animal must choose the red light, that is, not match the original sample to one of the subsequent choices.

Oddity problems   Tasks in which an animal is presented with at least three sample items, one of which differs from the others in some dimension (e.g., color, shape, material). To be correct, the animal must choose this differing (odd) sample.

Phoneme   The smallest bit of sound that causes two sets of utterances to mean different things. Thus the difference between "pea" and "tea" (in the International Phonetic Alphabet, /pi/ and /ti/) are the phonemes /p/ and /t/ (Wolfram and Johnson 1982).

Reversal learning   The ability of an animal to learn the reverse of the concept that it has already learned from a set of problems; the number of trials needed to learn to "reverse" itself is considered a measure of intelligence. Thus an animal is first taught to choose red when presented with red and green lights and is rewarded for this choice; the animal is then no longer rewarded for choosing red and the experimenter tracks how long it takes for the animal to switch to green to receive rewards.

Set learning   The ability of an animal to learn a concept from a set of problems, rather than the solution to a particular problem. An animal that is given the task of choosing the odd object from one collection is then given the same task for another collection of objects. Generally, the animal learns the correct response more quickly for the second collection. If the animal learns to respond correctly more quickly after experiencing each such collection, experimenters say that the animal has learned that *set* of problems.

CREDITS

Full citations can be found in the References.

Tables

Tables 3.1, 3.2, 7.1, and 7.2: *Ethology*, 55: 144, 55: 147, 75: 44, and 75: 52-53. Reprinted with the permission of Blackwell Wissenschafts-Verlag, Berlin.

Tables 4.1, 4.2, 5.1, 5.2, 5.3, and 5.4: *Animal Learning & Behavior*, 11: 181, 11: 182-183, 15: 427, 15: 428, 15: 431, and 15: 429. Reprinted with the permission of the Psychonomic Society, Inc.

Tables 6.1 and 6.2: *Journal of the Experimental Analysis of Behavior*, 50: 560 and 50: 561. Copyright © 1988 by the Society for the Experimental Analysis of Behavior. Reprinted with permission.

Tables 7.3, 7.4, 8.1, 8.2, 8.3, 8.4, 9.1, 9.2, and 10.1: *Journal of Comparative Psychology*, 108: 44, 108: 40, 104: 46-47, 104: 48, 106: 303-304, 106: 300, 105: 289, 105: 292, and 111: 68. Copyright © 1990, 1991, 1992, 1994, 1997 by the American Psychological Association. Reprinted with permission.

Tables 2.1, 11.1, 11.2, 11.3, 12.1, 13.1, and 13.2: *Applied Psycholinguistics*, 9: 63, 9: 66, 9: 67-68, 9: 69, 12: 160-161, 11: 32, and 11: 33. Reprinted with the permission of Cambridge University Press.

Table 14.1: Snowdon and Hausberger 1997: 163. Reprinted with the permission of Cambridge University Press.

Table 16.1: *The Auk*, 113: 53. Reprinted with permission.

Tables 15.1 and 15.2: *Journal of the Acoustical Society of America*, 96: 639 and 103: 2207. Reprinted with permission.

Tables 16.2 and 16.3: Patterson, Pepperberg, Story, and Hoffman 1997: 20, 22. Reprinted with the permission of the Society for Optical Engineering.

Figures

All photographs copyright © by William Muñoz and reprinted with permission.

Figures 7.1 and 7.2: *Ethology*, 75: 49 and 75: 51. Reprinted with the permission of Blackwell Wissenschafts-Verlag, Berlin.

INDEX